SOUTH POLAR REGION
GREENLAND
NORWEGIAN SEA
BARENTS SEA
KARA SEA
LAPTEV SEA
EAST SIBERIAN SEA
EUROPE
ASIA
NORTH ATLANTIC
AFRICA
SAHARA
ARABIAN PENINSULA
ARABIAN SEA
BAY OF BENGAL
INDIA
NORTH PACIFIC OCEAN
PHILIPPINE SEA
SOUTH CHINA SEA
AUSTRALIA
INDIAN OCEAN
SOUTH ATLANTIC OCEAN
ANGOLA PLAIN
ENDERBY PLAIN
SOUTH INDIAN BASIN
WEDDELL SEA
Queen Maud Land
ANTARCTICA
COPYRIGHT © 1996 NATIONAL GEOGRAPHIC SOCIETY
WORLD CLIMATE
CLIMATIC ZONES BASED ON KÖPPEN SYSTEM
TROPICAL
DRY
MILD
CONTINENTAL
POLAR
HIGH ELEVATIONS

Fundamentals of World Regional Geography

3e

Joseph J. Hobbs
University of Missouri, Columbia

Cartography by Andrew Dolan

BROOKS/COLE
CENGAGE Learning

Australia • Brazil • Japan • Korea • Mexico • Singapore • Spain • United Kingdom • United States

BROOKS/COLE
CENGAGE Learning

Fundamentals of World Regional Geography, Third Edition, International Edition
Joseph J. Hobbs

Sr. Acquisitions Sponsoring Editor: Aileen Berg

Developmental Editor: Liana Sarkisian

Editorial Assistant: Margaux Cameron

Media Editor: Alexandria Brady

Marketing Manager: Jack B. Cooney

Marketing Program Manager: Darlene Macanan

Marketing Coordinator: Lianne Ramsauer

Content Project Manager: Hal Humphrey

Art Director: Pam Galbreath

Manufacturing Planner: Rebecca Cross

Production Service: Teresa Christie, MPS North America

Rights Acquisitions Specialist: Roberta Broyer

Photo Researcher: Jeremy Glover, Bill Smith Group

Text Researcher: Sue C. Howard

Cartographer: Andrew Dolan

Cover Image: © Peter Adams/Corbis

Library of Congress Control Number: 2012937193

For product information and technology assistance, contact us at **Cengage Learning Customer & Sales Support, 1-800-354-9706.**

For permission to use material from this text or product, submit all requests online at **www.cengage.com/permissions.**
Further permissions questions can be e-mailed to **permissionrequest@cengage.com.**

International Edition:

ISBN-13: 978-1-133-11390-4

ISBN-10: 1-133-11390-7

Cengage Learning International Offices

Asia
www.cengageasia.com
tel: (65) 6410 1200

Australia/New Zealand
www.cengage.com.au
tel: (61) 3 9685 4111

Brazil
www.cengage.com.br
tel: (55) 11 3665 9900

India
www.cengage.co.in
tel: (91) 11 4364 1111

Latin America
www.cengage.com.mx
tel: (52) 55 1500 6000

UK/Europe/Middle East/Africa
www.cengage.co.uk
tel: (44) 0 1264 332 424

Represented in Canada by Nelson Education, Ltd.
www.nelson.com
tel: (416) 752 9100 / (800) 668 0671

Cengage Learning is a leading provider of customized learning solutions with office locations around the globe, including Singapore, the United Kingdom, Australia, Mexico, Brazil, and Japan. Locate your local office at: **www.cengage.com/global**

For product information and free companion resources: **www.cengage.com/international**

Visit your local office: **www.cengage.com/global**

Visit our corporate website: **www.cengage.com**

Printed in Canada
1 2 3 4 5 6 7 15 14 13 12

For Mom
Who has been with me for every step
On every journey

With love

Brief Contents

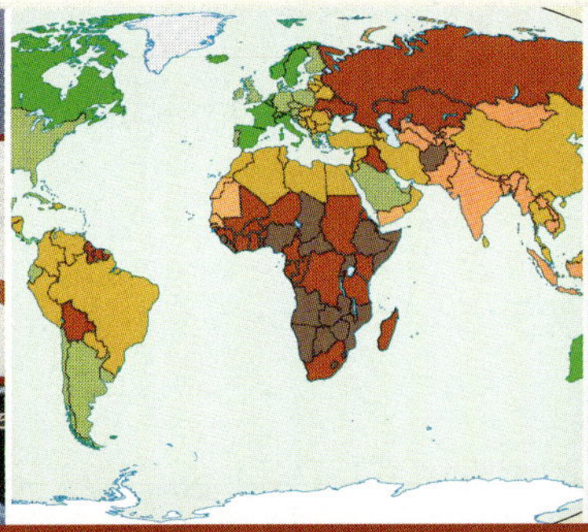

Contents

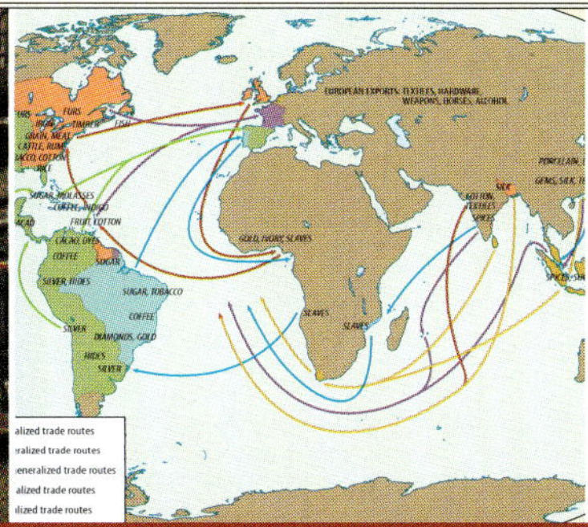

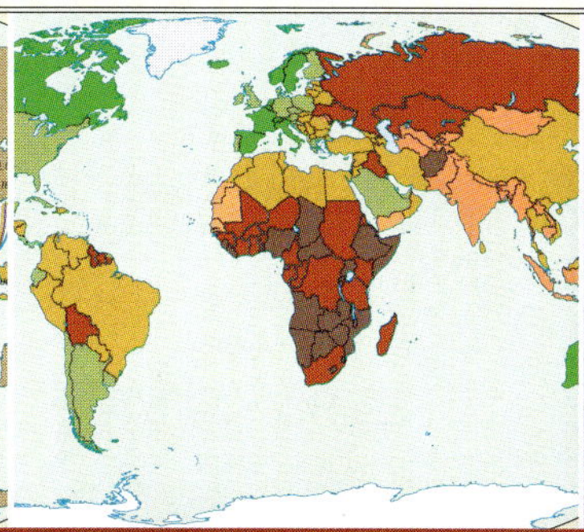

INTERNE
e-mail
copias
papeleria
renta
venta
asesorias
software
cursos
accesorios
mayar

الاختيار لك
تنظيم الأسرة
من أجل حياة أفضل

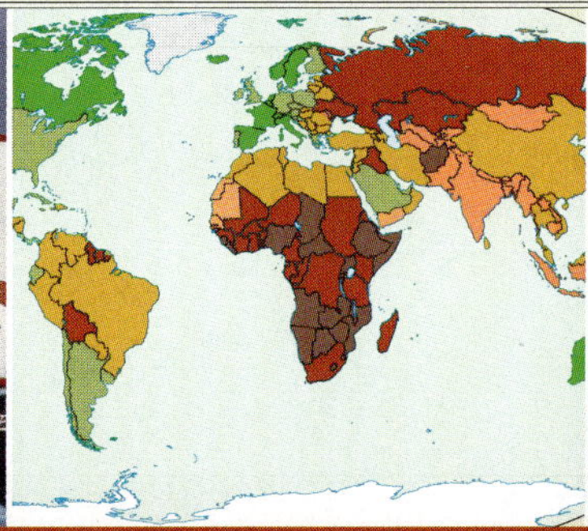

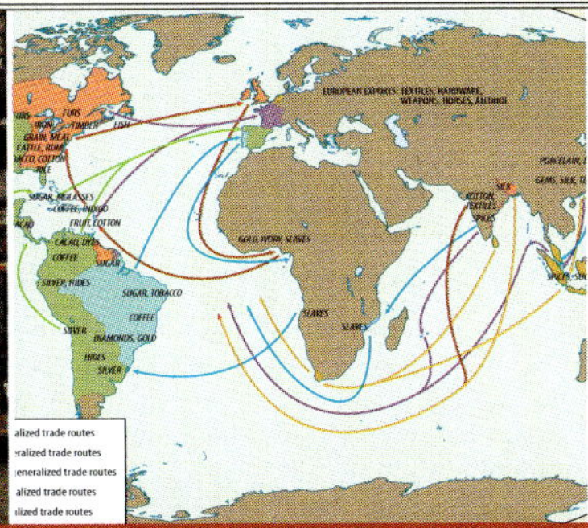

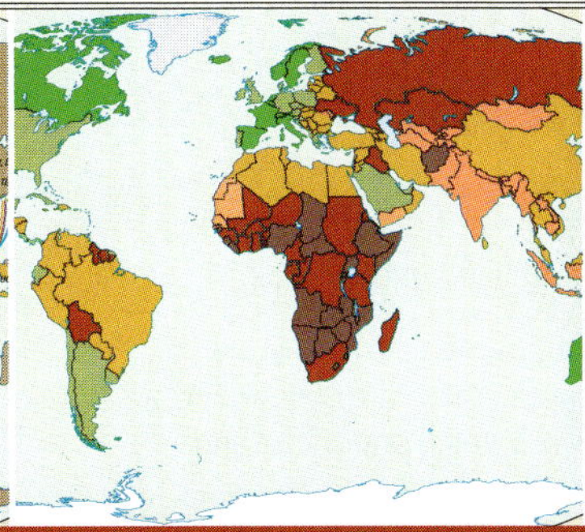

INTERNE
e-mail
copias
papelería
renta
venta
asesorias
software
cursos
accesorios
mayar

Maps

Preface

To appreciate how the world works today, you will benefit from a solid grounding in the environmental, cultural, historic, economic, and geopolitical contexts of the world's regions and nations. *Fundamentals of World Regional Geography* establishes that foundation for you and offers you an opportunity to explore in more detail the events, issues, and landscapes of your world.

My aim in writing the book is to make the world feel *accessible* to you. I hope that no place or issue will seem too far away or strange. I encourage you to think and talk more about what you read here, including through social media. I've started a Twitter site for us all to communicate with one another about what's happening in the world: follow me on Twitter: *@worldprof*

Chapters 1 through 3 provide the basic concepts, tools, and vocabulary for world regional geography. In the first chapter, geography's uniquely spatial approach to the world is introduced, along with some of the discipline's milestone concepts and its considerable career possibilities—especially those growing from the "geospatial revolution." The second chapter covers the essential characteristics of the world's physical processes and how some of them have been altered by human activity. Climate change and the treaties to control it have a prominent role in that chapter. Chapter 3 traces the modification of landscapes by human actions, describes trends and projections of population growth, and considers efforts to slow destructive trends in resource use.

Then come eight chapters exploring the world's regions through a consistent, thematic approach focusing in turn on five elements: "Area and Population," "Physical Geography and Human Adaptations," "Cultural and Historical Geographies," "Economic Geography," and "Geopolitical Issues." The final section of each chapter, entitled "Regional Issues and Landscapes," contains a selection of short studies of critical problems in global affairs—for example, the Palestinian-Israeli conflict—and exemplary or important problems in human or physical geography, such as ethnicity and petroleum in Nigeria, and the sources and impacts of deforestation in the Amazon Basin. The book is built for a one- or two-semester course. Instructors and students have the freedom to use some or all of content in a given chapter. If time is limited, you can concentrate on the five thematic elements of the chapter and limit the case studies selected from the Regional Issues and Landscapes section. It does not matter what order you read the book in; after the three introductory chapters, no chapter presumes that you have read any other chapter.

New to the Third Edition

Both longtime and first-time users of *Fundamentals of World Regional Geography* should be pleased with this text. I have tried to make the writing more *accessible* than ever. I use first person fairly often so that students can recognize the geographer behind the writing.

- I have written some new *Perspectives from the Field* essays, describing a meeting with the leader of a "Cargo Cult" in Vanuatu and observing landscape changes in Southeast Asia over a period of decades, for example.
- The visual layout is appealing. Cartographer Andrew Dolan builds on his reputation as the best cartographer of the world regional geography texts. For this edition's world maps he uses the Winkel-Trippel projection for the first time, to excellent effect. Also for the first time, I have dipped into my extensive collection of digital photographs. Since the second edition, I have researched, travelled, and photographed across Oceania, South and East Asia, and Latin America.
- There is more *geospatial* content throughout, with extensive use of remotely-sensed imagery and practical applications of GIS.
- For the first time I will correspond regularly with students and professors, especially to track issues related to the book. With the twitter handle *@worldprof*, I will share events and insights with readers.
- The book has, of course, been thoroughly updated. Since the second edition was published, the world has experienced the Great Recession. The economic fallout is a common thread throughout the book. The continued ascendance of China is another pervasive theme.
- I have made it easier to understand and use the important "margin cross-reference" feature which stitches common strands together all across the world.
- A new feature called "Try It" encourages students to use their own critical thinking and problem-solving skills to address a wide range of issues. The student may do these individually or in group exercises assigned by the instructor.
- More attention has been given to solutions, jumping from the principles of sustainable development to specific applications, especially in the developing world.
- I work purposefully to be objective, always trying to present pros and cons of controversial issues (e.g., of large dams) and multiple perspectives of complex political problems (e.g., of the Arab/Israeli conflict).
- Major metropolitan area populations are in tables rather than in the text, where they would interrupt the flow of prose.
- The Arctic Ocean appears to be opening up to trade and minerals exploration as seasonal and permanent sea ice coverage shrinks. In several chapters, we follow the race for Arctic routes and resources.
- I continue using the pattern of five geographic themes for each region: Area and Population; Physical Geography and Human Adaptations; Cultural and Historical Geographies; Economic Geography; and Geopolitical Issues. The book's signature emphasis on geopolitics is as strong as ever. This

edition has an enhanced focus on Economic Geography, introducing students to such blocs as the "PIGS" of Europe and the emerging market "BRICS." Conventional and alternative energy sources receive considerable attention in each region.

Chapter-Specific Changes

- Chapter 1, "Objectives and Tools of World Regional Geography," introduces "the Geospatial Revolution," with discussion of GIS and Remote Sensing. Students are encouraged to use Google Earth habitually. The AAG's exciting forecasts about job growth in Geography are shared.
- Chapter 2, "Physical Processes and World Regions," introduces the concept of the Anthropocene and includes a useful new map on the human "footprint." The term *global warming* is largely replaced by the more nuanced and variable *climate change*. Measures for adaptation to and mitigation of climate change are reviewed. There is more information on marine resources. I have removed the detailed information on weather fronts; profs can introduce these with other materials if they wish.
- Chapter 3, "Human Processes and World Regions," digs more deeply into the paradoxes of population, including how "underpopulation" as well as overpopulation can threaten economic well-being.
- Chapter 4 on Europe helps students make sense of the "Euromess:" the continent's sovereign debt crisis, problems with the Euro, and threats to the integrity of the European Union itself. Europe's growing immigration problems, especially in the south, are examined in detail. How can Europe restrict movement while widening Schengenland?
- Chapter 5, "Russia and the Near Abroad," sees Russia deepening its dependence on petroleum while Vladimir Putin consolidates his hold on power. The two are closely tied in what is described as "Putinomics." Russia continues to use oil and natural gas as bludgeons in foreign policy, through a geographic cat's cradle of pipelines.
- Chapter 6 is about the Middle East and North Africa. The major innovation here is a thorough analysis of the Arab Spring, including both the broad underlying themes and fallout, and an individual country survey showing how the revolution played out differently. We examine the near-death and resurrection of Iraq's southern marshes. Iran's prospective nuclear weapons development and the potential for preemptive military action are described. By popular demand, Afghanistan has been moved out of the Middle East and into Asia.
- Chapter Seven, the book's longest (because it covers the most populous region), is now "South and East Asia" rather than "Monsoon Asia." Villagers still outnumber urbanites here, but in anticipation of a rapid change in that equation there is attention to Asia's megacities. I discuss the phenomenon of "language death" due partly to computerized communication, and also the potential of computers to save endangered languages.
- Chapter 8 takes us across Oceania and Antarctica. The "Lucky Country" of Australia continues on it lucky streak on the back of ever-growing commodities demand in China. The Alliance of Small Island States is more alarmed than ever over rising sea levels.
- Chapter 9 on Sub-Saharan Africa portrays the continent at its most hopeful state is decades. Conflicts have subsided. Ironically, some of the greatest threats come from having too much wealth in oil and other natural resources: this is the so-called "resource curse." Technologies are growing quickly in Africa. Mobile phones are providing a shortcut to the information superhighway, and African software developers have contributed an important crowdsourcing tool, Ushahidi.
- Chapter 10 on Latin America describes the unspeakable tragedy of Haiti's 2010 earthquake and the world's other seismically prone cities like Port-au-Prince that are "rubbles in waiting." There are new discussions of religious movements, including those of Santeria and Maria Lionza. Brazil's impressive recent economic growth, based in part on newly discovered offshore oil reserves, is described.
- Chapter 11's disccusion of United States and Canada depicts two developed countries that still have their doors open to immigrants. In the United States, we had a 2010 census data form for 63 distinct ethnic/racial groups. The recent booms and busts of the Great Recession are depicted. We look at the phenomenon of "birth tourism" and the slowdown in illegal immigration due to economic recession. We see the Arid West becoming more arid, with some authorities saying this is the old and the new normal. Meteorological natural hazards of many kinds seem to be on the rise. Canada's tar sands and natural gas released by "fracking" have given North America more energy independence than they have enjoyed in decades. The *Deepwater Horizon* accident and its impacts on the Gulf are here. Outsourcing has begun to give way to insourcing (onshoring) and new life in American manufacturing.

Acknowledgments

I am grateful to everyone who encouraged and helped me with the book, especially Cindy, Katie, Lily, and Mom. Andy Dolan is the best cartographer there is. Andy wrote much of glossary and, as always, provided a lot of guidance on what he thought should be added to and removed from the text. The Cengage Learning team was headed up by Acquisitions Editor Aileen Berg and included Development Editor Liana Sarkisian, Project Manager Teresa Christie of MPS North America, Editorial Assistant Margaux Cameron, Media Editor Alexandria Brady, Marketing Manager Jack B. Cooney, Marketing Coordinator Lianne Ramsauer, Marketing Communications Manager Darlene Macanan, Rights Acquisitions Specialist Roberta Broyer, Photo Researcher Jeremy Glover of Bill Smith Group, Text Permissions Researcher Sue C. Howard, and Art Director Pam Galbreath.

Ancillaries

This text is accompanied by a number of ancillary publications to assist instructors and enhance student learning.

Instructor Resources

PowerLecture with JoinIn and ExamView

This one-stop digital library and presentation tool includes preassembled Microsoft® PowerPoint® lecture slides. In addition to a full Instructor's Manual and Test Bank, PowerLecture also includes an image library, animations, videos, blank maps, Google Earth activities, JoinIn™ Student Response System, and ExamView® testing software with all the test items from the printed Test Bank in electronic format, enabling you to create customized tests in print or online.

Online Instructor's Manual and Test Bank

By Chad Garick of Jones County Junior College. Streamline and maximize the effectiveness of your course preparation using such resources as *chapter summaries and outlines, learning objectives, lecture suggestions, key terms, and answers to in-text questions*. This time-saving resource also includes a Test Bank that offers over 70 questions. Available for download on the instructor's companion Web site.

ExamView Computerized Testing Software

Featuring automatic grading, ExamView® allows you to create, deliver, and customize tests and study guides (both print and online) in minutes.

WebTutor Toolbox™

WebTutor™ Toolbox for WebCT™ or Blackboard® provides access to all the content of this text's rich Book Companion Web site from within your course management system. Robust communication tools—such as course calendar, asynchronous discussion, real-time chat, a whiteboard, and an integrated e-mail system—make it easy for your students to stay connected to the course.

Student Resources

Geography CourseMate

The more you study, the better the results. Make the most of your study time by accessing everything you need to succeed in one place. Geography CourseMate includes:

- An interactive eBook with highlighting, note taking, and an interactive glossary
- Interactive learning tools, including:
  - Quizzes
  - Flashcards
  - Videos
  - Animations
  - ... and more!

About the Author

Katie Hobbs

Joe Hobbs received his B.A. at the University of California Santa Cruz and his M.A. and Ph.D. at the University of Texas Austin. He is a professor and chairman of the Department of Geography at the University of Missouri. He is a geographer of the Middle East with many years of field research on Bedouin peoples and natural environments in the deserts of Egypt. Joe's interest in the region grew from his boyhood in Saudi Arabia. His profession in Geography has roots in all the travels and time abroad with his mom and dad. His research in Egypt has been supported by grants from Fulbright, the American Council of Learned Societies, the American Research Center in Egypt, and the National Geographic Society Committee for Research and Exploration. In the 1990s, he served as the team leader of the Bedouin Support Program, a component of the St. Katherine National Park project in Egypt's Sinai Peninsula. In 2007, he led an effort to establish a national plan for environmental management in the United Arab Emirates. From 2009–2012 he worked with a team funded by the Norwegian Research Council, studying the interactions between nomadic pastoralists and acacia trees in Egypt and the Sudan. His current research interests include indigenous peoples' participation in national parks and other protected areas, and educational and diplomatic relations between Vietnam and the United States.

Joe is the author of the books *Bedouin Life in the Egyptian Wilderness* and *Mount Sinai* (both University of Texas Press), co-author of *The Birds of Egypt* (Oxford University Press), and co-editor of *Dangerous Harvest: Drug Plants and the Transformation of Indigenous Landscapes* (Oxford University Press).

Joe has taught graduate and undergraduate courses in world regional geography, environmental geography, the geography of the Middle East, the geography of caves, and the geography of global current events, the geographies of drugs and terrorism, and a field course on the ancient Maya geography of Belize. He has received the University of Missouri's highest teaching award, the Kemper Fellowship, as well as MU awards for leadership in international education. He has led adventure tours to remote areas in Latin America, Africa, the Indian Ocean, Asia, Europe, and the High Arctic. Joe lives in Missouri with his wife Cindy, daughters Katie and Lily, and lots of animals.

"If a man takes no thought about what is distant, he will find sorrow near at home."

—CONFUCIUS

The Earth at night. In general, the prosperous countries are lit up, while poor countries are darker. The differences between prosperous and poor countries are discussed often in the book.

C. Mayhew & R. Simmon (NASA/GSFC), NOAA/ NGDC, DMSP Digital Archive

Joe Hobbs

1 Objectives and Tools of World Regional Geography

Before you begin reading, there is an important feature of the book that you need to know about: *cross-referencing*. The book is written with global interconnections in mind. Globalization is understandable as a concept, but how exactly does it work? The page and figure numbers in the book's margins serve to tie the diverse strands of global issues together. For example, when you read in Chapter 4 that Europeans generally dislike genetically modified foods, page numbers in the margin lead you to discussions of other countries' policies on GM foods and how genetic modification can affect ecosystems. Likewise, the discussion of how genetic modification of food crops can affect ecosystems has a page number in the margin, leading you back to the discussion of Europeans and GM foods. As you read about China's economic growth and appetite for raw materials, margin numbers will take you to places all around the world where these forces come into play. I hope you will use this feature often and learn a lot from it.

chapter outline

chapter objectives

This chapter should enable you to

- Appreciate the richness of geography
- Identify the six essential elements and the five themes of geography
- Learn some key concepts in geography
- Appreciate the book's overall objectives
- Learn the basic language of maps
- Understand the "geospatial revolution," geographic information systems (GIS), and remote sensing
- See how geographic knowledge is put to work in the job market

1.1 Welcome to World Regional Geography

About every five years, the National Geographic Society does a survey to see how much Americans between the ages of 18 and 24 know about the world. The findings can be discouraging, especially when it comes to the big issues; take knowledge of Afghanistan, for example. From a base inside Afghanistan, al-Qa'ida planned its attacks against the United States on September 11, 2001. The United States responded by opening a war in Afghanistan that would become the longest in American history. This conflict often dominated the news, and almost every American knew or had seen someone deployed to Afghanistan. But where is Afghanistan? In National Geographic's most recent survey, 90 percent of the Americans polled could not locate Afghanistan on a map.

So what? Does it matter if you can't find Afghanistan on a map? Does it matter if you can't even find the United States on a map? Years ago, geography earned a reputation for forcing students to memorize lists of places and facts (•**Figure 1.1**). And the truth is, by themselves, those pieces of knowledge probably mean little. But geography is all about context and connections. Understanding *where* things are makes it much easier to appreciate and answer the *who*, *what*, *when*, *why*, and *how* questions in life, at every scale—from your daily activities to world affairs. Geography always starts with the "where" question, but it is far more interesting and important than its old reputation for memorizing places suggests. Helping you to understand contexts and relationships, geography can help you make better informed judgments and decisions. See the Try It feature on page 3.

Geographic knowledge has the power to transform our lives and contribute to the welfare of our communities and our countries. By the end of this chapter, you will know what geography is, recognize the benefits you might gain from learning world regional geography, understand the organization and objectives of this book, and learn some of the key concepts and tools of geography.

What Is Geography?

Geography, a term first used by the Greek scholar Eratosthenes in the third century B.C.E.,[1] literally means "description of the Earth" but is probably best characterized as "the study of the Earth as the home of humankind." Focusing on interactions between people and the environments in which we live, the modern academic discipline of geography has its roots in the Greek and Roman civilizations and the Scientific Revolution in Europe.

• **Figure 1.1** Geography used to be associated with memorizing mind-numbing facts. Not any more!

Geography has an important niche in modern science. Its role is well summarized by the National Geographic Society's **"six essential elements of geography."** They are as follows:

1. *The World in Spatial Terms.* Geography studies the relationships among people, places, and environments by mapping information about them into a spatial context (*spatial* means "of or relating to space").
2. *Places and Regions.* The identities and lives of individuals and peoples are rooted in particular places and in human constructs called regions.
3. *Physical Systems.* Physical processes shape the Earth's surface and interact with plant and animal life to create, sustain, and modify ecosystems.
4. *Human Systems.* People are central to geography; human activities, settlements, and structures help shape the Earth's surface, and humans compete for control of the Earth's surface.
5. *Environment and Society.* The physical environment is influenced by the ways in which human societies value and use the Earth's physical features and processes.
6. *Uses of Geography.* Knowledge of geography enables people to develop an understanding of the relationships among people, places, and environments over time—that is, of the Earth as it was, is, and might be.

Source: Geography for Life: National Geography Standards 1994, National Geographic Research and Exploration, Washington, DC, pp. 34–35.

Think about U.S. Involvement in Afghanistan

Try It

Why has it been difficult for the United States to be involved in Afghanistan? Suggest two or three reasons.

In trying to answer this important question (important partly because of the sacrifices, costs, and risks that came with U.S. involvement), it may be difficult for you *not* to use a map. You will probably consider Afghanistan in its regional context, looking at its neighbors and how it is positioned relative to major oceans and mountain ranges. You may think about some of the connections and interactions between places, cultures, beliefs, natural resources, politics, and other factors related to Afghanistan.

If it is a tough question to answer now, be patient; this book will help you answer it (on pages 240–244)—and many other questions that ask you to think critically about the world today.

After you have considered this question, you may have more empathy for a man who had to make tough choices about U.S. engagement in Afghanistan. Here is what the former U.S. Secretary of Defense Robert Gates told cadets at the U.S. Military Academy at West Point in 2011:

> Any future defense secretary who advises the president to again send a big American land army into Asia or into the Middle East or Africa should "have his head examined," as General [Douglas] MacArthur so delicately put it . . . Just think about the range of security challenges we face right now beyond Iraq and Afghanistan: terrorism and terrorists in search of weapons of mass destruction, Iran, North Korea, military modernization programs in Russia and China, failed and failing states, revolution in the Middle East, cyberpiracy . . . , natural and man-made disasters, and more. And I must tell you, when it comes to predicting the nature and location of our next military engagements, since Vietnam, our record has been perfect. We have never once gotten it right.

You may have to read those last two sentences more than once to appreciate Gates' purposeful irony. He might have wished that American presidents were advised by geographers! Geography is all about "getting it right" when it comes to understanding how the world works. Geographic knowledge of the *where, who, what, when, why,* and *how* can help guide informed decision making at all scales, from how you might get from point A to point B in town, to whether the United States should commit troops to a ground war in Asia.

Another summary of what geography is all about is known as the **Five Themes of Geography**. The National Council for Geographic Education and the Association of American Geographers developed this list. Because of its effective simplicity, some geographers prefer using it:

1. Location
2. Place
3. Human–Environment Interaction
4. Movement
5. Region

Although brief, these six elements and five themes cover a lot of ground. Geography is probably the most all-encompassing of the social sciences (a point of pride for geographers). Broadly, the discipline has two major branches, **physical geography** and **human geography**, each of which has roots and relationships with other disciplines in the social and physical sciences (•Figure 1.2). Geographers often bridge the social and natural sciences in their research, publication, and teaching (another point of pride for them). As you can see in the center of Figure 1.2, where all the components of the discipline converge, geography is almost always concerned with the theme of **human–environment interaction**. This concern has put geographers at the cutting edge of science and policy in the twenty-first century, because so many of the Earth's most pressing problems—climate change, population growth, and hunger, for example—involve the coupling of human and environmental systems.

Geographers' interests in human–environment interaction, and especially in the ways in which people are changing the face of the Earth, go way back. The great German geographer Alexander von Humboldt (1769–1859) began geography's modern era in a series of classic studies on this theme. From field observations in Venezuela, he concluded, "Felling the trees which cover the sides of the mountains provokes in every climate

Try It The Geography of Anyplace

Try using the "five themes of geography" to characterize any place. Staying with the U.S.–Afghanistan connection for now, here is an example to work from, using Ground Zero in Manhattan.

Geographic Characteristics of Ground Zero

Location: Lower Manhattan, New York City (with an exact location of north latitude: 40 degrees, 42 minutes, 43 seconds, and west longitude: 74 degrees 00 minutes, 49 seconds (later in the chapter, we will look at latitude and longitude).

Place: Formerly, office buildings and firms at the heart of one of the world's great financial centers (a reason it was targeted for destruction); now, a place of historical significance and collective grief for people of the United States.

Human–Environment Interaction: Lower Manhattan occupies low-lying ground that once was marshy swampland. Construction of the twin towers of the World Trade Center, as well as the buildings erected after the 9/11 attacks, required special foundations to keep the Hudson River's water from pouring in.

Movement: Before 9/11, the daily comings and goings of office workers in the World Trade Center; on 9/11, the diversion of airplanes to target the buildings; after 9/11, the flow of mourners, tourists, and construction crews to the site.

Region: Situated in region of the United States known as the Northeast, in a humid continental climate region (in the next chapter we will look at such physical regions).

You can use these themes to appreciate any place geographically, from the Great Pyramids of Egypt to where you are now. Try it.

two disasters for future generations: a want of fuel and a scarcity of water."[2] A century and a half later, we are von Humboldt's future generations. Look at some of the most pressing global environmental issues that concern us today: they include deforestation and shortages of fresh water.

In Humboldt's wake, other geographers in Europe and the United States wrote about environmental changes due to deforestation and the expansion of agriculture and industry. The American geographer Carl Sauer (1889–1975) wrote, "We have accustomed ourselves to think of ever expanding productive capacity, of ever fresh spaces of the world to be filled with people, of ever new discoveries of kinds and sources of raw materials, of continuous technical progress operating indefinitely to solve problems of supply. Yet our modern expansion has been affected in large measure at the cost of an actual and permanent impoverishment of the world."[3] These words have a modern ring to them, but Sauer, a geographer at the University of California–Berkeley, wrote them in 1938. Sauer is credited with founding the **landscape perspective** in American geography, based on the method of studying the transformation through time of a **natural landscape** to a **cultural landscape.** Essentially, Sauer challenged us 16
to think of what the world would look like without people and then understand what people have done to reshape the world through time. Geographers are interested in how the forces of nature and culture have been at work in the creation of the **landscape**—the collection of physical and human geographic features on the Earth's surface—and in particular the roles that human ideas, activities, and cultures play in modifying the landscape. **Culture**—the system of values, beliefs, and attitudes that shapes and influences perception and behavior[4]—underlies many of our decisions about how to use and modify the landscape. That is why geographers are so concerned with cultural features such as ethnicity, language, and religion, and why you will learn much about these issues in this book.

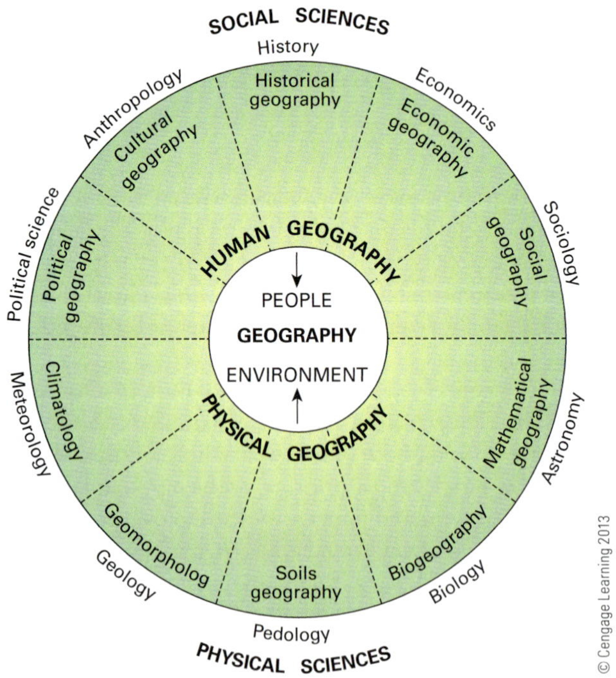

• **Figure 1.2** Selected subfields of geography. These are the main subject areas in human geography and physical geography and their links with the most closely related disciplines in the social and natural sciences.

The World Regional Approach to Geography

The world regional approach to geography ranges across the human and physical subfields of geography, synthesizing, simplifying, and characterizing the human experiences of Earth as home. It is impossible to deal with something as large and diverse as our planet without an organizing framework. World regional geography simplifies the task by dividing the world into **regions** (•**Figure 1.3** and •**Table 1.1**). These subdivisions of space are human constructs, not "facts on the

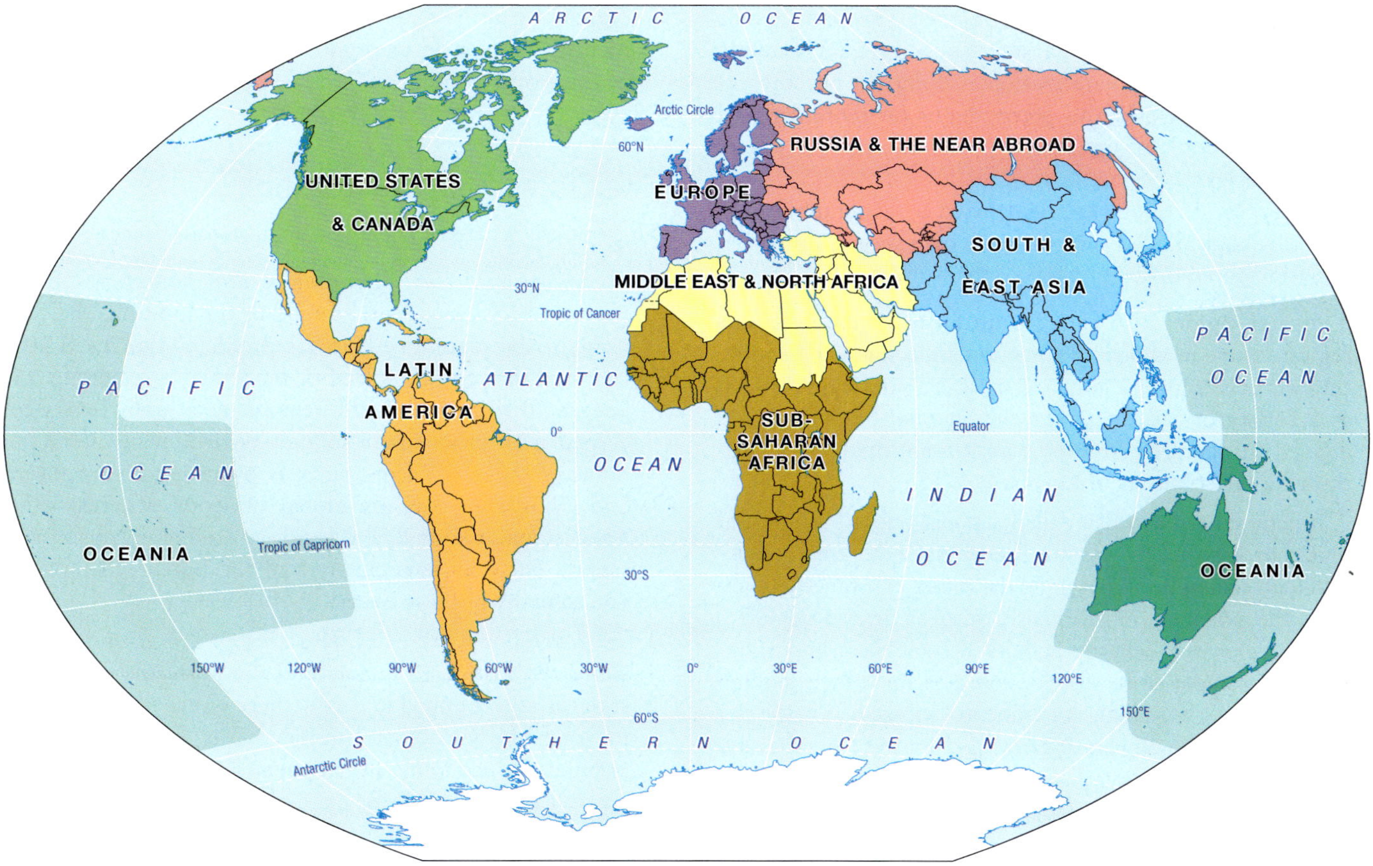

• **Figure 1.3** World regions as identified and used in this book.

Table 1.1 The Major World Regions: Basic Data

Region	Area (thousands; sq mi)	Area (thousands; sq km)	Estimated Population (millions)	Estimated Population Density (sq mi)	Estimated Population Density (sq km)	Annual Rate of Natural Increase (%)	Human Development Index	Urban Population (%)	GNI PPP Per Capita ($US)
Europe	1,959.3	5,072.0	532.3	272	105	0.0	0.873	72	31,070
Russia and the Near Abroad	8,533.2	22,100.8	282.0	33	13	0.2	0.726	64	12,300
Middle East and North Africa	5,416.1	14,027.8	503.4	93	36	1.8	0.650	62	9,460
Monsoon Asia	8,265.3	21,407.1	3,829.2	463	179	1.0	0.621	42	6,080
Oceania	3,306.8	8,564.5	37.3	11	4	1.1	0.822	67	27,440
Sub-Saharan Africa	8,655.2	22,417.2	846.5	98	38	2.5	0.423	32	2,080
Latin America	7,946.2	20,580.7	595.8	75	29	1.2	0.724	79	10,560
North America	8,403.8	21,765.8	346.3	41	16	0.5	0.909	79	44,800
World Regional Totals	**52,485.9**	**135,938.5**	**6,972.8**	**133**	**51**	**1.2**	**0.647**	**51**	**10,240**
United States (for comparison)	3,717.8	9,629.1	311.7	84	32	0.5	0.910	79	45,640

Sources: World Population Data Sheet, Population Reference Bureau, 2011; U.N. Human Development Report, United Nations, 2010; World Factbook, CIA, 2011.

ground." People create and draw boundaries around regions that share relatively similar characteristics. A region is simply a convenience and a generalization, helping us become acquainted with the world and preparing us for more detailed insights. This book recognizes eight world regions; others have more or less.

Three types of regions are recognized by geographers and used in this book. Each is helpful in its own way in conveying information about different parts of the world:

- A **formal region** (also called a **uniform** or **homogeneous region**) is one in which all the population shares a defining trait or set of traits. A good example is a political unit such as a county or a state, where the regional boundaries are defined on a map. Figure 4.1 on page 66 is a formal region map showing the countries of Europe.
- A **functional region** (also called a **nodal region**) is a spatial unit characterized by a central focus on some kind activity (often an economic activity). At the center of a functional region, the activity is most intense, whereas toward the edges of the region the defining activity becomes less important. A good example is the distribution area for a metropolitan newspaper, with the highest numbers of subscribers in the city and diminishing numbers at growing distances from the city.
- A **vernacular region** (or **perceptual region**) is a region that exists in the minds of many people. This region may play an important role in cultural identity, but does not necessarily have official or clear-cut borders. Good examples are the South, the Bible Belt, and the Rust Belt in the United States (•Figure 1.4). These regional terms have economic and cultural connotations, but 10 people might have 10 different definitions of the qualities and boundaries of these regions. Vernacular or perceptual regions, created by individuals and cultures, represent the regional identities that help us organize, simplify, and make sense of the world around us.

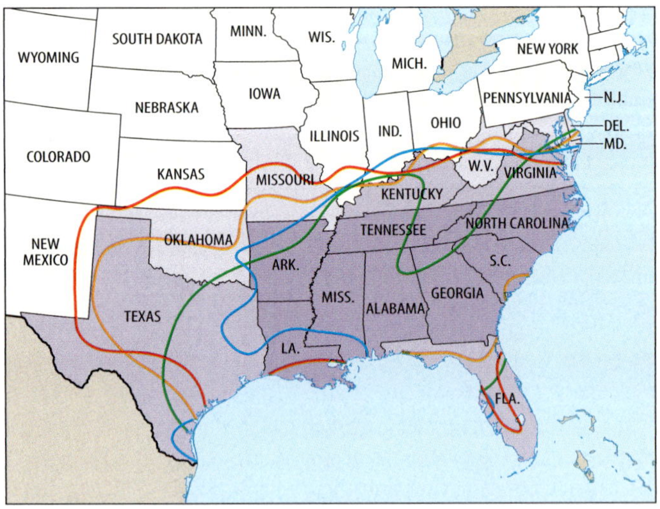

• **Figure 1.4** Definitions of a vernacular region, the American South. Purple shading represents three state-based delineations; colored lines delimit various religious, linguistic, and cultural "Souths." These are just a few of the many different interpretations of the region.

The Objectives of This Book

This book is written to help you achieve four objectives:

1. *To understand Earth's problems and potential solutions to these problems.* As the home of humankind, the Earth needs to be maintained. It has problems that need attention and that can be solved or made less threatening. Like geography in general, world regional geography is very concerned with problems in human–environment interaction. Some of these problems, such as overpopulation, poverty, and climate change, are global in scope, whereas others are national, regional, and local, or are manifested at these scales. The next two chapters of the book will introduce you to many of these issues.
2. *To develop a habit of synthesizing information to understand the world.* To understand the problems of Afghanistan (or any country) and why these issues may be relevant to your life, you must consider many factors: resources, population, economic development, ethnicity, history, and geopolitical interests, for example. Is that too much information for you to take in? Not if you have the right tool for filtering and synthesizing the information. World regional geography is that tool. Again, geography is all about context and connections. Using geography's holistic and integrative approach in a regional framework, you will look at relationships between places and people. You will be using information, techniques, and perspectives from both the natural sciences and the social sciences. You will become familiar with issues in political science, history, economics, anthropology, sociology, geology, atmospheric science, and other areas. In pulling these issues and perspectives together and finding the links among them, you will be doing geography.

   This approach will be rewarding for you in both tangible and intangible ways. Your overall university experience should become richer as you become better able to link the various courses you are taking. In the future, you will be more likely to interact with the world with new wisdom that can be rewarding both professionally and personally. More complete knowledge of the world—good geography—is also good business. In the competitive environment of the global economy, better understanding of cultures and environments throughout the world helps boost the "bottom line." You may be surprised how much your geographic knowledge will help you in your career, whatever it turns out to be.
3. *To understand current events.* This book is carefully written to set the stage of world events for you. With this book and your professor's guidance, you should become able to read or view news with a much better understanding of the issues underlying world events. Incidents like violence in

Joe Hobbs

• **Figure 1.5** Where is this place?[5] What clues on the landscape or in the man's appearance might tell you where you are?

the Middle East, earthquakes and tsunamis in the western Pacific, and disease epidemics are not random, unpredictable events. They are rooted in consistent, recognizable problems that have geographic dimensions. You should find it satisfying to feel so well informed about a problem that "suddenly" appears in the news.

4. *To develop the ability to interpret places and "read" landscapes.* As you saw in the exercise on Ground Zero, in
16 doing geography you are concerned both with **space** (the exact placement of locations on the face of the Earth) and with **place** (the imprecise but important physical and cultural contexts of a location). Place is much more subjective than space because, like a vernacular region, it often is defined by the meanings of a particular location. For example, your perceptions of New York City may be very different from those of your friend and may be shaped by personal experience in the "Big Apple" or by photographs or movies you have seen. In this book, there is much discussion of the "sense of place" that individuals and groups have about locations and regions. Perception of place can have a very strong influence on how we make decisions and interact with other people. Perception of place can even have a strong impact on world events. For example, in Chapter 6 on the Middle East, you will see how Jewish and Muslim perceptions of sacred places located within a few meters of each other in Jerusalem play crucial roles in conflict and peacemaking in the Middle East and beyond.

As an example of how you can define and identify place, study the photograph in •Figure 1.5, and try to name the place—or at least the region—where this photograph was taken. What clues in the physical and human geographies of this place help you locate it? As you use this book and progress through your course, you should be increasingly skilled in looking at images of a place with insight into many of the features that make up its identity. These include the climate, vegetation, and landforms of the physical environment and the language, religion, history, and livelihoods of the people who live in that environment. The skill of interpreting place will also serve you well when you travel. For this place, see note 5 on page 19.

1.2 The Language of Maps

How skilled are you at reading maps? What about imagining or drawing them? (See the Insights feature on page 8.) Did you ever wish you were better at these skills? This section should help you—and it will be useful as you navigate this book.

We turn now to the most important tool that geographers use to explore and explain relationships on our planet: the map. As geographers study people, places, and environments, they usually (but not always) collect and depict information that can be mapped. In other words, they are interested in the **spatial** context of things. As noted in the first essential element of geography, *spatial* means "pertaining to space."

A **map** is a representation of various phenomena over all or a part of the Earth's surface, usually rendered on a flat surface such as paper or a computer monitor. The science of making maps is called **cartography.** There are two basic types of maps: reference maps and thematic maps. **Reference maps** are concerned mainly with depicting the locations of various features, both natural and human-made, on the Earth's surface (road atlas maps are a good example, as are the opening maps for each regional chapter, such as Europe in Figure 4.1). **Thematic maps** show the spatial distribution of one or more attributes across

INSIGHTS | Mental Maps

When someone asks you, "Could you draw me a map of how to get there?" you might quickly draw some lines, write down some street names, talk about some familiar landmarks, and apologize for how crude your map is. Your map would probably end up looking very different from that of another person asked the same question. Our understanding of location is not completely objective. Each of us has a personal sense of space and place and associations with them.

A **mental map,** like a vernacular region, is a collection of personal geographic information that each of us uses to spatially organize the images and facts we have about places, both local and distant. We constantly draw upon that geographic information to make our way through daily life, and are always revising and updating that information as we succeed or fail on our way. Sometimes we use that information to create actual maps. These maps are not accurate, precise, or scientific, but they portray useful information and tell us much about the individuals and cultures that create them (• **Figure 1.A, B**).

Joe Hobbs

A

Joe Hobbs

B

• **Figure 1.A, B** The man in Figure 1.A is Saleh Ali, a Bedouin tribesman from the Eastern Desert of Egypt. During his 75 years of living, he has never used a map, but has drawn on a wealth of personal experience and oral traditions—his mental maps—to find his way across one of the world's most rugged natural environments, the Sahara. On this day, he used his "mental maps" to take the author to a valley called Wadi Abu Haadh, which, he said, was "a place of many vipers." You can see him probing the gravel under an acacia tree—using his mental map of that habitat—looking for one of these poisonous snakes. In Figure 1.B he has found one of the creatures. "Don't step there," he warned me. Can you see the head of this deadly animal?[6] See note 6 on page 19.

a given area. Thematic maps can be divided into two categories: quantitative and qualitative. Quantitative thematic maps show the spatial distribution of numerical information (such as population density or income levels, as in Figure 3.14 on page 55), whereas qualitative thematic maps display nonnumeric data (such as the distribution of climates or languages, as in Figure 2.4 on page 24.

As maps are an essential tool in the study of world regional geography, it is important that you know how to read them. The main elements of the "language of maps" are *scale, coordinate systems, projections,* and *symbolization.*

Scale

A map is a reducer; it shrinks an area to the manageable size of a chart, piece of paper, or computer monitor. The amount of reduction appears on the map's **scale,** which shows the actual distance on Earth as represented by a given linear unit on the map. A common way to depict scale is with a fraction or ratio, such as 1:10,000 or 1:10,000,000. In the fraction, one linear unit on the map (for example, 1 inch or 1 centimeter) represents 10,000 or 10,000,000 such real-world units on the ground. A **large-scale map** is one with a relatively large representative fraction (for example, 1:10,000 or even 1:100) that portrays a relatively small area in more detail. A **small-scale map** has a relatively small representative fraction (such as 1:1,000,000 or 1:10,000,000) that portrays a relatively large area in more generalized terms. Compare the two maps in •Figure 1.6. Figure 1.6a is a small-scale map showing San Francisco and surrounding parts of the Bay Area. Figure 1.6b is a large-scale map that zooms in on part of San Francisco. Remember this inverse relationship: A small-scale map shows a large area, and a large-scale map shows a small area.

Coordinate Systems

Maps cannot convey the subjective meanings associated with place, but they are very effective in conveying information about space and location. In this book, you will be concerned with two kinds of **location**: relative location and absolute location.

Relative location defines a place in relationship to other places and is information you can derive from many maps. It is one of the most basic reference tools of everyday life; you might say you live south of the city, just west of the shopping mall, or next door to a good friend. Relative location will become part of your basic geographic knowledge and your critical thinking about geography. You might look at Figure 10.1. on page 351 to see, for example, how tantalizingly close the landlocked country of Bolivia is to the Pacific Ocean, and wonder, "Why on earth is Bolivia landlocked?!" Understanding the implications of relative location will be quite useful for you in following world affairs; keep your eye on Bolivia's efforts to recover in international courts the sea access it lost to Chile through warfare in 1879.

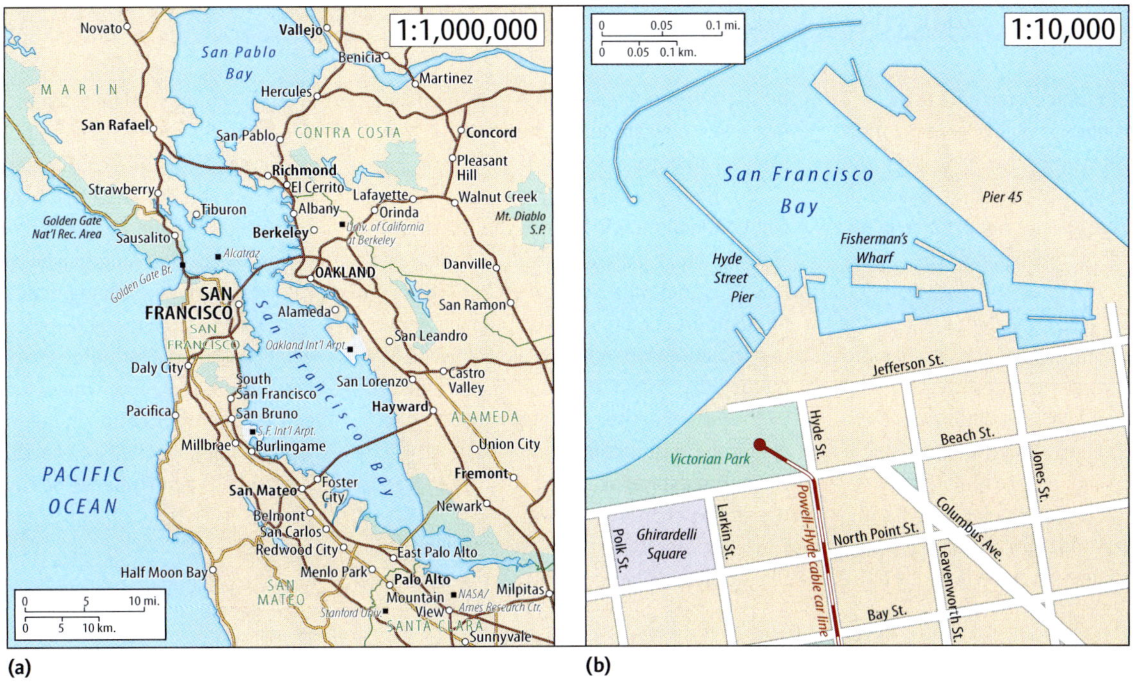

• **Figure 1.6** **(a)** Small-scale and **(b)** large-scale maps of San Francisco and environs.

Absolute location, also known as **mathematical location,** is crucial for reference maps but is not always necessary for thematic maps. **Coordinate systems** are used to determine absolute location. These coordinate systems use grids consisting of horizontal and vertical lines covering the entire globe. The intersections of these lines create addresses in a global coordinate system, giving each location a specific, unique, and mathematical placement (as appears, for example, as a waypoint in the common GPS device).

The most common coordinate system uses **parallels** of **latitude** and **meridians** of **longitude.** The term *latitude* denotes position with respect to the equator and the poles (see •Figure 1.7). Latitude and longitude are measured in **degrees** (°), **minutes** (′), and **seconds** (″). Each degree of latitude, which is made up of 60 minutes, is about 69 miles (111 km) apart; these distances vary a little because Earth is a slightly flattened (oblate) sphere or ellipsoid. Each minute of latitude, which is made up of 60 seconds, is therefore roughly a mile apart. The **equator,** which circles the globe east and west midway between the poles, has a latitude of 0°. All other latitudinal lines are parallel to the equator and to each other, which explains why they are called *parallels*.

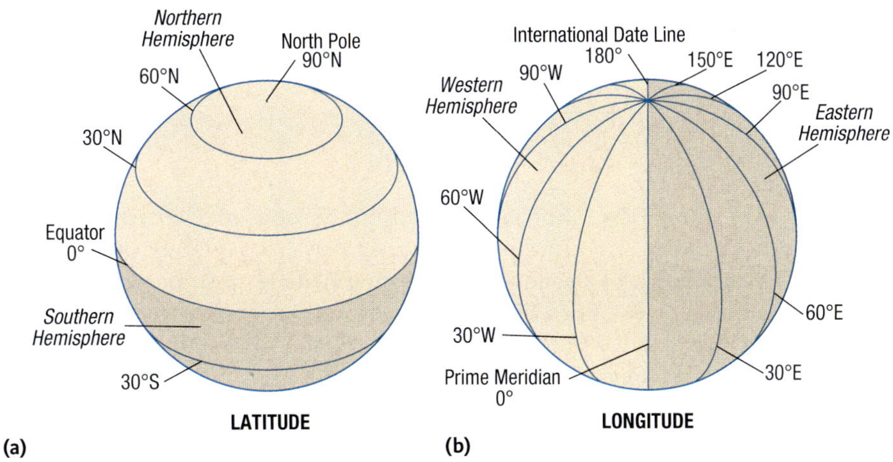

• **Figure 1.7** **(a)** Earth's lines of latitude (parallels) in increments of 30 degrees, from the equator (0 degrees) to the North Pole (90 degrees north latitude). **(b)** Earth's lines of longitude (meridians) in increments of 30 degrees.

INSIGHTS | Core Location and Peripheral Location

Among the geographical concepts used throughout this book is that of core location and peripheral location (the region of Europe is subdivided along these lines, for example). 65 Some locales have greater importance in local, regional, or world affairs because they have a central, or core, location relative to others. Other locales are less important because they are situated far from "where the action is." A comparison of two countries, the United Kingdom (U.K.) and New Zealand, provides a good example (• **Figure 1.C**). Both are island countries. Their climates are remarkably similar, although they are in opposite **hemispheres** and are about as far apart as two places on Earth can be. Westerly winds blow off the surrounding seas, bringing abundant rain and moderate temperatures throughout the year to both.

But there are important differences. The United Kingdom is located in the Northern Hemisphere, which has the bulk of the world's land (it is sometimes described as Earth's **land hemisphere**) and most of its principal centers of population and industry; New Zealand is on the other side of the equator, surrounded by the vast expanses of water in the Southern Hemisphere (sometimes known as Earth's **water hemisphere**) and off the beaten track of the globe's economic activity.

As Figure 1.C illustrates well, the United Kingdom is located near the center of the world's landmasses. Only a narrow channel separates it from the densely populated industrial areas of western continental Europe. Many major oceanic commercial routes converge on this western seaboard area of Europe. For centuries, the United Kingdom has played a major role in the economic and political development of northwestern Europe. New Zealand, meanwhile, has been a far outpost of that development and history. The United Kingdom has a central or **core location** in the modern framework of human activity on Earth, whereas New Zealand has a **peripheral location**.

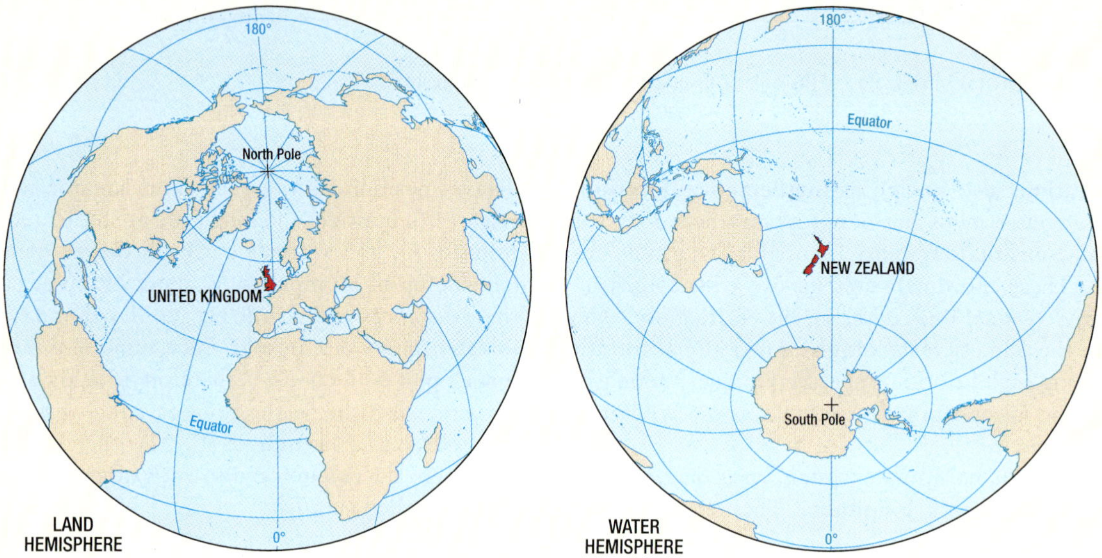

• **Figure 1.C** In the left map, note how the major landmasses are grouped around the margins of the Atlantic and Arctic Oceans. The British Isles and the northwestern coast of Europe lie in the center of the "land hemisphere," which constitutes 80 percent of the world's total land area and has about 90 percent of the world's population. In the map on the right, New Zealand lies near the center of the opposite hemisphere, or "water hemisphere," which has only 20 percent of the land and about 10 percent of the population.

Every point on a parallel has the same latitude (for example, places on the Equator in both South America and Africa are located at 0° latitude). Places north of the equator are in **north latitude.** Places south of the equator are in **south latitude.** The highest latitude a place can have is 90°N (the **North Pole**) or 90°S (the **South Pole**). Places located between the **Arctic Circle** at 65.56°N and the North Pole, and between the **Antarctic Circle** at 65.56°S and the South Pole, form the most commonly recognized boundaries of the **high latitudes**. Places located between the **Tropic of Cancer** and the **Tropic of Capricorn,** at 23.44°N and 23.44°S, respectively, are said to be in **low latitudes.** Places occupying an intermediate position with respect to the poles and the equator are said to be in the **middle latitudes.** Incidentally, there are no universally accepted definitions for the boundaries of the high, middle, and low latitudes. The northern half of the Earth between the equator and the North Pole is called the **Northern Hemisphere,** and the southern half between the equator and the South Pole is the **Southern Hemisphere.**

Meridians of longitude are straight lines connecting the poles (see Figure 1.7b). Every meridian runs due north–south. All the meridians converge at the poles and are farthest apart at the equator. Lines of longitude are not the same distance from one another across the globe, so their values vary. At the equator, the distance between lines of longitude is about 69 miles (111 km), whereas at the Arctic Circle it is only about 28 miles (45 km). The meridian most commonly used as a starting point is the one running through the Royal Astronomical Observatory in Greenwich (pronounced "Grenich"), England. Known as the Greenwich Meridian, or **prime meridian,** it has a longitude of 0°. Places east of the prime meridian are in **east longitude;** places west of it are in **west longitude.**

The meridian of 180°, exactly halfway around the world from the prime meridian, is the other dividing line between places east and west of Greenwich. All of the Earth's surface eastward from the prime meridian to 180° is in the **Eastern Hemisphere,** and all of the Earth's surface westward from the prime meridian to 180° is in the **Western Hemisphere.** This meridian of 180° has another purpose in addition to complementing the prime meridian. It serves as the **International Date Line**, where the beginning of one day and the end of another day meet. The Earth has 24 time zones, and there must be a line where the Earth's clock begins and ends. The line has a few zigzags in it for political and practical reasons (especially to fit a country or part of a country in a single time zone; see Figure 8.1, page 285). The date west of the line is one day ahead of the date east of the line. The person traveling west across the International Date Line gains a day (crossing from Monday to Tuesday, for example); and traveling east, loses a day (crossing from Monday to Sunday).

You now have the ability to create and interpret the absolute location of any spot on Earth. Try It—see the feature on this page.

Try It

Latitude and Longitude

Here is a useful exercise to ensure that you understand how latitude, longitude, and absolute location work.

On the map in • **Figure 1.D**, the latitude of Madrid, Spain, is approximately 41 degrees north latitude, 4 degrees west longitude (41°N, 4°W).

What are the approximate latitude and longitude coordinates of Oslo, Norway? In which hemispheres (north/south; east/west) is Oslo located[7]?

Understanding absolute location is a simple and indispensible part of the language of maps.

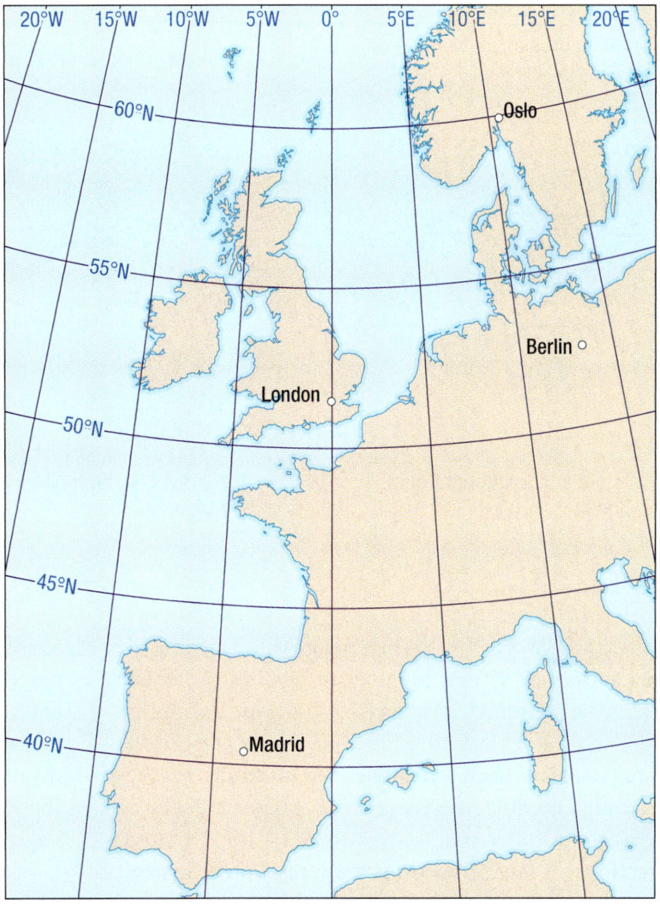

• **Figure 1.D** What are the approximate latitude and longitude coordinates of Oslo, Norway? The answer is in footnote 7 on page 19.

Projections

Because the Earth is spherical, any map created on a flat surface will inevitably have some distortion. **Map projections** are mathematical attempts to minimize this distortion. There are thousands of projections, but there is no single "correct" or "perfect" projection for any particular map. All flat maps have varying amounts of distortion among the five basic properties of a globe: area, shape, scale, distance, and direction. On large-scale maps (such as a map of a city) the distortions are small enough that they may be disregarded for most purposes, but on maps of smaller scales (countries, continents, the entire world) distortion is an increasing problem as scale becomes smaller.

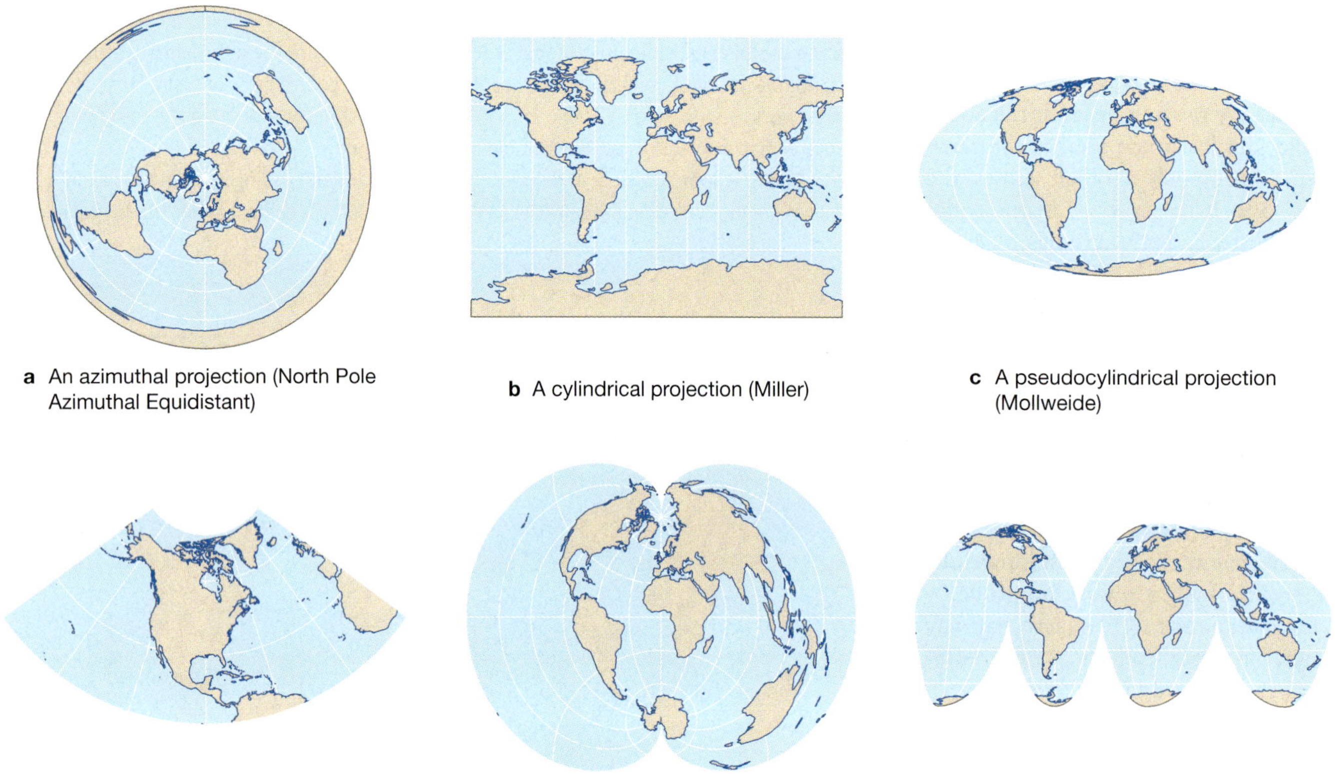

a An azimuthal projection (North Pole Azimuthal Equidistant)

b A cylindrical projection (Miller)

c A pseudocylindrical projection (Mollweide)

d A conic projection (Albers Equal-Area Conic)

e A polyconic projection (American Polyconic)

f An interrupted projection (Goode Homolosine)

• Figure 1.8 Examples of different map projections.

Most projections work by transferring the features of the spherical Earth onto a "developable surface" (a geometric surface that can be flattened into a plane without tearing or stretching) such as a plane, cylinder, or cone. These projections are referred to as **azimuthal**, **cylindrical**, and **conic**, respectively (there are also notable subsets of these categories, such as pseudocylindrical and polyconic.) See **•Figure 1.8** for examples of these projections. A few projections are interrupted, instead of contiguous; the most famous of these is Goode's Homolosine, which has a "peeled orange" look.

Projections are also classified by which metric property they preserve (or distort the least). **Conformal projections**, which include the famous Mercator Projection discussed in the feature on page 13, preserves shapes well. **Equal-area projections**, as the name suggests, preserve area. But no equal-area projection can preserve shapes, and no conformal projection can preserve areas. **Equidistant projections** preserve distance from a specific point to all other points. Map projections that do not preserve any one metric, or try to distort all properties about equally for aesthetic purposes (making the map "look right"), are called **compromise projections.** The Winkel Tripel projection used for the world maps in this book is a compromise projection.

Another important property of a map is its **orientation**, the relationship between the direction on the map and the corresponding compass directions in reality. Almost all maps place north at the top, but there are sometimes reasons to orient a map differently, and it is possible to present a different perception of geographical space by changing a map's orientation. There are some radical orientations that literally turn the world upside down—see A Different Viewpoint feature on page 14.

Symbolization

Maps allow us to get information, to see patterns of distribution, and to compare these patterns with one another. But no map can be a complete record of a given area. In a process called *cartographic abstraction*, important details are chosen by the cartographer to convey the map's information, while less relevant details are often not shown. Understanding the abstraction process is an important skill to have when using maps: the user knows that no map can be "complete," and many details must be simplified or omitted to keep a map legible. Ironically, all maps must "lie" to some degree to inform their readers. There is a fine book about the use and abuse of maps entitled *How to Lie with Maps*.

The geographic features shown on maps are represented by symbols, such as lines, fills, shapes, colors, and type. For example, reference maps can show different types of roads by using lines of differing width, and make different countries easier to tell apart by using different fill colors for each.

Thematic maps often (but not always) use just one type of symbol to display their data, and they can be classified by which symbol they use. **Choropleth maps**, the most common type of thematic map in this book, display their data by filling in political units with differing colors. A good example of a choropleth map is, again, the map of Europe in Figure 4.1 on page 66. **Isarithmic maps** do not use political units, but instead use lines or bands of color to join points of equal value across the mapped area. A map showing high temperatures or a topographic map showing contour lines are both examples of isarithmic maps, as is the map of vernacular map of the South in Figure 1.4. **Graduated symbol maps** use simple symbols such as circles or bar graphs, scaled proportionally to the quantity of the attribute being mapped. **Dot maps** use dots to represent a stated amount of some phenomenon within a political unit (for example, if one dot equaled 1,000 people, 12 dots in an area would indicate 12,000 people). **Flow maps** use arrows of various widths to indicate the movement of people or goods from one area to another (see the map of people on the move in Figure 11.3 on page 394).

1.3 Geographic Technologies and Careers

To close this chapter, we turn to some of the most innovative tools and breakthroughs in geography and consider how you (or someone you know) might become part of the action in this growing field. This section should also open your eyes to some of the newest tools that will help you have an easier and yet deeper understanding of the world's regional geographies. The opportunities that may open to you by studying geography are abundant and exciting. "We are living in the era of the geographer," Stanford ecologist Hal Mooney wrote.[8]

The Geospatial Revolution

In recent decades, technological advances in the field of geography and related sciences have accelerated so quickly that they can truly be called revolutionary. These advances are not confined to the laboratory or the library. Like information on the Web, they have been democratized: you can use the power of geographic tools on your own computer, tablet, or smartphone.

A relatively new term relates to most of these technological advances: **geospatial**. This adjective means "pertaining to the geographic location and characteristics of natural or constructed features and boundaries on, above, or below the Earth's surface, especially referring to data that is geographic and spatial in nature."[9] Even with simple paper maps, geographers have always worked with geospatial information.

A closer LOOK

The Mercator Projection

Arguably the most well-known map projection is the one developed by Gerardus Mercator in 1569. Mercator (1512–1594) developed his cylindrical, conformal projection mainly for navigational use. The **Mercator Projection**, which you can see in • **Figure 1.E**, was designed to show lines of constant compass bearing (known as rhumb lines) as straight lines, which was very important to assist sailing vessels in charting their courses. This was a significant improvement over previous projections, and Mercator projections are still used for water navigation today. However, the projection's usefulness at sea makes it largely unsuitable for other purposes, including reference world maps. In order for the rhumb lines to be shown straight, the projection must continually increase the spacing between the parallels away from the Equator. This results in enormous distortions of size approaching the polar areas (in fact, the poles themselves cannot be shown on a Mercator map as they lie at infinity). On a Mercator map, Greenland and Africa appear similarly sized, whereas in reality Africa is about 14 times larger than Greenland!

Despite the objections of cartographers for many decades, the Mercator projection is still a common sight on world maps in school classrooms, TV newscasts, and online mapping services such as Google's. Its straight lines and convenient rectangular shape make it attractive for such uses. The Mercator Projection also makes it possible to show the United States or Europe, despite their northern locations, as being essentially at the "center of the world." This is done by removing most of Antarctica but keeping much of the Arctic in the frame. Indeed, the Mercator projection's exaggeration of mid-northern latitudes made it very popular in the West during the age of European colonialism. The 45
Mercator Projection was gradually accepted as being "authoritative," and some people are still slow to abandon it for less distorted projections. Most atlases and textbooks no longer use Mercator. But it is still prevalent enough in other media that it is important to recognize this projection's drawbacks and to think critically about how a map projection can influence our perceptions of the world.

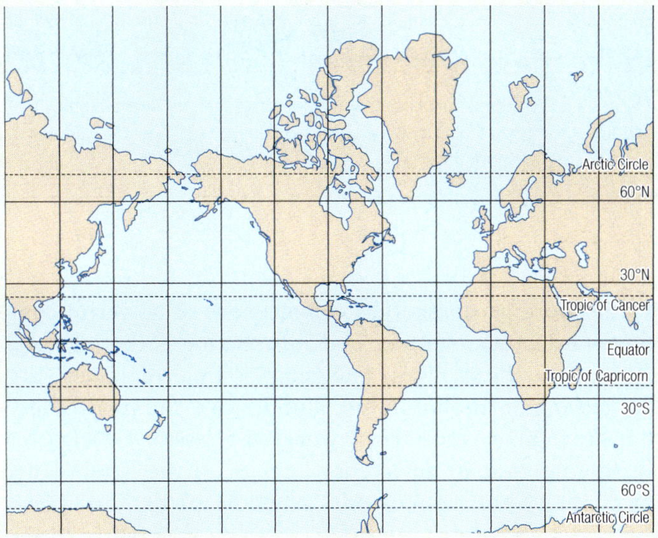

• **Figure 1.E** The Mercator Projection.

A DIFFERENT **viewpoint**

"What's Up: South!"

Now you know how world maps can look different, depending on what projection is used, and how they can be used to produce certain impressions (such as Europe or North America being at the "center of the world"). But imagine *not* being from North America or Europe and getting very tired of how your part of the world is depicted on world maps. Some Australians from "Down Under" have produced a map that challenges the basic premise that "Up Over" people have about the world: that north is at the top. These Aussie cartographers have produced a map called "What's Up: South!" with the world turned upside down (See •**Figure 1.F**). They argue that, like any ball, the Earth has no definitive top and bottom. The conventional choice to put north at the top is just that, a convention. The "South Up" map causes us to see the world in a different way. For one thing, our eyes see a world that is more watery. And we see that Australia and the other "southern" continents now dominate the world! Incidentally, there has been a long history of alternative map orientations. Classical Arab cartographers, producing the world's finest maps of their time, routinely placed south at the map's top (a good example is al-Idrisi's twelfth-century world map). For our purposes in this book, however, unless otherwise noted north is always at the top. This is true of most maps.

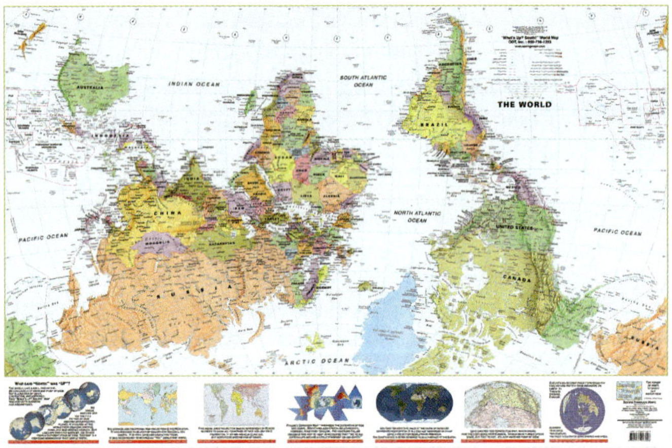

• **Figure 1.F** Who says north has to be at the top of the map? Here is an Australian projection that is literally meant to make you see the world differently.

But the word geospatial has connotations of powerful computer hardware, software, and information-gathering tools including satellites. These technologies are evolving rapidly. The geospatial revolution is still underway, and it will touch your life in many ways. Have you seen or used a GPS (global positioning system) device, in a car or in the hand? Have you looked up directions on Google Maps or MapQuest? These devices and applications provide you with simple information, but underlying them are enormous and sophisticated technologies. Two of the most important geospatial technologies are geographic information systems (GIS) and remote sensing.

GIS—the familiar acronym for **geographic information systems**—represents the cutting edge of geography today. This is one of geography's most powerful tools, and is the field's leading area of growth and employment because of its value in addressing so many real-world problems. GIS is quite complex, but in essence it is a computer-based tool that allows people to create, view, manipulate, analyze, and store geospatial data. These data can be far more than just the latitude–longitude coordinates of a feature on the Earth's surface—a mouse-click on a road, for example, will bring up a box displaying the road's name; its length and width; when it was last repaved; its speed limit; whether it is owned by the city, county, or state; the amount of traffic it receives in a certain time frame; and so forth. These different types of information, called attributes, do not magically appear, but must be created by people who spend an enormous amount of time and energy to compile them from a variety of sources.

GIS data are created and displayed in "layers"—databases storing the locations and attributes of features belonging to a single theme (for example, with layers showing potential customers, the streets on which those customers live, their land ownership, the land's elevation, and land uses in their community, as in •**Figure 1.9**). A GIS can have dozens of layers displaying all kinds of geospatial data for a given area. GIS users can select and highlight certain attributes in various layers that relate to their interests or analysis, while turning off those attributes that are not needed. The ability to selectively

• **Figure 1.9** Geographic information systems (GIS) create and use layers of spatial data. GIS data, images, and models have an enormous range of applications.

INSIGHTS | GIS in Action

GIS makes it possible to see geographic patterns, problems, and connections easily and efficiently—a capability that Esri, the world's largest GIS software maker, calls "The Geographic Advantage."[10] One of the best ways to see what GIS can do is to visit Esri's Web site, http://www.gis.com, where you will see an amazing number of ways that GIS is being used all around you and all around the world. Here are a few examples of the geographic advantage:

- *Natural Resource Management.* In forestry, caring for existing and future trees ensures a steady supply of wood for the world's building needs. GIS provides tools to help determine where to cut today and where to seed tomorrow while minimizing negative impacts.
- *Business.* Every day, businesses deliver goods and services to clients all around a city. Each truck driver needs a route of how to most efficiently visit each client. GIS provides tools to create efficient routes that save time and money and reduce pollution.
- *Defense.* In the military, leaders need to understand terrain to make decisions about how and where to deploy their troops, equipment, and expertise. They need to know which areas to avoid and which are safe. GIS provides tools to help get personnel and materials to the place where they can best do their job.
- *Emergency Preparedness and Response.* During floods and hurricanes, emergency response teams save lives and property. GIS provides tools to help locate shelters, distribute food and medicine, and evacuate those in need.
- *Communications and Media.* In telecommunications, when phone service is out, it means part of the network may be disconnected. GIS provides tools to help find out what part of the network is affected and brings that information to the field so workers can get everyone talking again.
- *Planning.* Planners of all kinds—business analysts, city planners, environmental planners, and strategists from all organizations—use GIS to lay out a framework so that growth can occur in a managed way. As you will see in Chapter 7, GIS can even help us plan for a warmer world in which rising sea levels threaten to flood coastal cities and farms. 260

view certain types of geospatial data, as well as to add other data (such as from a handheld GPS device, from a digitized paper map, or from an image taken by an orbiting satellite), allows users to see spatial relationships with extraordinary ease and clarity. The knowledge they gain from understanding geographic relationships is extremely valuable for decision making and explains why GIS is known as a "critical thinking technology" (see Insights, "GIS in Action" on this page). GIS, in other words, is a high-tech example of what geography is all about.

Remote sensing, also known as *Earth observation*, is the science of acquiring information about the Earth's surface without being in direct contact with it. Most remote sensing data is obtained by aerial photography (cameras mounted on airplanes taking pictures of the ground) or by sensors on Earth-orbiting satellites. Remote sensing is not limited to cameras that capture visible light; much important information about processes and features on the surface or in the atmosphere is gleaned from satellite sensors that can "see" other parts of the electromagnetic spectrum such as infrared and microwave wavelengths. Radar (which measures the reflection of satellite-emitted radio waves bouncing off of ground features) and light detection and ranging ([LIDAR], which uses light in the same way) are also examples of remote sensing.

Remote sensing is an exceptionally good tool for helping geographers understand how people and natural processes modify the Earth. This book uses remote sensing to introduce you to many places, patterns, and problems. "A picture is worth a thousand words," the adage says, and a remote sensing image often goes far in describing a landscape. For example, •**Figure 1.G** (a remotely sensed image taken on April 4, 2010, by a satellite 425 miles above the Earth), is a fine depiction of the cultural landscape of Sendai, in northeast Japan. Just inland from the concrete wave breakers and scalloped shoreline, people have built homes between a greenbelt of trees and an irrigation canal. The canal is a straight and uniform feature—a good indication that it is the handiwork of people rather than of nature. Bridges and streets also stand out as built features. Continuing westward (left) across the image, you see many more homes, including some nestled against a patch of trees (the Japanese prize their forests, and go to great lengths to protect them). Finally, at the far left of the image the long brown rows of empty land are fields for wet rice, which would become green when planted and irrigated from the nearby canal. This satellite image provides an excellent snapshot of how the people of Japan modify nature to suit their needs and desires.

Illustrating dramatically how remote sensing allows us to appreciate landscape change, the next image (Figure 1.H) shows how nature reclaimed Sendai. On March 11, 2011, a magnitude 9.0 earthquake thrust a section of the Earth's crust upward from the seafloor about 80 miles from Sendai. The displaced water roared ashore as devastating

Try It Reading the Landscape with Remote Sensing

Remote sensing images have a way of bringing out the detective in all of us. Compare Figures G and H, examining and describing the details of this landscape's transformation. What changes has this natural disaster wrought on Sendai?

GeoEye/EyeQ

G

GeoEye/EyeQ

H

• **Figure 1.G, H** The shoreline of Sendai, Japan, before and after the gigantic earthquake and tsunami on March 11, 2011.

tsunami waves as high as 40 feet. The results are breathtaking: almost every feature on the landscape, both natural and cultural, has been destroyed or transformed. See the Try It feature on this page, which invites you to interpret these remote sensing images, and A Closer Look feature on
254 page 17.

Careers in Geography

This striking poster in •**Figure 1.10** was produced by the Association of American Geographers (AAG). The employment trends in favor of geography since 2000 have been remarkable. The U.S. Department of Labor has identified geospatial technology as one of the most important emerging and evolving fields in the technology industry. In 2010, this agency projected "much faster than average" growth—more than 20 percent—in jobs for geographers, geoscientists, cartographers, urban and regional planners, and other geographic professionals, through the year 2018.[11]

Much of this job growth will occur in professions in which people feel like they can have a positive impact. The Association of American Geographers had these observations of global trends that are contributing "to a renaissance of geography and its potential for making a difference in society and the world:"

> These include globalization at an increasing pace and scale, phenomena that compel greater understanding of the world, places, people, and natural systems that affect us as a planet and as global citizens and consumers. It includes a recent proliferation of geographic technologies, once fairly obscure and now pervasive in our daily lives, such as GPS in cell phones and cars, online mapping at your fingertips, cable news reports using spatial visualizations, and many more applications in modern business and government services that underlie operations, planning, and progress in all sectors everywhere we live and work. It also includes an academic trend toward greater interdisciplinarity, especially a renewed focus on big questions that matter but that require a breadth of knowledge and multiple fields to tackle. Geography's long-standing intellectual traditions in crossing those usual disciplinary boundaries are now better understood, increasingly seen as relevant and more widely respected in scholarly circles. These trends have produced unprecedented growth in the field.[12]

More and more students are graduating with geography degrees—bachelors, masters, and doctoral—and more of them are finding jobs. Here are just a few examples of starting jobs of recent graduates from the University of Texas-Austin Department of Geography. B.A. students became a Nature Conservancy preserve manager, a transportation planner, a natural science teacher, and a geospatial engineer. Masters students became a U.S. Geological Survey water specialist, the chief executive officer of an insurance company in the West Indies, and a vice president for market research in real

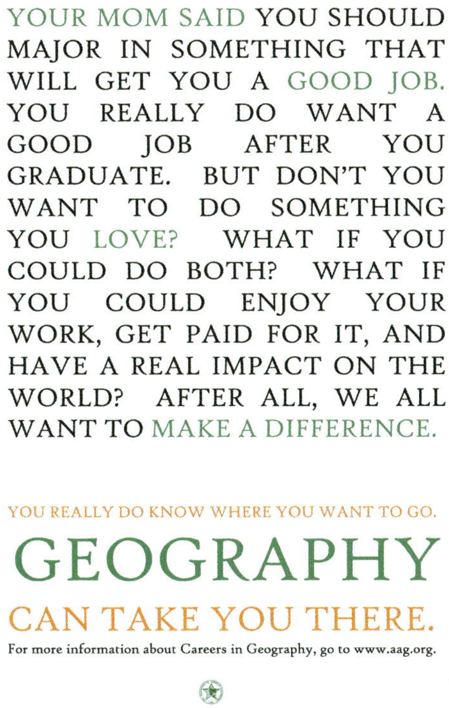

• Figure 1.10 Geography is awesome.

Source: Dr. Patricia Solís, Association of American Geographers, copyright registered 2004. Reprint permission granted for educational and dissemination purposes only; please do not reprint, translate, or otherwise alter without express written permission by the author.

estate.[13] Many Ph.D. graduates became geography professors (ask your prof which of the 60 specialty groups of the AAG she or he belongs to), whereas others took executive positions in GIS departments and companies or became decision makers in foundations and research institutions all around the world. But even if you are not a geography major or you have not yet seen an advertisement proclaiming "Geographer Wanted," chances are that many professional opportunities will arise if you use geography's tools to grow knowledge of the world in which you live. This book will help you do just that.

A closer LOOK

Google Earth

I would like to recommend two tools that can greatly enhance your understanding and enjoyment of world regional geography. One is a globe. As we saw earlier, flat maps distort representations of the Earth, but a globe puts a representation of the true world in your hands. If you keep a globe handy, and use it as you work your way through the book, it will help you learn more about the Earth and remember what you learn.

The other tool is a virtual globe, map, and geographic information software program like Google Earth or NASA's World Wind. If you have not already done so (and if you have permission to do so), on the PC or Mac, navigate to http://www.earth.google.com (apps are also available for most smartphones and tablets). Download the free Google Earth application, open it, bookmark it, and use it often—*PLEASE!* With Google Earth you are in control of an easy-to-use but incredibly powerful geospatial toolkit. It is so powerful, in fact, that it enables you to become a rather effective spy: you can see not only your home or apartment, but top secret North Korean and Iranian nuclear facilities. A Russian intelligence official grumbled of Google Earth, "Terrorists don't need to reconnoiter their target. Now an American company is working for them!"[14]

By the way, when I photograph a bridge, power plant, or dam (as I do often, thinking I might use the photo in one of my classes, or for this book), I am sometimes detained and questioned by security guards. That is ironic, considering what Google Earth can do.

Google Earth's base map is composed of many thousands of remotely sensed images, from both aerial photography and orbiting satellites, put together as a giant mosaic. And although Google Earth is not a full-fledged GIS, its "Layers" sidebar works in a similar fashion: you can turn on and off certain features like country boundaries and road networks. With the "Weather" layer, you can get real-time views of storm systems and temperatures. How hot or cold is it in Arabia or Siberia right now? It's easy to find out. You might also be interested in the "Global Awareness" layer that lets you explore some of the world's environmental problems. And then there is the "Street View" layer for many locations, infamous for news stories like this: "A woman, checking out a female friend's house on Google Maps, was surprised to see her husband's Range Rover out front. A divorce is underway."[15]

summary

- A recent study showed that Americans generally have poor knowledge of world geography.
- There are six essential elements of the national geography standards: the world in spatial terms, places and regions, physical systems, human systems, environment and society, and the uses of geography.
- Geography means "description of the Earth" and is also defined as "the study of the Earth as the home of humankind."
- The four main objectives of the text are for readers to understand Earth's problems and potential solutions for these problems; develop a habit of synthesizing information to understand the world; understand current events; and develop the ability to interpret places and "read" landscapes.
- Maps are the geographers' most basic tools. The basic language of maps includes the concepts and terms of scale, coordinate systems, projection, and symbolization. Maps can depict spatial data in a variety of ways.

- Individuals and cultures generate their own unique "mental maps." Regions are in effect mental maps that help us make sense of a complex world. Modern geographic thought derives from a long legacy of interest in how people interact with the environment. The dominant approach has been to understand how people have changed the landscape or the face of the Earth.
- The discipline of geography may be divided into regional and systematic specialties, with the systematic fields having the most followers. Their concerns overlap many disciplines in the natural and social sciences. Geographers are employed in many private and public capacities. The strongest growth area with the most jobs is in geographic information systems (GIS). The "geospatial revolution" employs many geographic tools and touches many parts of our daily lives.

Key Terms + Concepts

absolute location (p. 9)
Antarctic Circle (p. 10)
Arctic Circle (p. 10)
cartography (p. 7)
choropleth map (p. 13)
coordinate systems (p. 9)
core location (p. 10)
cultural landscape (p. 4)
culture (p. 4)
degrees (p. 9)
dot map (p. 13)
equator (p. 9)
five themes of geography (p. 3)
flow map (p. 13)
formal region (p. 6)
functional region (p. 6)
geographic information systems (GIS) (p. 14)
geography (p. 2)
geospatial (p. 13)
graduated symbol map (p. 13)
hemisphere (p. 10)
  Eastern Hemisphere (p. 11)
  land hemisphere (p. 10)
  Northern Hemisphere (p. 10)
  Southern Hemisphere (p. 10)
  water hemisphere (p. 10)
  Western Hemisphere (p. 11)
homogeneous region (p. 6)
human–environment interaction (p. 3)
human geography (p. 3)
International Date Line (p. 11)
isarithmic map (p. 13)
landscape (p. 4)
landscape perspective (p. 4)
large-scale map (p. 8)
latitude (p. 9)
  high (p. 10)
  low (p. 10)
  middle, (p. 10)
  north (p. 10)
  south (p. 10)
location (p. 8)
  absolute location (p. 9)
  relative location (p. 8)
longitude (p. 9)
  east (p. 11)
  west (p. 11)
map (p. 7)
map projection (p. 11)
  azimuthal (p. 12)
  compromise (p. 12)
  conformal (p. 12)
  conic (p. 12)
  cylindrical (p. 12)
  equal-area (p. 12)
  equidistant (p. 12)
  Mercator (p. 13)
mathematical location (p. 9)
mental map (p. 8)
meridian (p. 9)
  prime meridian (p. 11)
minutes (p. 9)
natural landscape (p. 4)
nodal region (p. 6)
North Pole (p. 10)
orientation (p. 12)
parallel (p. 9)
perceptual region (p. 6)
peripheral location (p. 10)
physical geography (p. 3)
place (p. 7)
reference map (p. 7)
region (p. 4)
remote sensing (p. 15)
scale (p. 8)
seconds (p. 9)
six essential elements of geography (p. 3)
small-scale map (p. 8)
South Pole (p. 10)
space (p. 7)
spatial (p. 7)
symbolization (p. 12)
thematic map (p. 7)
Tropic of Cancer (p. 10)
Tropic of Capricorn (p. 10)
uniform region (p. 6)
vernacular region (p. 6)

Review Questions

1. What is geography? What are some of its characteristic approaches?
2. What are the six essential elements of geography, as defined by the National Geographic Society? What does each element indicate about geography's concern with space, place, or the environment? What are the five themes of geography? Are you comfortable applying them to your knowledge of places?
3. What does *spatial* mean, and how does geography's interest in space differentiate it from other disciplines?
4. What geographic features make the United Kingdom and New Zealand different?
5. What are the major terms and concepts associated with scale, coordinate systems, projections, and symbolization?
6. Why is a map made with the Mercator projection more suitable for navigation than a map made with other projections?
7. What is the difference between a dot map and a choropleth map?
8. What is a mental map?
9. What is GIS, and what typically makes it different from old-fashioned manual cartography? What are some applications of remote sensing? How can modern geospatial tools help us with longstanding concerns in geography?
10. What do geographers do for a living?

Notes

1. The abbreviation B.C.E. stands for "before the Common Era," which is a reference to the dating system invented by European Christians that sets the birth of Jesus Christ as year 1. In Christian cultures, dates before that year are expressed as B.C., meaning "before Christ," and later years are identified as A.D., which stands for *anno Domini* (Latin, "in the year of our Lord"). Religion-neutral dating systems such as the one used in this book employ B.C.E. (before the Common Era) and C.E. (Common Era), respectively, but the years are numbered the same in the two systems.
2. Quoted in Geoffrey J. Martin and Preston E. James, *All Possible Worlds: A History of Geographical Ideas* (New York: Wiley, 1993), p. 150.
3. Quoted in Andrew Goudie, *The Human Impact on the Natural Environment* (Oxford: Blackwell, 1986), p. 6.
4. The anthropologist Kathleen A. Dahl authored this definition of culture.
5. You are in Monument Valley, northeastern Arizona near the Utah border, in the southwestern region of the United States. Place clues include an arid landscape, sandstone buttes that you have probably seen in American Western movies, and a Native American of the Navajo group, wearing typical Navajo turquoise and silver jewelry.
6. The viper's triangular head is just left of the center of the photo. This snake has buried himself in the shallow sand and gravel, a "virtual land mine" ready to explode on its prey or foe: a mouse, a Bedouin, or a geographer.
7. Using this map, you should determine Oslo's location as about 60 degrees north, 11 degrees east (60°N, 11°E). Oslo is therefore in the Northern Hemisphere, the Eastern Hemisphere, and the land hemisphere.
8. Committee on Strategic Directions for the Geographical Sciences in the Next Decade, National Research Council, 2010, Understanding the Changing Planet: Strategic Directions for the Geographical Sciences, Washington, DC: The National Academies Press.
9. http://www.dictionary.com.
10. Esri, at http://www.gisday.com/cd2009/fliers/what_is_gis.pdf.
11. http://www.aag.org/cs/careers
12. http://www.aag.org/cs/jobs_and_careers/geography_can_take_you_there
13. Information courtesy of Bill Doolittle, University of Texas at Austin.
14. Lieutenant General Leonid Sazhin, Federal Security Service, quoted in Katie Hafner and Saritha Rai, "Google Offers a Bird's-Eye View, and Some Governments Tremble," *New York Times*, December 20, 2005, p. 1.
15. http://gawker.com/5191459/cheating-husband-said-caught-via-google-street-view.

Online Resources

CourseMate: Make the most of your study time by accessing everything you need to succeed in one place. Read your textbook, take notes, review flashcards, watch videos, complete activities, take practice quizzes, and more—online with CourseMate. Log in at **www.cengagebrain.com**.

"The acquisition of wisdom is above that of pearls. The topaz of Ethiopia cannot equal it, Nor can it be valued in pure gold. Where then does wisdom come from? And where is the place of understanding?"

—Job 28

Joe Hobbs

Joe Hobbs

Russia's Arctic wilderness of Franz Josef Land, not far from the North Pole. The dark patches of ground are permafrost, or "permanently" frozen ground. But as the Arctic warms, permafrost is melting and adding greenhouse gases to the Earth's atmosphere. The Arctic is the "canary in the coal mine" of climate change.

2

Physical Processes and World Regions

Many issues in world regional geography relate to the interaction of people and the natural environment, and how these relationships have changed the face of the Earth. Physical processes are the main focus in this chapter, and human processes the focus in Chapter 3. But these processes are so intertwined that human forces receive much attention in this chapter, and natural forces much attention in the following chapter.

This discussion of physical geography introduces you to the four "spheres" that make up the Earth's habitable environment: the **lithosphere**, its outer "rind" of rock; the **hydrosphere**, made up of all the world's water features; the **atmosphere**, the layer of gases surrounding the Earth; and the **biosphere** (also known as the ecosphere), which is the global ecological system including all the relationships played out among the lithosphere, hydrosphere, and atmosphere. Earth's lithosphere is a work in progress, and you will see how its changing surfaces provide both opportunities and threats to people. We will consider the climate and vegetation types that play such large roles in human activities, and appreciate the rich diversity of wild plant and animal species. We will look briefly at the planet's often-overlooked oceans and the resources they hold. Finally, we will examine how the climate may be changing, what these changes may mean for life on Earth, and what people can do to prevent or lessen some of the most dangerous climatic changes.

chapter outline

chapter objectives

This chapter should enable you to

- Understand the tectonic forces behind some of the world's major landforms and natural hazards
- Recognize consistent global patterns in the distribution of vegetation types and climates
- Identify the natural areas most threatened by human activity and explain how natural habitat loss may endanger human welfare
- Appreciate the important roles of the world's oceans
- Describe the potential impacts of global climate change and international efforts to prevent them

2.1 Geologic Processes

The Earth's surface is in motion. Those of you living on the West Coast of the United States have probably had some personal experience of the unforgettable sensation of the ground moving beneath you in a mild "temblor" or even a large earthquake. These events reflect processes that are constantly
16, changing the face of the Earth. Here we explore just one set
254, of geological processes—arguably the most important of them
261 all because of its role in shaping global landforms—plate tectonics.

Plate Tectonics

About a century ago, a German geologist named Alfred Wegener came up with what seemed to many like an outlandish theory known as **continental drift.** Pointing to such things as the jigsaw puzzle–like geometry of Africa's west coast and South America's east coast, he proposed that the continents were once joined in a supercontinent (which he named Pangaea) but that they drifted apart over time. However, Wegener was not able to explain the forces behind these movements, and he was derided by a scientific community that seemed unwilling to accept anything other than a *terra firma* that was truly firm. Peers pronounced his conclusions "utter damned rot" and "mere geopoetry."[1] But later discoveries in deep-sea science established Wegener's basic proposition as factual, and today a good deal is known about how continental drift occurs.

The Earth's lithosphere, which varies from 50 to 125 miles in thickness, is made up of about a dozen giant and several smaller sections of rock called **plates.** These plates move in various directions in processes known collectively as **plate tectonics** (•Figure 2.1). On the ocean floors in places such as the Mid-Atlantic Ridge and the East Pacific Rise, new lithosphere is "born" as molten material rises from the Earth's mantle and cools into solid rock (•Figure 2.2). Plate tectonics are often explained by the useful analogy of a conveyor belt (the convection cell in Figure 2.2) in constant motion. On either side of the long, roughly continuous ridges, the two young plates move away from one another, carrying islands with them; this process is called **seafloor spreading.**

Seafloor spreading has few impacts on people, but when the Earth's plates collide, there is cause for great concern: such **tectonic forces** are among the planet's greatest natural hazards. The most dangerous tectonic forces are related to **seismic activity** (*seismic* refers to Earth vibrations, mainly earthquakes) that causes earthquakes and **tsunamis,** and the **volcanism** (movement of molten earth material, especially in volcanoes) often associated with plate movements. The globe's most active and deadly realm of tectonic activity is the so-called "Ring of Fire" on the rim of the Pacific Ocean (see Figure 2.1 and page 358).

Plates collide and converge in different ways and with different consequences; among the most notable consequences are the
building of Earth's great mountain ranges. In some parts of the 16,
world—off the east coast of Japan, as we saw in Chapter 1, and 254, 358,
off the coast of the U.S. Pacific Northwest, for example—one 396
plate "dives" below another in a process known as **subduction** (see Figure 2.2). The descending lithosphere is melted again as it dives into the Earth's superheated mantle along a deep linear feature known as a **trench** (for example, the Mariana Trench off Japan). Subduction is another stage along the "conveyor belt" process that will eventually see this material recycled as newborn lithospheric crust.

This subduction process releases enormous amounts of energy. Periodically, the great stress of one plate pushing beneath another is released in the form of an earthquake. The world's largest recorded earthquakes—registering 9.5 (Chile, 1960), 9.2 (United States, 1964), 9.1 (Indonesia, 2004), and 9.0 (Japan, 2011), respectively, on the **Richter scale,** which measures the strength of the earthquake at its source—have
struck along these subduction zones. This sudden displace- 214,
ment of a section of oceanic lithosphere is also what triggers 358,
a tsunami, one of nature's most powerful and destructive 396
processes. See pages 17, 254, and 288 for more insight into tsunamis.

In other places where they meet, lithospheric plates grind and slide along one another, as in California and Turkey. The processes of rock crowding together or pulling apart along these fracture lines is known as **faulting,** with movement along various kinds of **faults** causing earthquakes, the emergence of new landforms, and other consequences.

Volcanism generally takes place along and near **subduction zones** (see Figure 2.2) and also in the world's several dozen geologic hot spots where molten material has broken through the crust as a "plume" (as in Yellowstone National Park and in Hawaii—see page 288). Despite posing a host of natural hazards

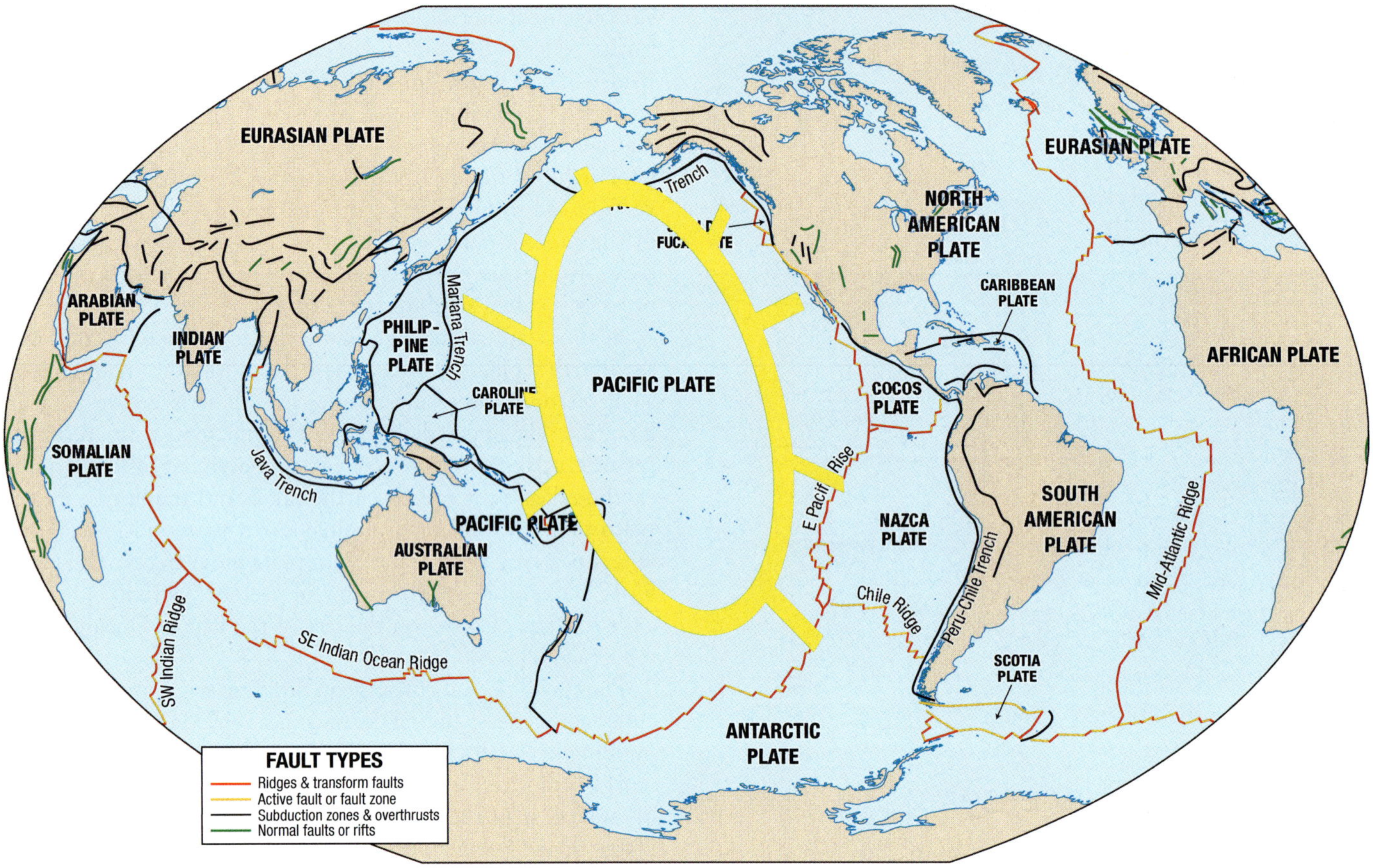

• **Figure 2.1** Major tectonic plates and their general direction of movement. Earthquakes, volcanoes, and other geologic events are concentrated where plates separate, collide, or slide past one another. Where they separate, rifting produces very low land elevations (well below sea level at the Dead Sea of Israel and Jordan, for example) or the emergence of new crust on the ocean floor (in the middle of the Atlantic Ocean, for example). The yellow ring points to the rough outline of the "Ring of Fire."

Source: Adapted from NASA, "Global Tectonic Activity Map of the Earth," DTAM-1, 2002.

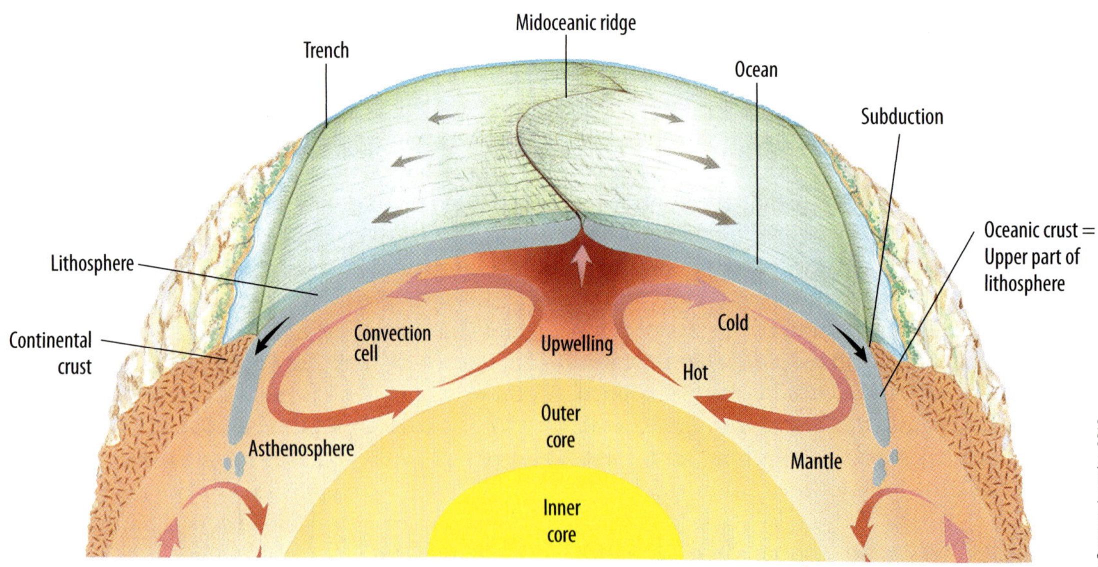

• **Figure 2.2** Generalized cross section of the Earth, showing its main concentric layers and the process by which its lithosphere is recycled.

to people living on their slopes or downwind—including pyroclastic flows (fast-moving currents of rock fragments, hot gases, and ash), lava flows, ashfalls, and other dangers—volcanoes have beneficial qualities. Volcanic rock usually breaks down to form fertile soils (as in Ethiopia and the Nile Valley), and the range of climate and vegetation types on their slopes creates op-
181, portunities for growing crops and raising livestock.
311 We turn now to global patterns of climate and vegetation.

2.2 Patterns of Climate and Vegetation

"Everybody talks about the weather, but nobody does anything about it," Mark Twain reportedly said. We will see in this chapter that we have actually done a lot about the weather. First, we look at natural patterns in vegetation and climate that present different opportunities for people around the world. Once you recognize these patterns, you will be more prepared to explore all the world's regions.

As you experience a warm, dry, cloudless summer day or a cold, wet, overcast winter day, you are encountering the **weather**—the atmospheric conditions occurring at a given time and place. **Climate** is the average weather of a place over a long time period. Along with surface conditions such as elevation and soil type, climatic patterns have a strong correlation with patterns of natural vegetation and in turn with human opportunities and activities on the landscape.

Precipitation and temperature are the key variables in weather and climate. Warm air holds more moisture than cool air, and **precipitation**—rain, snow, sleet, and hail—results from processes that cool the air to release moisture. Water is, of course, essential for life on Earth. In this book, you will see many examples of how the precious resource of fresh water is increasingly
imperiled and contested. Some geographers would tell you that 180,
the map of the world's precipitation—•**Figure 2.3**—is the most 250, 259,
important of all maps in understanding life on Earth. 265,
Combinations of precipitation, temperature, latitude, and 436
elevation produce a great variety of local climates. Geographers group these local climates into a number of major climate types, each of which occurs in more than one part of the world (•**Figure 2.4**) and is associated closely with other types of natural features, especially vegetation. Geographers recognize 10 to 20 major types of terrestrial ecosystems, called **biomes,** which are categorized by dominant types of natural vegetation (•**Figures 2.5** and **2.6**). When you are using maps of biomes, it is important to keep in mind that these depict vegetation that would dominate in the absence of human activity. For example, Figure 2.5 depicts a vast, unbroken area of temperate mixed forest in the eastern United States, but of course much of this region has been

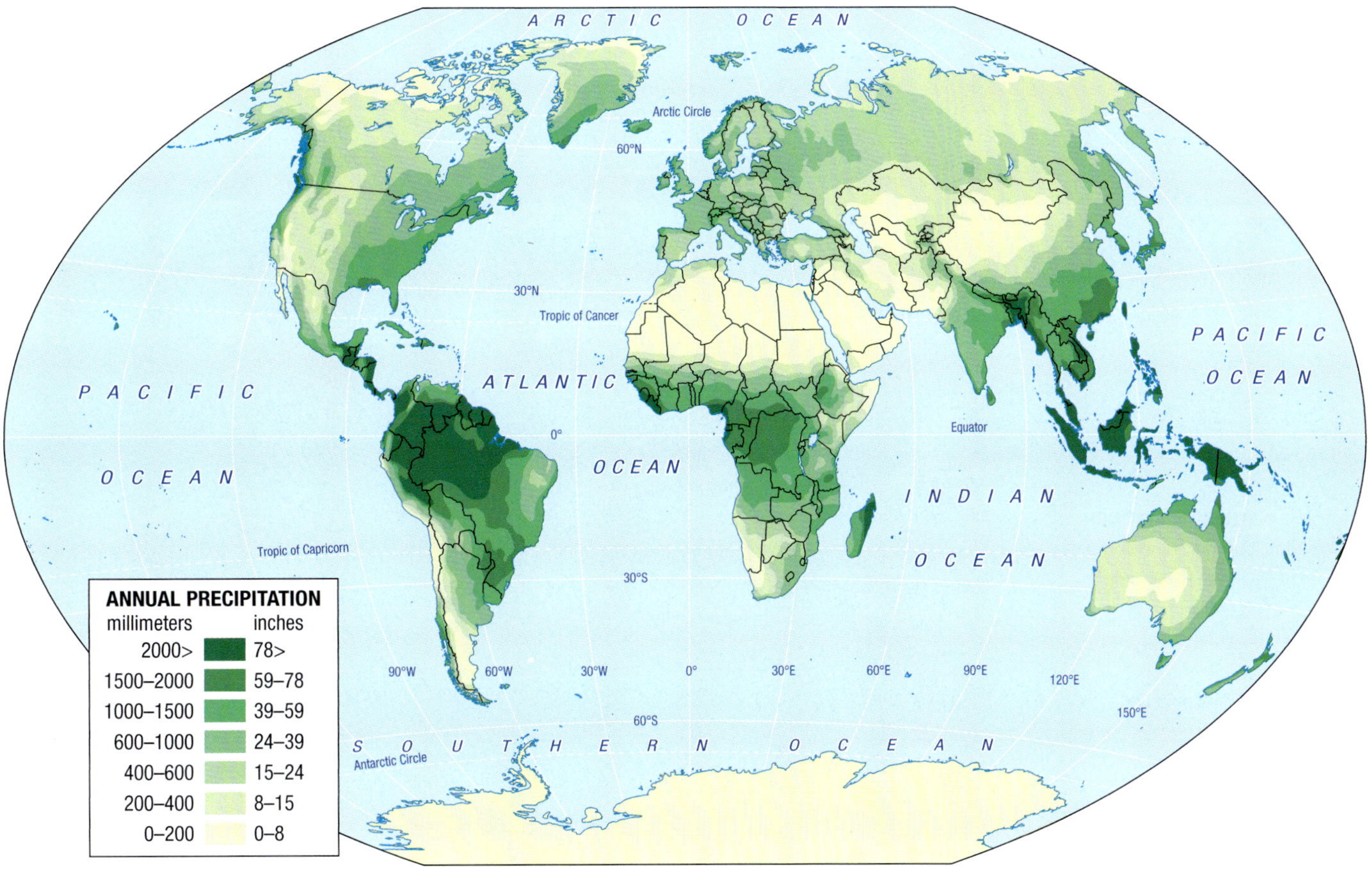

• **Figure 2.3** World precipitation map. Is this, as some authorities say, the most important of all world maps?
Source: Adapted with permission from UNEP/GRID and UEA/CRU, "Mean Annual Precipitation," GRID - Geneva Data Sets - Atmosphere, GNV174.

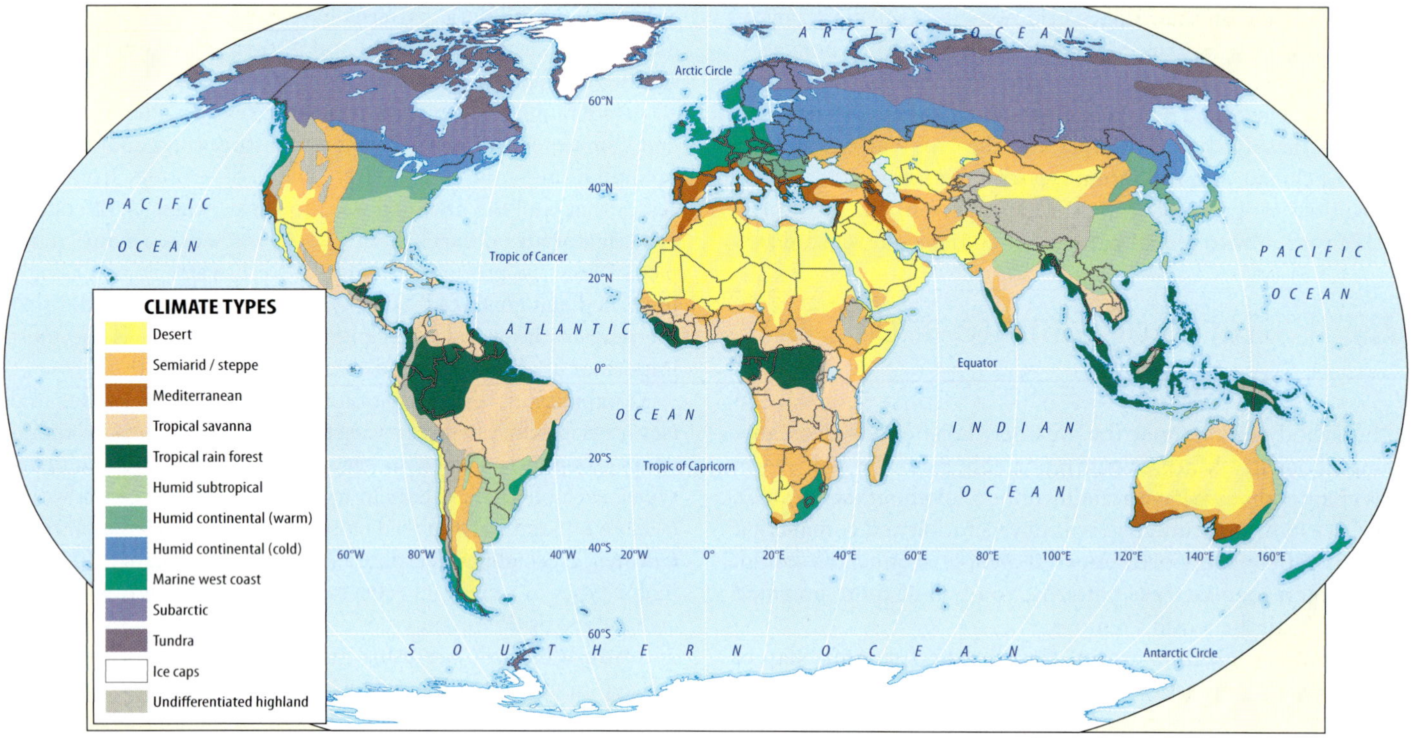

• **Figure 2.4** World climates.
Source: Based on Rand McNally's Classroom Atlas, 2003.

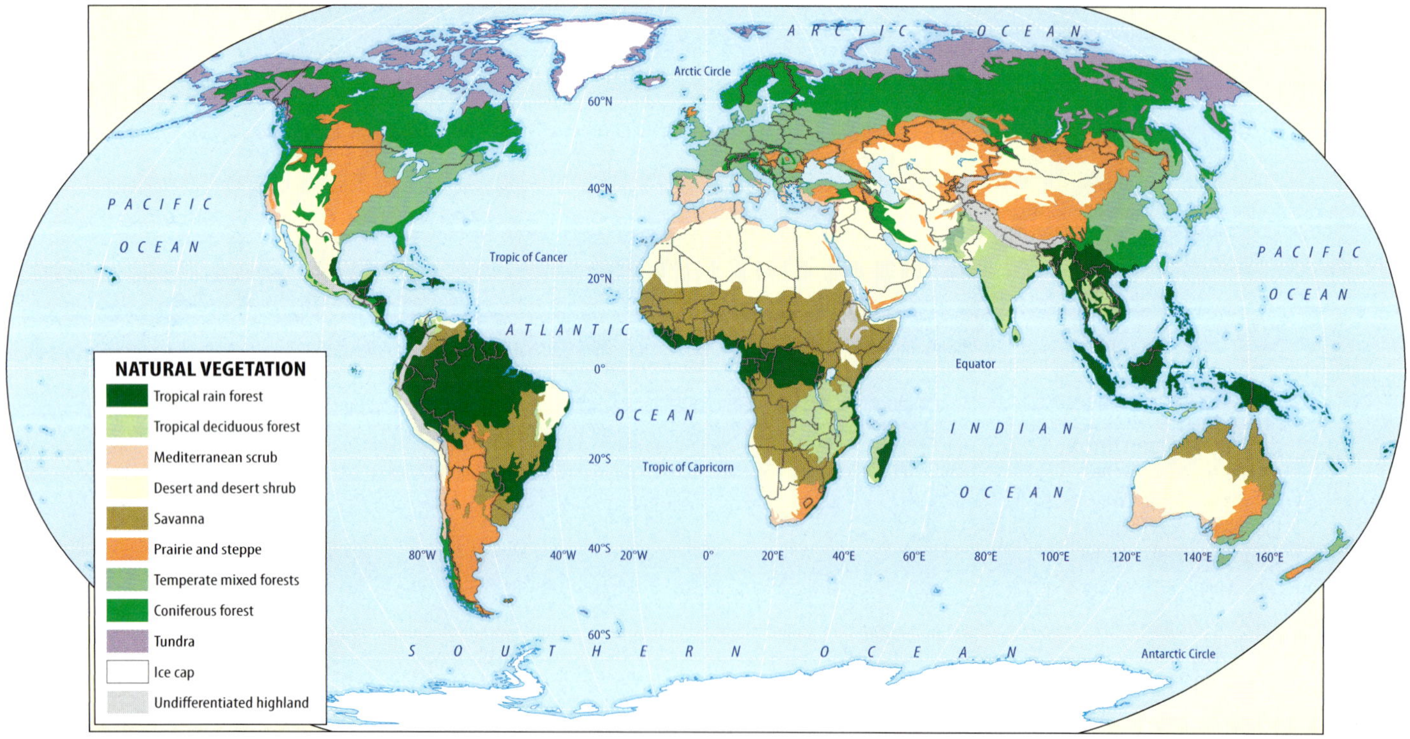

• **Figure 2.5** World biomes (natural vegetation) map.
Source: Based on World Wildlife Fund ecoregions data, 1999.

transformed by agriculture and urbanization. The map of the "human footprint" on page 42 in the next chapter shows what people have done to change the face of the Earth.

Your Global Environmental Counterparts — Try It

Examine the maps in Figures 2.4 and 2.5 (by the way, are these large-scale or small-scale maps?). In each map, find approximately where you live. What kind of climate region is this? What is the characteristic biome or natural vegetation there? Now find one or more other areas in the world that have the same climate and biome types—what places are most like where you live? Remember, these maps are greatly generalized, so you may need to be forgiving about the details.

Climate plays the main role in determining the distribution of biomes, but differing soils and landforms may promote different types of vegetation where the climatic pattern is essentially the same. Vegetation and climate types are sufficiently related that many climate types take their names from vegetation types—for example, the tropical rain forest climate and the tundra climate. You may easily see the geographic links between climate and vegetation by comparing the maps in Figures 2.4 and 2.5. The spatial distributions of climate and vegetation types do not overlap perfectly, but there is a high degree of correlation (see Try It on this page).

In the **ice cap, tundra,** and **subarctic** climates, the dominant feature is a long, severely cold winter, making agriculture difficult or even impossible. Summers are short and cool. Perpetual ice in the form of glaciers may be found at very high elevations in the lower latitudes (even in equatorial regions) and in the polar realms. The ice cap biome (•Figure 2.6a) has no plant life, except in those very few spots where enough ice or snow melts in the summer to allow tundra vegetation to grow. Tundra vegetation (•Figure 2.6b) is made up of mosses, lichens, shrubs, dwarfed trees, and some grasses. Needleleaf evergreen **coniferous trees** can stand long periods when the ground is frozen, depriving them of moisture. **Coniferous forests** (often called **boreal forests** or *taiga*, their Russian name; •Figure 2.6c) thrive in areas of subarctic climate,
121 including large parts of Russia's Siberia region.

In **desert** and in **semiarid** or **steppe** climates, the dominant feature is aridity (extreme dryness) or semiaridity (mainly dry conditions, with sparse precipitation). Deserts and steppes occur in both low and middle latitudes. Agriculture in these areas usually requires irrigation. The Earth's largest dry region extends in a broad band across northern Africa and southwestern and central Asia. The deserts of the 151, 160
middle and low latitudes are generally too dry for either trees or grasslands. They have **desert shrub** vegetation (•Figure 2.6d), often only in dry riverbeds, and in many places have no vegetation at all. Grasslands dominate in the moister steppe climate, which represents a transitional zone between very arid

Joe Hobbs

• **Figure 2.6 (b)** Tundra, northern Norway.

Joe Hobbs

• **Figure 2.6 (a)** Ice cap biome, Franz Josef Land, Russian Arctic region.

Antropov/Shutterstock.com

• **Figure 2.6 (c)** Coniferous forest.

Joe Hobbs

• **Figure 2.6** **(d)** Desert shrub, southern Sinai Peninsula, Egypt.

Joe Hobbs

• **Figure 2.6** **(f)** Tropical rain forest.

deserts and more humid areas. The biome composed mainly of short grasses is also called the steppe, or **temperate grassland** (•**Figure 2.6e**). The temperate grassland region of the United States and Canada originally supported both tall-grass and short-grass
399 vegetation types, known in those countries as **prairies.**

The **tropical rain forest** and **tropical savanna** climates are rainy, low-latitude climates. The key difference between them is that the tropical savanna type has one or more prolonged dry seasons, whereas the tropical rain forest climate has a steadier rainfall regime. Heat and moisture dominate in the tropical rain forest biome (•**Figure 2.6f**), where broadleaf evergreen trees (which lose their leaves, but not seasonally or in tandem) dominate the vegetation. In dry tropical areas that have enough moisture to support trees, **tropical deciduous forest** replaces the rain forest. Here the broadleaf trees are not green throughout the year; they lose their leaves and are dormant during the dry season and then add foliage and resume their growth during the wet season. The tropical deciduous forest approaches the luxuriance of the tropical rain forest in wetter areas but thins out to low, sparse scrub and thorn forest (•**Figure 2.6g**) in drier areas. **Savanna** vegetation (•**Figure 2.6h**) has taller grasses than the steppe and usually includes some trees, if found in areas of greater overall rainfall and more pronounced wet and dry seasons. You have seen much savanna vegetation in documentaries
about the great animal herds of East Africa. 314

In the **marine west coast** climate, occupying the western sides of continents in the higher middle latitudes (the Pacific Northwest region of the United States, for example), warm ocean currents moderate the winter temperatures, and summers tend to be cool. Coniferous forest dominates in some cool, wet areas of marine west coast climate; a good example is the redwood and sequoia forest of northern California (•**Figure 2.6i**). These humid middle-latitude regions have mild to hot summers, and winters ranging from mild to cold, with several other types of climate interspersed.

The **Mediterranean** climate (named after its most prevalent area of distribution, the lands around the Mediterranean Sea) is typically found between a marine west coast climate and a

Joe Hobbs

• **Figure 2.6** **(e)** Steppe, eastern Turkey.

Joe Hobbs

• **Figure 2.6** **(g)** Scrub and thorn forest, northern Zimbabwe.

Joe Hobbs

• **Figure 2.6** **(h)** Savanna, southern Kenya.

76 lower-latitude steppe or desert climate. In the summer high-sun period, it lies under high atmospheric pressure and is typically rainless. In the winter low-sun period, it usually receives precipitation from low-pressure systems carried by westerly winds. **Mediterranean scrub forest** (•Figure 2.6j), known locally by such names as *maquis* and *chaparral*, characterizes Mediterranean climate areas. Many of these shrubs contain oils that are highly aromatic, repelling wild grazing animals but attracting human culinary interest: sage and thyme

Joe Hobbs

• **Figure 2.6** **(i)** Marine west coast forest type; giant sequoias in King's Canyon National Park, central California.

Joe Hobbs

• **Figure 2.6** **(j)** Mediterranean scrub forest (dark area in left foreground) east of Los Angeles. The mountains produce a "rain shadow" effect: the light area in the right and background is desert.

are good examples. Much of California has Mediterranean climate and vegetation.

The **humid subtropical** climate occupies the eastern portion of continents between approximately 20 and 40 degrees of latitude and is characterized by hot summers, mild to cool winters, and ample precipitation for agriculture. The **humid continental** climate lies poleward of the humid subtropical type; it has cold winters, warm to hot summers, and enough rainfall for agriculture, with the greater part of the precipitation falling in the summer. This climate type is often subdivided into warm and cold subtypes, indicating the greater severity of winter in the zones closer to the poles. In middle-latitude areas with these humid subtropical and humid continental climate types, a **temperate mixed forest** with mostly broadleaf but also coniferous trees (as in the U.S. Midwest and Northeast; •Figure 2.6k) is found. As cold winter temperatures freeze the water within reach of plant roots, broadleaf trees shed their colorful leaves and cease to grow, thus reducing water loss. Then they produce new foliage and grow vigorously during the hot, wet summer.

Joe Hobbs

• **Figure 2.6** **(k)** Temperate mixed forest, central Missouri, United States.

Joe Hobbs

• **Figure 2.6** **(l)** Undifferentiated highland vegetation, San Juan Mountains, Colorado, United States. Note the "timber line" above which trees do not grow. Here, biomes change with elevation, and so the vegetation types have to be characterized as "undifferentiated" on small scale maps.

Coniferous forests can thrive in some hot and moist locations where porous, sandy soil allows water to escape downward, giving conifers (which can withstand drier soil conditions) an advantage over broadleaf trees. Pine forests on the coastal plains of the southern United States are an example.

Undifferentiated highland climates have a range of conditions according to elevation and exposure to wind and sun. The vegetation in these climates (•**Figure 2.6l**) differs greatly, depending on elevation, degree and direction of slope, and other factors. These climates are "undifferentiated" in the context of world regional geography in that a small mountainous area may contain numerous biomes, and it would be impossible to map them on a small scale. The world's mountain regions have a complex array of natural conditions and opportunities for human use. With increasing elevation, temperatures fall by a predictable **lapse rate,** on average about 3.6°F (2.0°C) for each increase of 1,000 feet (305 m) in elevation. In climbing from sea level to the summit of a high mountain peak near the equator (for example, in western Ecuador), a person would experience many of the major climate and biome types found in a
356 sea-level walk from the equator to the North Pole!

2.3 Biodiversity

All of Earth's natural regions have remarkable life forms. But scientists recognize the exceptional importance of some biomes because of their **biological diversity** (or **biodiversity**)—the number of plant and animal species present and the variety of genetic materials these organisms contain.

The most diverse biome is the tropical rain forest. From a single tree in Peru's Amazon region, the entomologist Terry Erwin recovered about 10,000 insect species. From another tree several yards away, he counted another 10,000, many of which differed from those of the first tree. Alwyn Gentry of the Missouri Botanical Garden recorded 300 tree species in a 1-hectare (2.47-acre) plot of the Peruvian rain forest. Less than 50 years ago, scientists calculated that there were 4 million to 5 million species of plants and animals on Earth. Now, however, their estimates are much higher, in the range of 40 million to 80 million. This startling revision is based on research, still in its infancy, on species inhabiting the rain forest. Many of these are small insects. Incidentally, the number of animal species far exceeds that of plant species.

The Importance of Biodiversity

Such diversity is important in its own right, but it also has vital implications for nature's ongoing evolution and for people's lives on Earth. Humankind now relies on a handful of crops as staple foods. In our agricultural systems, the trend in recent decades has been to develop high-yield varieties of grains and to plant them as vast **monocultures** (single-crop plantings). This trend, which is the cornerstone of the so-called Green Revolution, is controversial. On the one hand, it puts more food on the
global table. But on the other, it may render agriculture more 249
vulnerable to pests and diseases and thus pose long-term risks of famine. In evolutionary terms, the Green Revolution has reduced the natural diversity of crop varieties that allows nature and farmer to turn to alternatives when adversity strikes. At the same time, while we remove tropical rain forests and other natural ecosystems to provide ourselves with timber, agriculture, or living space, we may be eliminating the foods, medicines, and raw materials of tomorrow, even before we have collected them and assigned them scientific names. "We are causing the death of birth," lamented biologist Norman Myers.[2]

Regions where human activities are rapidly depleting the rich 290,
variety of plant and animal life are known as **biodiversity hot** 344
spots, which scientists believe deserve immediate attention for study and conservation. Thirty-four priority regions identified by Conservation International are depicted in •**Figure 2.7**. By referring to the map of the Earth's biomes in Figure 2.5, you will see that most of these hot spots are in tropical rain forest areas. Many are islands that tend to have high biodiversity because species on them have evolved in isolation to fulfill special roles in these ecosystems and because human pressures on island eco-
systems are particularly intense. Efforts are under way in most 290,
of these hot spots to establish national parks and other protected 344
areas. Conservationists are also turning their attention increasingly to the state of the world's oceans, which play critical roles in the Earth's physical processes and sustain many people.

2.4 The World's Oceans

Life on Earth would be impossible without the roles and resources of the hydrosphere, which includes both oceans and freshwater sources such as lakes and rivers. This book deals often with issues of freshwater, which is appropriate considering the projection by the World Commission on Water that by 2025, half of the world's population will live under conditions of severe water stress.[3] In this section, the focus is on the world's oceans.

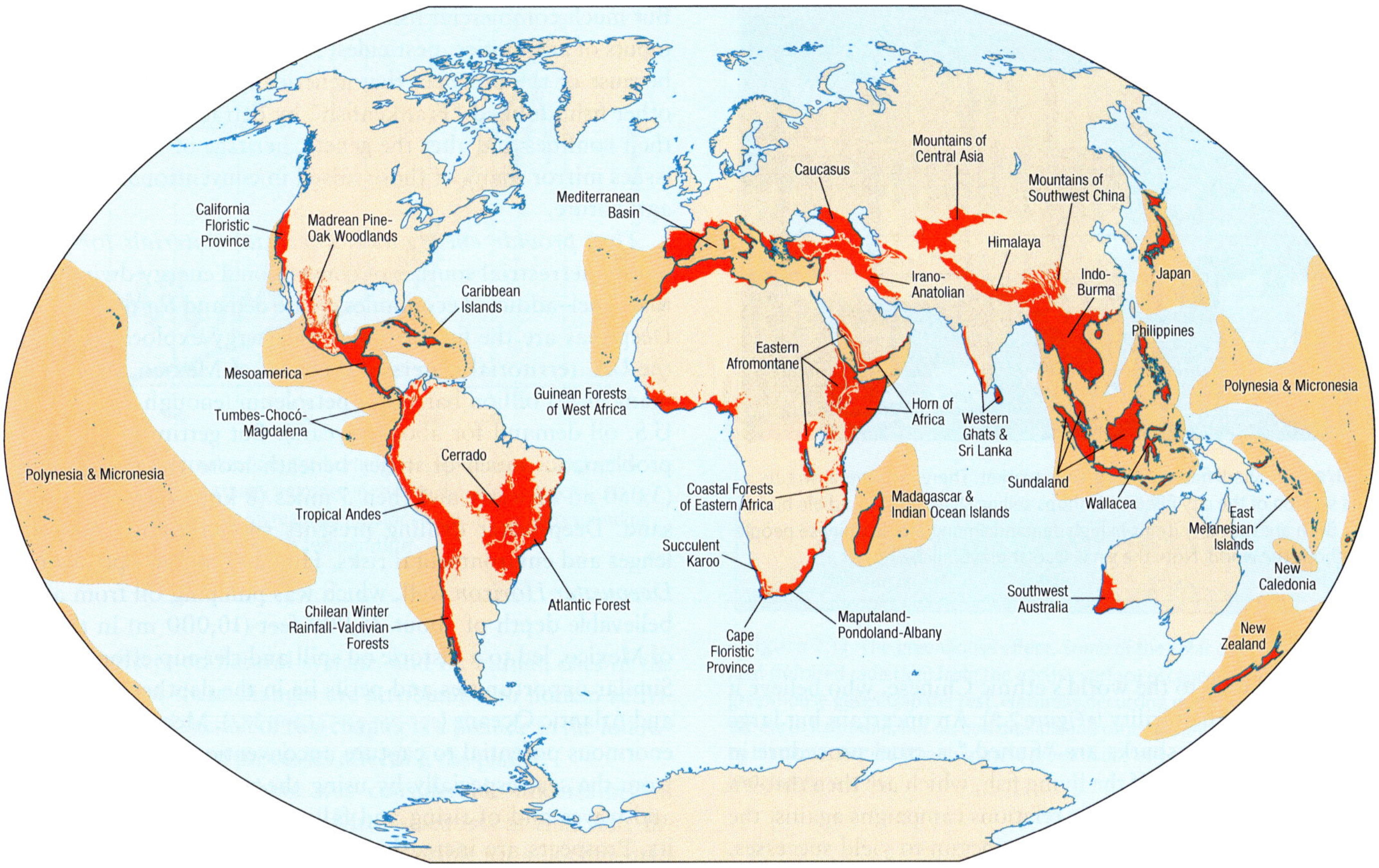

• **Figure 2.7** World biodiversity hot spots.

Source: Adapted from "Threatened Ecoregions" map, National Geographic Family Reference Atlas of the World, p. 40. Copyright © 2002 National Geographic Society. Source data provided by Conservation International.

Why Should We Care about Oceans?

I live in Missouri, where a day at the beach or a plate of fresh seafood is just a dream. Even those of you living close to the sea may have very little interaction with it. Why bother thinking about the world's oceans, much less worrying about them?

It's a watery world. About 70 percent of the world's surface is comprised of water. That is a huge proportion, and it is very important to consider how water shapes life on the planet. Oceans have the largest role in the **hydrologic cycle,** as the sun evaporates seawater into vapor that is then released as freshwater precipitation. Without the seas, the Earth's usable freshwater resources would be extremely limited.

The oceans feed us. About a billion people, or 15 percent of the world's population, rely mainly on fish for their protein (•**Figure 2.8**). Growing human populations, increased buying power among the middle classes in developing countries, and technological improvements in the fishing industry are putting these resources under unprecedented pressure.[4] Since 1980, there has been a spectacular 40 percent growth in worldwide demand for seafood. Some scientists have issued a stern warning on the future of the world's fisheries: Unless major steps are taken quickly, there will be a "global collapse" of all species currently fished by 2050.[5]

Defining "collapse" as a population less than 10 percent of its previous levels, these authorities noted that 30 percent of the fish species currently exploited have already collapsed. The declines are most extreme among stocks of large predatory fish like sharks, tuna, and swordfish. An estimated 75 million sharks are killed each year (mainly by Spanish and Indonesian fishermen) to provide shark fin soup (which can cost up

Joe Hobbs

• **Figure 2.8** Seattle's Pike Street Market. Humanity's demand for fish and seafood is growing, but what about supplies? Stocks are falling, and in many species plummeting.

Table 3.2 Top and Bottom Countries in Per Capita GNI-PPP (U.S. Dollars, 2011)*

Top Countries	GNI-PPP Per Capita	Bottom Countries	GNI-PPP Per Capita
Luxembourg	59,590	Afghanistan	860
Norway	55,420	Togo	850
Singapore	49,780	Sierra Leone	790
Switzerland	47,100	Malawi	780
United States	45,640	Central African Rep.	750
Netherlands	39,740	Niger	680
Denmark	38,780	Eritrea	580
Australia	38,510	Burundi	390
Austria	38,410	Congo, Dem. Rep.	300
Sweden	38,050	Liberia	290

*This table excludes city-states, territories, colonies, and dependencies. Only countries with available data are listed.

Source: 2011 World Population Sheet, Population Reference Bureau.

of improvement in the material conditions of people through diffusion of knowledge and technology.

The Human Development Index. Definitions like that, and statistics like per capita GDP and per capita GNI PPP, reveal little about measures of well-being such as income distribution, gender equality, literacy, and life expectancy. Recognizing the shortcomings of strictly economic definitions, the United Nations Development Programme created the **Human Development Index (HDI)**, a scale that considers attributes of quality of life. This book uses HDI in the Basic Data tables of the chapter that introduce each region (for example, Table 4.1 on page 67). On the HDI scale, a measure of 1.0 means "perfect." According to this index, with a rating of 0.938, Norway is "the world's best place to live," although it ranks second in per capita GNI PPP. In Chapter 4, we will see why the Norwegians and other Scandinavians are living so well. Following Norway, in descending order, are Australia, New Zealand, the United States, and Ireland. In HDI terms, with a measure of just 0.140, Zimbabwe is the world's worst place to live; the Democratic Republic of the Congo, Niger, Burundi, and Mozambique fare only slightly better. Note that all five of these low-rated countries are in Africa, for reasons discussed in Chapter 9.

On the basis of per capita GNI PPP, 1.2 billion, or 18 percent, of the world's people, inhabit the MDCs. Most citizens of these countries, such as the United States, Canada, Japan, South Korea, Australia, New Zealand, and the nations of Europe, enjoy an affluent lifestyle with freedom from hunger. Employed in industries or services, most of the people live in cities rather than in rural areas. Disposable income, the money that people can spend on goods beyond their subsistence needs, is generally high. There is a large middle class. Population growth is low as a result of low birth rates and low death rates (these demographic terms are explained in Section 3.3 on population). Life expectancy is long, and the literacy rate is high.

Life for a very large number of the planet's other 82 percent, or about 5.7 billion people, is quite different. In the LDCs, including many countries in Latin America, Africa, and Asia, poverty and often hunger prevail. The leading occupation is subsistence agriculture, and the industrial base is small. Most people live in rural areas, but that statistic will change soon; 46 percent of the LDCs' population is urban. The middle class tends to be small but growing, with an enormous gulf between the vast majority of poor and a very small wealthy elite, who own most of the private landholdings. With high birth rates and falling death rates, population growth is high. Life expectancy is short, and the literacy rate is low.

LDCs are not predestined to be in poverty forever. Effective development policies and practices have changed the outlook for many countries. Doomsayers predicted for decades that the colossal country of India would one day collapse under the weight of overpopulation and poverty. As we will see in Chapter 7, India is instead emerging as a global economic power to be reckoned with, for example, by using its human capital to generate wealth from software and outsourced services (see Insights feature on page 235). A few decades ago, China was a desperately poor country, but today it has the world's second largest economy. Still, because of their large populations, both countries face many challenges—including big gaps between their haves and have-nots—and neither ranks as a more developed country. 232

With four-fifths of the world's people living in the poorer, less developed countries, it is important to understand the root causes of underdevelopment and to appreciate how wealth and poverty affect the global environment in very different but equally profound ways. With this knowledge, it becomes possible to envision solutions.

Why Are Some Countries Rich and Others Poor?

Many theories attempt to explain the disparities between MDCs and LDCs. It is important to recognize that there is no single, widely accepted explanation about development in general or in a particular region or country. Probably the best thing you can do with your critical thinking skills is to weigh the various explanations and see which one, or which combination of them, seems to fit the situation of a given country or region. Here are the main explanations you are likely to come across.

Embraced most strongly in the LDCs, **dependency theory** argues that the worldwide economic pattern established by the Industrial Revolution and by the forces of colonialism that accompanied it persists today. In his book *Ecological Imperialism*, the historian and geographer Alfred Crosby explains how dependency led to the rich–poor divide, depicting the two very different ways in which European powers used foreign lands during the Industrial Revolution.[3] In the pattern of **settler colonization**,

Europeans sought to create new Europes, or **neo-Europes**, in lands much like their own: temperate middle-latitude zones with moderate rainfall and rich soils where they could raise wheat and cattle. And so, between 1630 and 1930, more than 50 million Europeans emigrated from their homelands to create European-style settlements in what are now Canada, the United States, Argentina, Uruguay, Brazil, South Africa, Australia, and New Zealand. These lands were destined to become some of the world's wealthier regions and countries.

In contrast to their preference to settle familiar mid-latitude environments, Europeans viewed the world's tropical lands mainly as sources of raw materials and markets for their manufactured goods. The environment was too different from home to make settlement attractive. In establishing a pattern of **mercantile colonialism**, Crosby explains, Europeans were less inhabitants than conquering occupiers of the colonies, overseeing indigenous peoples and resettled slaves in the production of primary or unfinished products: sugar in the Caribbean; rubber in Latin America, West Africa, and Southeast Asia; and gold and copper in southern Africa, for example. Colonialism required huge migrations of people to extract the Earth's resources, including 30 million slaves and contracted workers from Africa, India, and China to work mines and plantations
294, 322, 365, 409 around the globe.

In the mercantile system, the colony provided raw materials to the ruling country in return for finished goods; this meant that people in India would purchase clothing made in England from the raw cotton they themselves had harvested. The relationship was most advantageous to the colonizer. England, for example, would not allow its colony India to purchase finished goods from any country but England. It prohibited India from producing any raw materials the empire already had in abundance, such as salt (India's Mohandas Gandhi defiantly violated this prohibition in his famous "March to the Sea"). Finished products are **value-added products**, meaning they are worth much more than the raw materials they are made from, so manufacturing in the ruling country concentrated wealth there while limiting industrial and economic development in the colony.

The colony was obliged to contribute to, but was prohibited from competing with, the economy of the ruling country—a relationship that dependency theorists insist continues today. Dependency theory asserts that to participate in the world economy, the former colonies, now independent countries, continue to depend on exports of raw materials to and purchases of finished goods from their former colonizers and other MDCs, and this disadvantageous position keeps them poor. Dependency theorists call this relationship **neocolonialism**. With independence, the former colonies needed revenue. To earn that money, they continued to produce the goods for which markets already existed—generally the same unprocessed primary products they supplied in colonial times. Dependency theorists argue that when the former colonies try to break their dependency by becoming exporters of manufactured goods, the MDCs impose trade barriers and quotas to block that development (see Insights, page 420). Dependency theory and neocolonialism also envision the world in terms of core and periphery—a concept of duality you have seen already in Chapter 1 and will encounter often in the book—with wealth flowing from the periphery of the world's poor countries to its core of wealthy countries.

Whether because of neocolonialism or a more complex array of variables, many developing countries continue to rely heavily on income from the export of a handful of raw materials. This makes them vulnerable to the whims of nature and the world economy. The economy of a country heavily dependent on rubber exports, for example, may suffer if an insect pest wipes out the crop or if a foreign laboratory develops a synthetic substitute. When demand for rubber rises, that country may actually harm itself trying to increase its market share by producing more rubber, because in doing so, it drives down the price. (Consider oil: members of the Organization of Petroleum Exporting Countries [OPEC] cartel of oil-producing countries drove oil prices to all-time lows in the mid-1980s when they overproduced oil in a bid to earn more revenue.) If the rubber-producing country withholds production to shore up rubber's price, it provides consuming countries with an incentive to look for substitutes and alternative sources. (Again, look at oil: after OPEC embargoed shipments of oil to the United States in the 1970s, the United States began developing domestic oil supplies and becoming more energy-efficient, thus temporarily reducing oil prices and OPEC's revenues.) The developing country is in a dependent and disadvantaged position—even if it produces something as vital as oil.

Many geographers view dependency theory and neocolonialism as simplistic and politically biased explanations of development. They consider a wider and more complex set of factors, including culture, location, and natural environment, to explain why some countries are wealthy and others are poor. These include the following factors:

Advantageous and disadvantageous location. Does a country's location play a role in how rich or poor it is? In some cases, yes. Location can influence a country's economic fortunes. For example, as discussed in Chapter 1, because it is situated close to a great mainland with which to trade, the island of Great Britain enjoys a core location favorable for economic development. Japan has a similar location relative to the Asian landmass. In contrast, landlocked nations such as Bolivia in South America and numerous nations in Africa have locations unfavorable for trade and economic development, and they have not overcome this disadvantage. But it is important to recognize that geographic location is never the sole decisive factor in development. Like Japan and Britain, Madagascar and Sri Lanka are island nations situated close to large mainlands, but for a variety of (mainly political) reasons discussed in the relevant chapters, neither has experienced prosperity.

Resource wealth or poverty. Having or lacking a diversity or abundance or **natural resources** (naturally occurring materials that people can use) plays a significant role in development.

INSIGHTS Globalization: "The World Is Flat"

One of the most remarkable international trends of recent decades has been **globalization**—the spread of free trade, free markets, investments, and ideas across borders and the political and cultural adjustments that accompany this diffusion. There has been much debate and sometimes-violent conflict over the pros and cons of globalization. For good or ill, globalization is a reality. *The World Is Flat*, journalist Thomas Friedman entitled his book on the subject. He observed, ironically, that in the search for India, Christopher Columbus proved the world was round; but in visiting India, Friedman himself found the world to be flat: the "playing field" of the international economy had been leveled: "It is now possible for more people than ever to collaborate and compete in real time with more people on more different kinds of work from more different corners of the planet and on a more equal footing than at any previous time in the history of the world—using computers, e-mail, fiber-optic networks, teleconferencing and dynamic new software."[4]

Supporters of globalization argue that the newly emerging global economy will bring increased prosperity to the entire world. Innovations in one country will be transferred instantly to another country, productivity will increase, and standards of living will improve. They propose that one obvious solution to the problem of the LDCs' inability to compete in the world economy, and therefore escape their dependency, is to reduce the trade barriers that MDCs have erected against them (some of these barriers are discussed on page 420 in Chapter 11). With free trade, free enterprise will prosper, pumping additional capital into national economies and raising incomes for all. Many of the inputs into globalization come from **multinational companies** (also called *transnational companies*, meaning companies with operations outside their home countries) as they increase their investments abroad. Most of the companies are based in the MDCs, but multinational corporations have also grown in LDCs such as China, India, and Mexico.

Opponents of globalization argue that the process will actually increase the gap between rich and poor countries, and even within countries; a selected few developing nations (or people within a country) will prosper from increased foreign investment and resulting industrialization, but the hoped-for global wealth will bypass other countries (or people within a country) altogether. And the increasing interdependence of the world economies will make all the players more vulnerable to economic and political instability. The multinational companies will recognize huge profits at the expense of poor wage laborers. In addition, environments will be harmed if environmental regulations are reduced to a lowest common denominator (for example, the high standards of air quality demanded by the U.S. Clean Air Act would be deemed noncompliant with World Trade Organization rules because they make it harder for countries with "dirtier" technologies to compete in the marketplace). Such concerns often lead to massive protests against "corporate-led globalization," especially at meetings of the World Trade Organization, the International Monetary Fund, and the World Bank. Many more confrontations like these can be anticipated. Typical protesters' demands are that working conditions be improved in the foreign "sweatshops," where textiles and other goods are produced at low cost for U.S. corporations, and that corporations like Starbucks should sell only "fair trade" coffee beans bought at a price giving peasant coffee growers a living wage rather than at the "exploitive" price typically paid. Corporations like Starbucks have often been quick to respond to the demands of the environmentally and socially conscious consumer.

It is often assumed that the rapid growth and spread of computer and wireless technologies are benefiting everyone, but in fact there is still a **digital divide** to overcome. The divide is between the handful of countries that are the leaders in **information technology (IT)** and the majority of nations that have little ability to create, purchase, or use new technologies. The fear in the technology-lagging countries is that the growth of e-commerce will concentrate the wealth generated by that commerce in the technology leadership countries. This would enable the leaders to make even further advances in technology and realize even bigger economic gains from a so-called **knowledge economy** based on innovation and services, while the technology laggards fail to catch up and simply become poorer (•**Figure 3.A**). There are many efforts underway to empower the world's poor through access to information technology; the "One Laptop per Child" program seeks to provide the world's children with rugged, low-cost, low-power, Internet-connected laptops (see http://.onelaptop.org). Rapid developments and falling prices in smartphone technology could also help level the playing fields of globalization. 331

Joe Hobbs

• **Figure 3.A** An Internet café in Mérida, Yucatán, Mexico. Information technology—especially in the form of computers, the Internet, and cellular telephones—is spreading rapidly around the 196
world. Some people believe that there is now an unfolding Information Revolution comparable to the Agricultural and Industrial Revolutions in its capacity to change humanity's relationship with the earth.

Superabundance of an especially valuable resource (for example, oil in the Persian/Arabian Gulf countries) or a diversity of natural resources has helped some countries become more developed than others. The former Soviet Union and the United States achieved superpower status in the twentieth century in large part by using the enormous natural resources of both countries.

Cultural and historical factors. In some cases, human industriousness has helped compensate for resource limitations and helped to promote development. For example, Japan has a rather small territory with few natural resources (including almost no petroleum). Yet in the second half of the twentieth 270
century, it became an industrial powerhouse largely because the

Japanese people united in a common purpose to rebuild from wartime devastation, placing priorities on education, technical training, and seaborne trade from their advantageous island location. Conversely, cultural or political problems like corruption and ethnic factionalism can hinder development in a resource-rich nation, as in the mineral-wealthy Democratic Republic of
340 Congo.

Incidentally, many students have told me they enjoyed reading the geographer Jared Diamond's book *Guns, Germs, and Steel* because he writes so engagingly about the forces that propelled countries forward or held them back. Diamond's book has come to be known as "A Short History of Everybody for the Last 13,000 Years"!

Environmental Impacts of Underdevelopment

Geographers are very interested in the environmental impacts of relations between MDCs and LDCs, especially on the poorer countries. LDCs generally lack the financial resources needed to build roads, dams, energy grids, and other infrastructure critical to development. Then they turn to the World Bank, International Monetary Fund (IMF), and other institutions of the MDCs to borrow funds for these projects. Many borrowers are unable to pay even the interest on these loans, which is sometimes huge; debtor nations have been known to spend as much as 40 percent of annual revenues on interest payments to the lenders, or more than they spend on education and health combined.

This debt cycle can be very harmful to natural environments. When lender institutions threaten to cut off assistance, borrowing countries often try to raise cash quickly to avoid this prospect, generally by one or two methods, or both. One method is to dedicate more land to the production of **cash crops** (also known as **commercial crops**). These are items such as coffee, tea, sugar, coconuts, and bananas exported to the MDCs, where they are luxuries or staples (illegal drug crops are also cash crops, of course, but fall outside most statistics and many of the general associations made here). Governments or foreign corporations often buy out, force out, or otherwise displace subsistence food farmers in the search for new lands on which to grow these commercial crops. In this process, known as **marginalization**, poor subsistence farmers are pushed onto fragile, inferior, or marginal lands that cannot support crops for long and end up depleted by cultivation. In Brazil's Amazon Basin, for example, peasant migrants arrive from Atlantic coastal regions, where government and wealthy private landowners cultivate the best soils for sugarcane, soybeans, and other cash crops. The newcomers to Amazonia slash and burn the rain forest to grow rice and other crops that exhaust the soil's limited fertility in a few years. Then they move on to cultivate new lands, and in their wake come cattle ranchers, whose land use further degrades the
221, 222, 384 soil. This is a very common, important problem in the LDCs, and you will see many examples of it in the following chapters.

The second way for the LDC debtor countries to raise cash quickly is to simply sell off their natural assets. Who is buying them? The wealthier countries, and increasingly and most strikingly, China—an issue that this book covers often. LDCs 329, 333, 367 often face difficult choices between using natural resources for immediate gain or for longer-term, more sustainable growth. In most cases, they feel compelled to take short-term profits, by cash cropping and other strategies, a choice that can have devastating environmental consequences. Most LDCs have resource-based economies that rely not on industry, but on stocks of productive soils, forests, and fisheries. The countries' long-term economic health could benefit through protection and careful use of these assets. But to pay off international debts and meet other needs, the LDCs often draw down their ecological capital faster than nature can replace it. In ecosystem terms, they exceed **sustainable yield** (also known as the 61
natural replacement rate), the highest rate at which a **renewable resource** can be used without decreasing its potential for renewal. One figure that comes up often in the literature is that in African countries as a whole, 28 trees are cut down for every tree that is planted.

A country where 28 trees are cut for every one planted is approaching or is in a state of **ecological bankruptcy**, the exhaustion of environmental capital. This environmental poverty plays itself out in several damaging ways. First, there are negative effects on the health and well-being of the people living in such countries. People in the LDCs feel the impacts of environmental degradation more directly than people in the MDCs (•**Figure 3.6**). Many of the world's poor drink directly from untreated water supplies; an estimated 1 billion people lack access to clean water. They tend to cook their meals with fuelwood rather than with fossil fuels. They are more dependent on nature's abilities to replenish and cleanse itself, and so they suffer more when those abilities are diminished (see Insights, page 48).

In addition, many political and social crises can emerge from the ecological bankruptcy process. Revolutions, wars, and refugee migrations in developing nations often have underlying

Joe Hobbs

• **Figure 3.6** The Thu Bon River in Vietnam. In the LDCs, it is very common for people to rely on polluted water sources for drinking, bathing, cooking, and washing their clothes and eating utensils. Health effects can be severe; one of the most tragic is infant diarrhea, typically a waterborne illness that kills an estimated 2 million youngsters worldwide every year.

INSIGHTS | Deforestation and The Fuelwood Crisis

Deforestation—the removal of tree cover—has many detrimental effects in the LDCs. You can see how these relationships play out in •**Figure 3.B**. As people remove trees to use as fuel, for construction, or to make room for crops, their existing crop fields lose protection against the erosive force of wind. Less water is available for crops because in the absence of tree roots to funnel water downward into the soil, it runs off quickly. Increased salinity (salt content) generally accompanies increased runoff, causing the quality of irrigation and drinking water downstream to decline. Eroded topsoil resulting from reduced plant cover can choke irrigation channels, reduce water delivery to crops, raise floodplain levels, and increase the chances that floods will destroy fields and settlements. As reservoirs fill with silt, hydroelectric generation, and therefore industrial production, is diminished. Upstream, where the problem began, fewer trees are available to use as fuel.

As they deplete nearby fuelwood supplies from trees and other vegetation, rural people make their own lives more difficult and have to change their activity patterns. Women and children, in almost every society the main fuelwood collectors, must walk farther, and then farther again, to gather fuel. Using animal dung and crop residues for fuel can solve one problem but create another: it deprives the soil of the fertilizers these poor people often use when farming. Reduced food output is the result. A family may eventually give up one cooked meal a day or put up with colder temperatures in their homes because fuel is lacking—steps that have a negative impact on the family's health. These impacts of fuelwood depletion are known collectively as the **fuelwood crisis**.

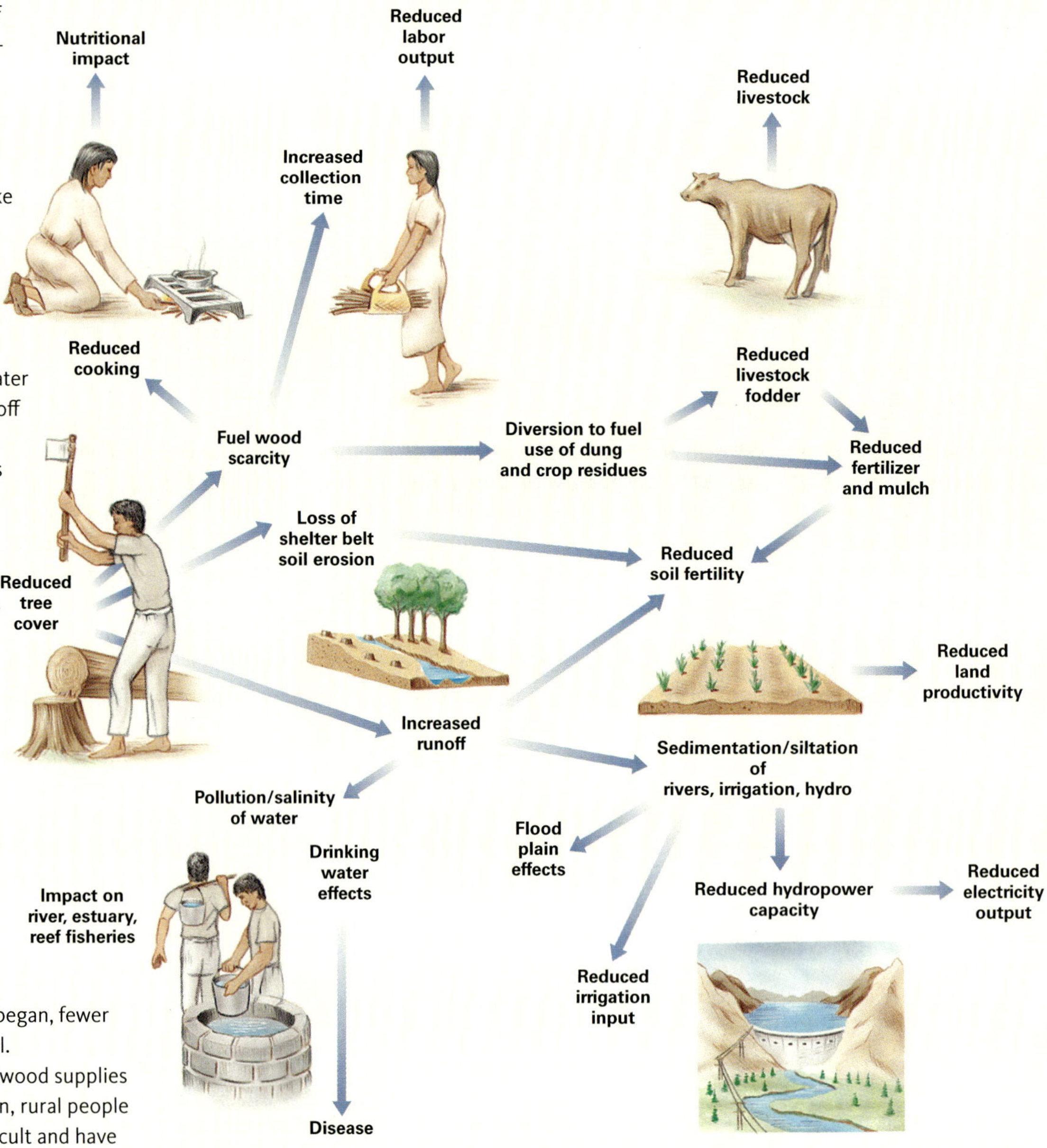

• **Figure 3.B** Impacts of deforestation in the LDCs.

The fuelwood crisis is a problem that governments and societies can confront successfully. A combination of new technologies, government policies, and sound economic growth can halt and even reverse deforestation. The biogas digester, a simple household device that can be used in warm climates, ferments animal manure for the production of methane cooking gas and thus spares fuelwood. Governments can protect forests while also allowing people to earn their livelihoods in them from **nontimber forest products (NTFPs)**, such as handicrafts made from bamboo (see www.ntfp.org). The process of economic development can itself lead to less pressure on forest resources. The world's wealthier countries have more forest now than they did in 1990, and forest cover has actually increased in about half of the world's 50 most forested countries since that time. A key challenge for the wealthier forested countries is to slow their demand for timber from the LDCs, and thus help break one of the cycles of deforestation, debt, and poverty.

environmental causes. Such problems, as related in the earlier discussion of climate change refugees in Chapter 2, are central to the national interests of the United States and other coun-
33 tries far beyond the affected nations. Deforestation in Mexico or Haiti, for example, is among the root causes why rural people from those countries try to move to the United States, legally or otherwise. When their home environments are so degraded that they simply cannot support farming as they once did, people will move.

3.3 The Geography of Population

Population may be the most critical issue in geography. It certainly is one of the most important issues for human life on Earth. The welfare of humanity and of the planet's other species and natural habitats is tied closely to two related issues: the number of people there are and the rates at which we consume resources. In some quarters, there are fears that the human **population explosion** the world has experienced since 1800 will lead to a precipitous crisis: a massive disease outbreak, famine, or some other kind of catastrophe triggered by too many people living too close together without sufficient food and other resources to sustain them. On the other hand, there is optimism that the human population growth rate is slowing, and our numbers are expected to stabilize—ideally at a population that can be sustained with minimal risk of famine or other suffering.

Meanwhile, large numbers of people move across and within political borders, usually by choice but in some cases by force. Migrants bring new cultures, ideas, and opportunities with them, but their reception is not always warm: tension and violence have characterized relations between majority and migrant minority groups in recent years in Europe and Australia, for example, and in the United States there is a heated debate about "immigration reform" directed at illegal migrants, especially from Latin America. Migration is one of the major themes discussed throughout this book.

Geography's interests in population are shared by many other fields, including biology, sociology, anthropology, and political science. The study of population is known as **demography**. The field of demography is most concerned with patterns of birth, death, marriage, and related issues in themselves, with less attention to issues of migration and population distributions. What most distinguishes population geography is its focus on spatial variations. This section of the book examines how many people have lived on Earth and how many we may be in the future, mindful of who these people are and why they have been so few or so many. It pays especially close attention to the fundamentally geographic issue of where they are. This discussion will be a very important reference for you as you work your way forward through the world regions. I hope you will find it interesting, and recommend that you explore the best Internet resource on world population (where you can also download a free world population data table): the Population Reference Bureau site at prb.org.

How Many People Have Ever Lived on Earth?

Around 100,000 years ago, our *Homo sapiens* ancestors came out of Africa across the land bridge of Suez and began to populate Eurasia. By around 10,000 years ago, about the time plants and animals began to be domesticated, there were probably about 5.3 million humans in the world—roughly the current number of residents of the Atlanta metropolitan area or the country of Finland. By 1 C.E., humans numbered between 250 and 300 million, about the population of the United States today. The first billion was reached around 1800. Then a staggering population explosion occurred in the wake of the Industrial Revolution. The second billion came in 1930, the fourth in 1975, the sixth in 1999, and the seventh in 2011 (•Figure 3.7). *Homo sapiens* is now by far the most populous large mammal on Earth and has succeeded where no other animal has in extending its range to the world's farthest corners. Using a 24-hour time period to represent the 200,000 years of our species' history, we reached our first billion within just the last three minutes. The people alive today represent about 6 percent of all the people who have ever lived on Earth!

Figure 3.7 illustrates very dramatically that right after about 1800, the human population surged. What happened after tens of thousands of years that made our numbers suddenly skyrocket? In addition to the factor, addressed earlier, of greater food surpluses that made it possible to feed more people, we answer that question by considering some of the basic vocabulary of population geography.

How Can We Measure Population Changes?

Excluding the issue of migration for now, two main variables determine population change in a given village, city, or country or the entire planet: birth rate and death rate. The **birth rate** is

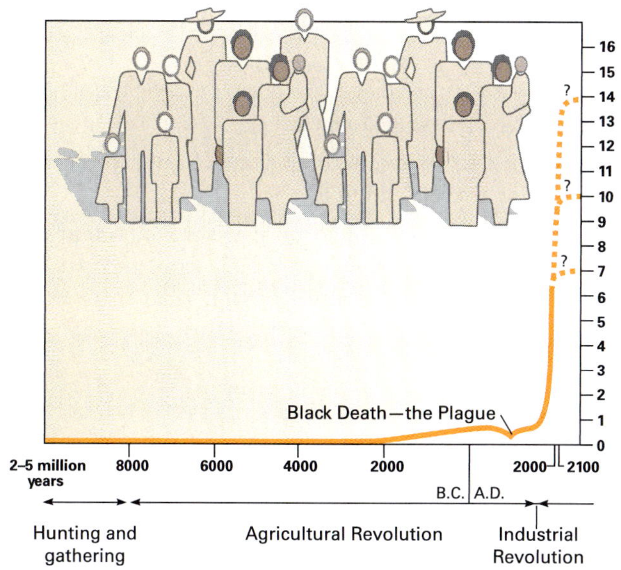

• **Figure 3.7** The human population has exploded since the Industrial Revolution, in a classic J-shaped curve of exponential growth.

the annual number of live births per 1,000 people in a population. The **death rate** is the annual number of deaths in that same sample population of 1,000. The **population change rate**—the figure that is often called the "population growth rate" but that may represent either growth or loss—is the birth rate minus the death rate in that population.

We can put these measures to work to appreciate the Earth's population now. Suppose there were a perfect sample of 1,000 people representing the world's population in 2011. Their birth rate was 20 per 1,000 and their death rate was 8 per 1,000. This means that by the end of that year, among the 1,000 people, 20 babies had been born and 8 people had died, resulting in a net growth of 12. That figure, 12 per 1,000, or 1.2 percent, represents the 2011 population change rate for the world.

We may now look at some of different combinations of birth rates, death rates, and population change rates and the various forces behind them.

What Determines Family Size?

Many factors affect birth rates, which tend to be much higher in the less developed countries of the world and among the poorer residents of more affluent countries. Some of the motivations and circumstances may seem unfamiliar at first, but you will see there are good reasons for them:

- Better-educated and wealthier people have fewer children. The parents of most of you reading this book probably considered the economic cost of raising you and sending you to college and made decisions about family size in part because of their education and yours.
- Conversely, less educated and poorer people generally want and have more children. One reason for this is economic: poorer parents are often convinced that "extra" children will help bring more income to the family by working in fields or factories and will help care for them in their old age.
- People in cities tend to have fewer children than those in rural areas.
- Those who marry earlier generally have more children (their reproductive life span is longer).
- Couples with access to and understanding of contraception generally have fewer children.
- Value systems and cultural norms play very important roles. Even where contraception is available and understood, a couple may decide not to interfere with what they perceive as God's will or may seek the social status associated with a larger family.

What Determines Death Rates?

Death rates are correlated mainly with health factors, particularly the level of nutrition and level of medical care available. Improvements in food production and distribution help reduce death rates. Better sanitation, better hygiene, and cleaner drinking water eliminate fatal diseases such as infant diarrhea, a common cause of infant mortality in the less developed countries. The availability of antibiotics, immunizations, insecticides, and other improvements in medical and public health technologies also correlate with death rates.

Human death rates overall have been on a steady trend of decline for decades. But death rates sometimes rise, of course, especially with the outbreak of epidemics such as HIV/AIDS. 316 Large proportions of the world's population have in fact been killed in disease epidemics and natural disasters. The "Black Death" (caused by bubonic plague) in Europe and Asia killed about 15 percent of the world's population between 1334 and 1349; at least 40 percent of Europe's population died. An influenza (flu) epidemic in 1918–19 killed as many as 100 million people. We cannot assume that such devastation will never happen again. Earlier this century, there were fears that the "bird flu" might kill as many as 150 million people if it mutated to a human strain.

Closely related to the measure of death rates is that of *life expectancy*, the number of years a person may expect to live in a given environment (typically defined as a country—see •**Figure 3.8**—and differentiated between women, who usually live longer, and men). As death rates fall, life expectancy increases, and the reverse is also true. In the United States in 2011, life expectancy for women was 80 years, and for men, 75 years. In Zimbabwe, a southern African country hit hard by the HIV/AIDS epidemic, a woman could expect to live 45 years and a man 46 years. But as we will see in Chapter 9, that is an improvement for Zimbabwe as it and other sub-Saharan African countries begin to make headway against the scourge of AIDS. 316

What Determines the Population Change Rate?

Throughout history, natural disasters, diseases, and wars have taken huge bites out of our numbers. Overall, however, with birth rates higher than death rates, the trend line has been one of growth—and since 1800, of spectacular growth. In 1968, the rate of population growth hit an all-time high of 2.0 percent. To appreciate how rapid that growth rate was, you can calculate the **doubling time**. Applying the often-used "Rule of 70," in which 70 is divided by the growth rate, the doubling time is the number of years required for the human population to double (assuming that the rate of growth would be unchanged over the entire period). In 1968, the human population was growing at a rate that would, if unchanged, have doubled in 35 years. That rate did slow down, however, illustrating the limited usefulness that doubling time has in projecting future growth. At the 2011 population change rate of 1.2 percent, our numbers would double in 58 years.

Doubling time may be only approximate, but it is a good tool for comparing among countries and also illustrates that human populations have the potential for exponential growth; that is, not an incremental or arithmetic increase from 1 to 2 to 3 to 4, and so on, but geometric growth from 2 to 4 to 8 to 16, for example. (Exponential growth in the context of the famous Malthusian scenario is discussed on page 58.) The annual rate of population change worldwide is depicted by country in •**Figure 3.9**.

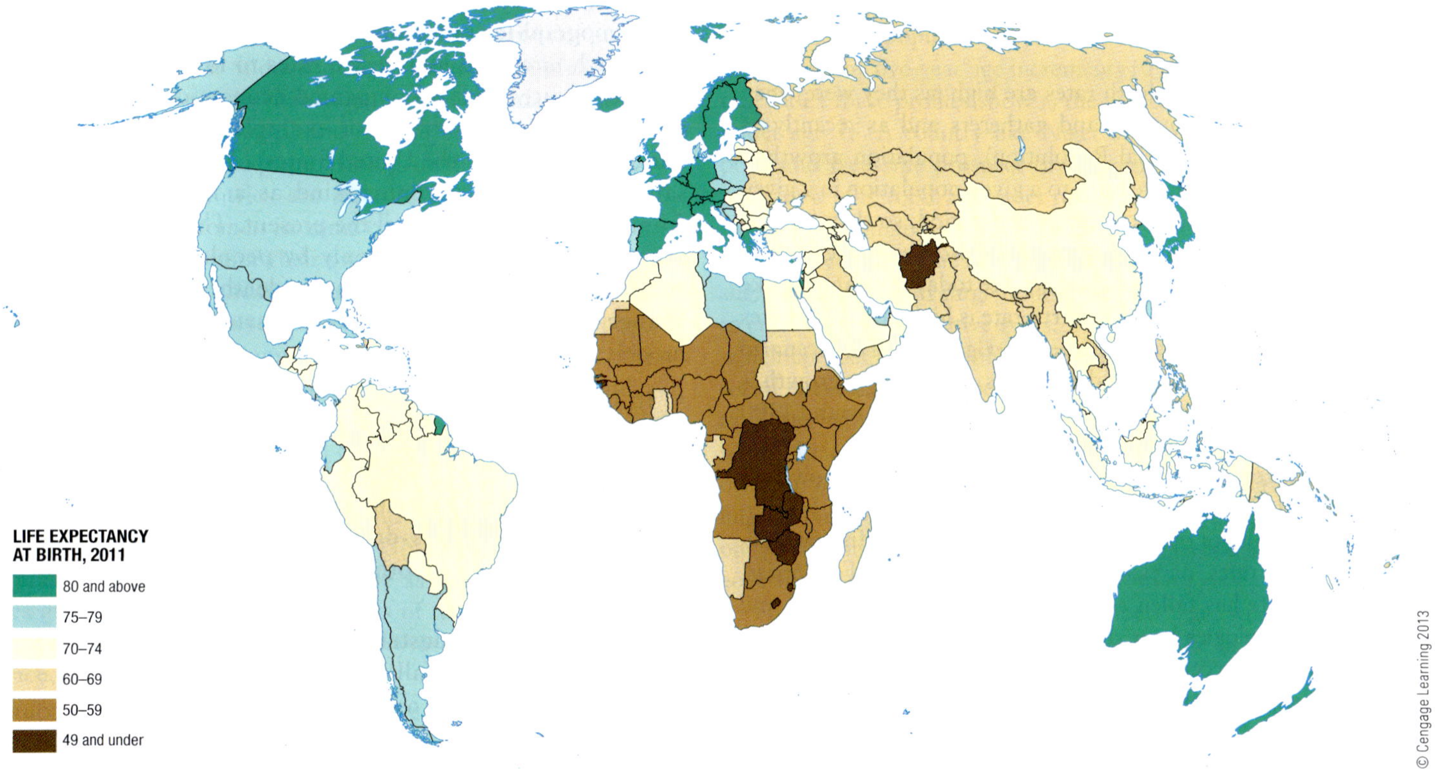

• **Figure 3.8** Life expectancy is closely tied to economic well-being; people live longer where they can afford the medicines and other amenities and technologies that prolong life.

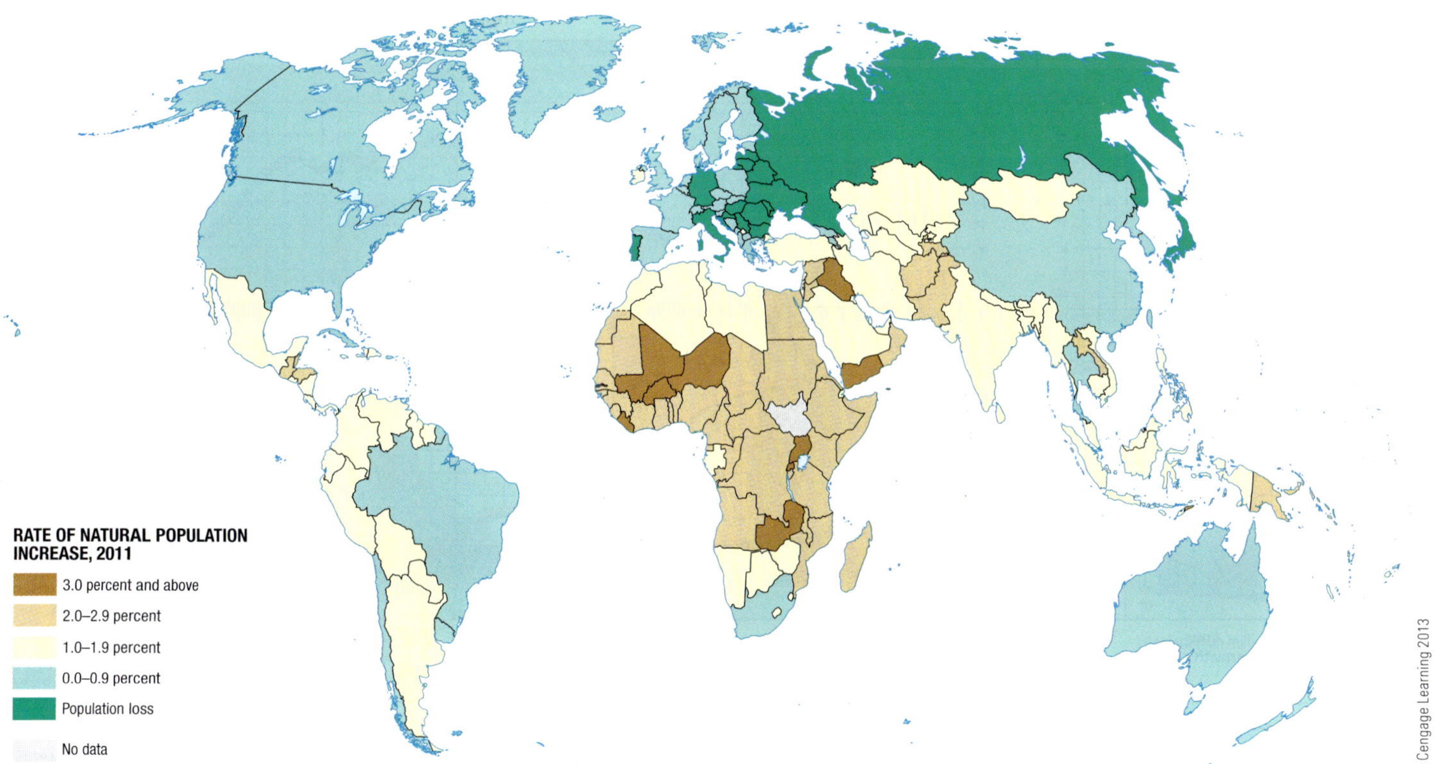

• **Figure 3.9** Population change rates are highest in the countries of Africa and other regions of the developing world and lowest in the more affluent countries.

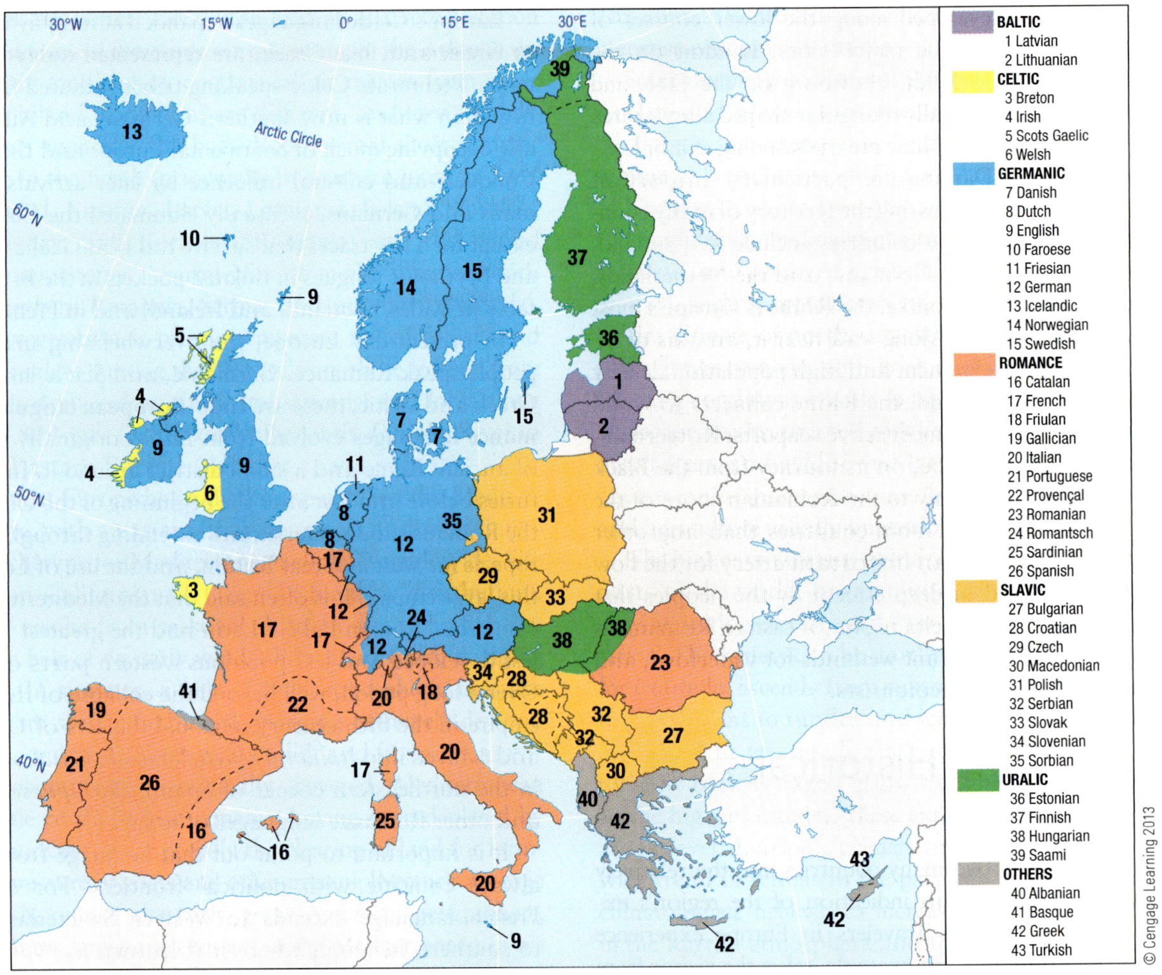

• **Figure 4.11** The languages of Europe.

English is a Germanic language, but it has many words and expressions derived from French, Latin, Greek, and other tongues. Originally, English was the language of the Germanic tribes known as Angles and Saxons who invaded England in the fifth and sixth centuries C.E. The Norman Conquest of England in the eleventh century established French for a time as the language of the English court and the upper classes. Modern English retains the Anglo-Saxon grammatical structure but borrows much vocabulary from French and other languages. English is now the principal language in most parts of the British Isles, having been imposed by conquest or spread by
93 cultural diffusion to areas outside England.

Although the European Union has 23 official languages, English is becoming the de facto *lingua franca* of this 27-nation organization, with about two-thirds of all EU documents drafted in English. By EU law, and at great expense, all those documents must be translated into all those other languages, which most recently have come to include Ireland's Gaelic. Although some of the key institutions of the European Union are in French-speaking cities, French usage across the bloc is declining. One reason is the recent accession of EU member countries in Eastern Europe, where about 60 percent of the population speaks English as a second language, compared with only 20 percent speaking French as a second language. English is by far the most widely taught second language in non-English-speaking European countries, where it is studied by over 90 percent of secondary school students (compared with 33 percent studying French and 13 percent German).

Slavic languages—also in the Indo-European language group—are dominant in most of Eastern Europe. These include *Russian* and *Ukrainian*, the primary languages of Europe's adjoining region of Russia and other former republics of the Soviet Union along Europe's edge; *Polish, Czech,* and *Slovak,* which are spoken in central Europe; and *Serbian, Croatian,* and *Bulgarian,* which prevail in the Balkan Peninsula. They apparently originated in Eastern Europe and Russia, spreading and differentiating from each other during the Middle Ages as people traded and migrated. Closely related to the Slavic languages are the **Baltic languages** of *Latvian* and *Lithuanian.*

A few languages in present-day Europe are not related to any of the groups just discussed. Some are ancient languages that have persisted from prehistoric times in isolated, usually mountainous, locales. Two outstanding examples are *Albanian* (of the Indo-European language family) in the Balkan Peninsula and *Basque* (an indigenous European language without Indo-European roots), spoken in and near the western Pyrenees

Mountains of Spain and France. Some languages unrelated to others in Europe have relatives in Russia and reached their present locales through migrations of peoples westward. The prime examples are *Finnish, Saami, Estonian,* and *Hungarian* (also called Magyar). These are all **Uralic languages,** thought to have originated in the Ural Mountain region of Russia and Kazakhstan. *Turkish,* an **Altaic language** that originated in central Asia, is spoken by the ethnic Turkish inhabitants of northern Cyprus. The **Roma (Gypsies)** speak *Romany,* in the
108, 225 **Indic** branch of the Indo-European languages.

What patterns can be seen in the rich and complex linguistic mosaic of Europe? Keeping in mind that language is a key component of ethnic identity, that question may be best answered by appreciating the seven ethnic patterns of Europe, as identified by David Levinson:[5]

1. Several nations are populated overwhelmingly by a single ethnic group—for example, the Germans of Germany, the Poles of Poland, and the Dutch of the Netherlands.
2. There are also substantial populations of ethnic groups that originated in other nations living in "foreign" countries (for example, the Poles of Lithuania), where they are known as **national minorities.**
3. There are groups distinguished more by culture than by ethnicity; they share a common culture but have many different ethnic origins (for example, the French and the Italians).
4. There are a few ethnolinguistic minorities, like the Saami (Lapps) of Scandinavia and the Ladin of Italy, who have resided in an area for a very long time but have retained cultural distinctiveness.
5. There are also regional ethnic minorities that have a major presence in a particular nation, where they are culturally very different from the country's other ethnic groups; examples are the Basques and Catalans in Spain and the Bretons in France. Such groups often call for the devolution of power from the host country to their ethnic subregion (see Insights, page 80).
6. There are recent immigrants from outside Europe, including guest workers mainly from former European colonies in Asia, Africa, and the Americas; examples are the Pakistanis and Indians of Britain and the North Africans of France.
7. Finally, there are ethnic groups with members all across Europe but without a homeland there; the Jews, who originated in modern Israel and adjacent areas (see Chapter 6, page 170) and the Roma (Gypsies), who probably originated in India, are the most prominent of these. The characteristics, contributions, and conflicts of Europe's ethnic groups are discussed in more detail later in the chapter.

Europeans' Religious Roots

Religion is one of the most important attributes of culture. On first glance, Europe's religious geography seems relatively straightforward: one of this region's principal culture traits is the dominance of **Christianity** (see the map of Europe's faiths in •Figure 4.12; the discussion of Christianity in Chapter 6, pages 172–173; and a note on the importance of religion in culture on page 4). But Christianity has a broad spectrum in Europe, where Christian encounters with other believers became part of the continent's troubled history. Initially outlawed by the Romans, Christianity was embraced as the faith of the realm by the Emperor Constantine in the fourth century, and it became a dominant institution in the late stages of the empire. It survived the empire's fall and continued to spread, first within Europe and then to other parts of the world. In total number of adherents, the **Roman Catholic Church,** headed by the pope in Vatican City in Rome, is Europe's largest religious group (with 280 million followers, or 45 percent of Europe's Christians), as it has been since the Christian church was first established (•Figure 4.13). Today, it remains the principal faith of a

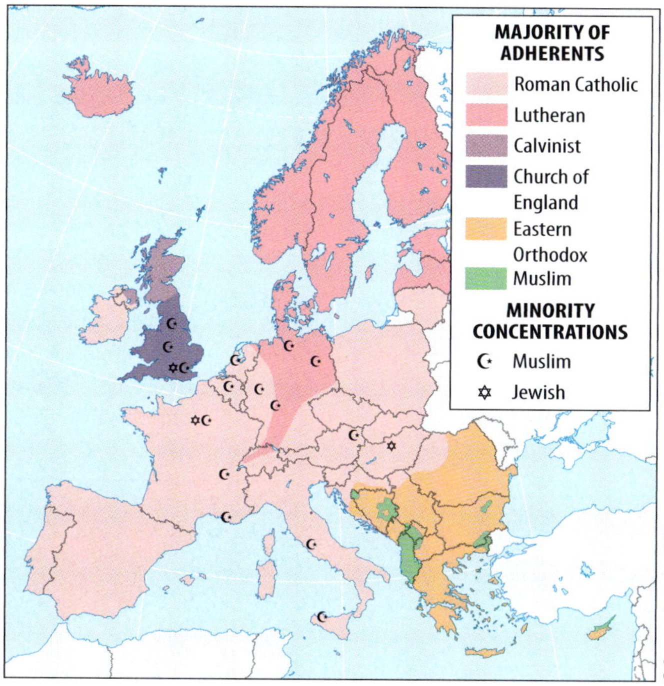

• **Figure 4.12** The religions of Europe. Sects of Christianity dominate, but Islam has a strong presence in the Balkans and, along with Judaism, in cities throughout Europe.

• **Figure 4.13** The Vatican in the "Eternal City" of Rome is the seat of the Roman Catholic Church.

INSIGHTS | Devolution

A notable political process in modern Europe, often related to issues of linguistic, religious, and ethnic identity, is **devolution**, the dispersal of political powers to ethnic minorities and other subnational groups. Like that of many
127, 174, 262, 404 other world regions, Europe's historical geography is rife with dominant groups occupying the traditional lands of smaller minority groups. Minorities are sometimes absorbed into the larger culture and political system, but more often than not, they retain some degree of distinctiveness and a desire to control their own affairs.

This book relates numerous examples from around the world of strife and war taking place because the dominant power refuses to yield any measure of authority to minority ethnolinguistic groups. But in Europe, the process of devolution is occurring peacefully for the most part and more energetically than ever. This may be because the European countries, after centuries of strife, are finally self-assured of their sovereignty and well-being and are so focused on international cooperation that demands for autonomy among their minorities are no longer perceived as threats. Devolution is an important issue in Wales, Scotland, Northern Ireland, Spain, and elsewhere; **•Figure 4.B** is a map of the major locales where devolution has occurred or is being appealed for in modern Europe.

• Figure 4.B Devolutionary areas of Europe. Power is being devolved from central governments to regional authorities in many parts of Europe. This map depicts places like Scotland and Wales, where the transfer has taken place, and others like Corsica, where devolution is anticipated, declared, or sought.

highly secularized Europe. The main areas that are predominantly Roman Catholic are Italy, Spain, Portugal, France, Belgium, the Republic of Ireland, large parts of the Netherlands, Germany, Switzerland, Austria, Poland, Hungary, the Czech Republic, Slovakia, Croatia, and Slovenia.

During the Middle Ages, a center of Christianity evolved in Constantinople (now Istanbul, Turkey) as a rival to Rome. From that seat of power, the **Eastern Orthodox Church** spread to
125, 172 Greece, Bulgaria, Romania, Serbia, Montenegro, and Macedonia, as well as to Ukraine and Russia in the east.

In the sixteenth century, the **Protestant Reformation** took root in various parts of Europe, and a subsequent series of religious wars, persecutions, and counterpersecutions left **Protestantism** dominant in Great Britain, northern Germany, the Netherlands, Denmark, Norway, Sweden, Iceland, and Finland. Except for the Netherlands, where a higher Catholic birth rate has since reversed the balance, these areas are still mainly Protestant, with the *Church of England* and *Calvinism* the strongest sects in Britain and *Lutheran Protestantism* dominant in Germany and Scandinavia. Among the three Baltic countries, Lutheran Protestantism is the dominant faith in Estonia and Latvia, while most Lithuanians are Catholic.

The fact that Christians form a vast majority is a somewhat misleading characterization of religion in Europe. Mainstream Christianity is in decline as Europeans have become increasingly secularized in recent decades. Historic legacy plays a strong role in the trend: over a period of centuries, Europeans paid a costly toll in lives by fighting for causes associated with religion, and they increasingly associate peace with secularism. Recent polls indicate that almost 75 percent of Europeans believe in God, but in most countries, even larger majorities shun traditions like attending church. In predominantly Catholic France, about one person in 20 attends a religious service every week (compared with one in three in the United States). In Poland, 95 percent of the people identify themselves as Catholic, but only about 40 percent attend church on Sunday. Various polls suggest that somewhere between 15 and 33 percent of Italians attend church in mainly Catholic Italy. Polls indicate that overall, about 20 percent of all Europeans regard religion as "very important" to them (compared to about 60 percent of Americans). The Christian churches that are growing—even surging—in attendance tend to be independent *Pentecostal churches* attended by immigrants from Nigeria, Sierra Leone, and other African countries.

The faith of **Islam** (practiced by **Muslims**) has the fastest-growing number of adherents of any world religion and is the fastest-growing religion in Europe (for a description of Islam, see Chapter 6, pages 173–175). Muslims are a majority in just three European countries: Albania, Kosovo, and Bosnia and Herzegovina on the Balkan Peninsula. Islam was once widespread in the Balkans (as that region is often called), where the Ottoman Turks established it during their period of rule from the fourteenth to nineteenth centuries. It was also the religion of the **Moors**, who ruled the Iberian Peninsula from the eighth century until the end of the fifteenth, when Spain's Catholic monarchy ousted them. Islam is growing throughout the continent, particularly among immigrant communities in Western Europe. Estimates vary, but France has as many as 4.7 million Muslims, followed by Germany with about 4.1 million, Britain with 2.8 million, and the Netherlands with almost 1 million.

As related in the earlier discussion of immigration, Muslims in Europe are often received with fear, suspicion, or hostility by "native" Europeans. But nothing can compare with the experience of the Jews as a minority in Europe. One of humankind's greatest tragedies is the story of their plight on the continent (see also Chapter 6, page 172). The Romans forcibly dispersed these followers of **Judaism** to Europe from Palestine in the first century C.E., and over time, they became more numerous in Europe than anywhere else. In 1880, before many emigrated to the United States, 90 percent of the world's Jews lived in Europe. In 1939, as World War II erupted, there were 9.7 million Jews, 60 percent of the world total, in Europe. By the end of the war, systematic murder by the Nazis in the **Holocaust** reduced their numbers to 3.9 million. About half of those subsequently emigrated to Israel and elsewhere, and there are about 1 million Jews in Europe today.

European Colonialism and Its Consequences

Far more than any other region, Europe has shaped the human geography of the modern world. Before the late fifteenth century, Europe played only a minor role in world trade patterns: goods moved from southern France and northern Italy northwest to Britain, for example, and more active trade connected Italy with the margins of the Mediterranean Sea and beyond to southwestern Asia. But the most important global trade routes were much farther east and south. The longest link was the 5,000-mile (8,000 km) **Silk Road** that moved goods overland (and on the Mediterranean Sea) from Chang-an (Xi'an) in China to Venice. 178,

The balance of world affairs started to shift to Europe with 230
the beginning of the Age of Discovery in the fifteenth century. European sailors, missionaries, traders, soldiers, and colonists burst onto the world scene. Although these agents had dif- 41
ferent motives, they shared a conviction that European ways were superior to all others, and by various means they sought to bring as much of the world as possible under their sway. Economic benefits for the European homelands were among their main motivations. By the end of the nineteenth century, Europeans had created a world in which they were economically dominant and exercised great influence on indigenous cultures.

The process of exploration and discovery by which Europeans filled in the world map began with Portuguese expeditions down the west coast of Africa. In 1488, a Portuguese expedition headed by Bartholomeu Dias rounded the Cape of Good Hope at the southern tip of Africa and opened the way for European voyages eastward into the Indian Ocean. In 1492, North America was brought into contact with Europe when a Spanish expedition commanded by a Genoese (Italian) man named Christopher Columbus (Cristóbal Colón) crossed the Atlantic to the Caribbean Sea. Less than half a century later, these early feats of exploration culminated in the first circumnavigation of the globe by a Spanish expedition commanded initially by a Portuguese man, Ferdinand Magellan. Worldwide exploration continued apace for centuries, with the Dutch, French, and English eventually wresting leadership from the Spanish and Portuguese.

The explorers were the vanguards of a global European invasion that would in turn bring the missionaries, soldiers, traders, settlers, and administrators. They gradually built the frameworks of European colonization of the Americas, Africa, Asia, and the Pacific. The colonial patterns are described on pages 44–45 in Chapter 3 and are not repeated in detail here. Briefly, their main manifestations were the transfer of wealth from the colonies to Europe, the movement of great numbers of indigenous laborers to serve European interests in other

Finally, Europe's leadership was weakened by a new dependence on outside sources of energy. The region's traditional reliance on coal was superseded by technological developments and environmental concerns, and the energy balance shifted to petroleum and natural gas. Despite the development of North Sea oil and gas resources (which mainly belonged to the United Kingdom and Norway), Europe became, and remains, highly dependent on imported Middle
97 Eastern oil.

An Imbalance of Wealth

A striking pattern in Europe's economic geography is that Western Europe is wealthier than Eastern Europe. This trend dates to at least the 1870s, when per capita incomes in the west were twice those in the east. The gap grew even wider throughout the twentieth century. After World War II, Eastern European countries were in effect colonized by the Soviet Union, serving as vassal states that gave up human and material resources to service the motherland. Only recently, with the dissolution of the Soviet Union in 1991 and the admission of Eastern European countries to the European Union, have the eastern countries been able to hope that they may eventually catch up with their western counterparts. Membership in the European Union is structured to help redistribute wealth from west to east, through mechanisms such as agricultural subsidies that could help to breathe new life into the faltering farms and factories of the east. As we will see however, the newer member states have not been automatically granted all the privileges enjoyed by the older, wealthier members.

Living off the Land and Sea

Despite the dominance of a postindustrial, urban economy, some European countries continue to rely significantly on the productivity of land and sea. Agriculture was the original foundation of Europe's economy, and it is still important. Food provided by the region's agriculture allowed Europe to become a populous area at an early time. After about 1500, a period of steady agricultural improvement began. Introduction of important new crops such as the potato played a part, but so did such practical improvements as new systems of crop rotation and scientific advances that produced better knowledge of the chemistry of fertilizers. The continuing expansion of industrial cities provided growing markets for European farmers, who received protection through **tariffs** (tax penalties imposed on imports) or direct **subsidies** (payments made to domestic producers) to encourage production and support rural incomes (these are also discussed on page 420 in Chapter 11, in the context of similar protective measures by the United States). As a result of European Union policies, surpluses of many products are common, including grains, butter, cheese, olive oil, and table wines. But Europe is not agriculturally self-sufficient and depends on trade for much of its foodstuffs.

© Jeffrey L. Rotman/CORBIS

• **Figure 4.15** Cod fishing in the North Atlantic Ocean.

Europe, like the United States, is finding itself increasingly at odds with the rest of the world as global free-trade agreements insist that its doors be open to more (and generally cheaper) food imports. Europe wants to protect its farmers, particularly those who raise what it calls the "sensitive" products (beef, chicken, and sugar), with high tariffs on imported foods and with subsidies that allow farmers to "dump" their surplus production onto the world market at very low prices. The European Union says it is doing enough to open its doors to imports because it imposes no tariffs on imports from the world's 50 poorest countries, but free-trade advocates insist that more trade barriers must fall.

Throughout history, fishing has been an important part of the European food economy, and the coasts still have many busy fishing ports (•**Figure 4.15**). At times, control of fishing grounds has been a major commercial and political objective of nations, even resulting in warfare. Fisheries are particularly important in shallow seas that are rich in plankton, the principal food of herring, cod, and other fish of commercial value. The Dogger Bank (to the east of Great Britain) in the North Sea and the waters off Norway and Iceland are major historical fishing areas, and Norway, Iceland, and Denmark are Europe's leading nations in total catch. Fishing has long been especially crucial to countries such as Norway and Iceland, which have little agriculturally productive land. However, overfishing of cod, the world's most popular food fish, sometimes known as the "beef of the sea," has grown so severe, particularly in the North Sea, that since 2006 the European Union and Norway have legislated cuts of cod fishing by as much as 25 percent. Environmentalists have cried foul; organizations, including the World Wildlife Fund and Greenpeace, have insisted on a complete ban on cod fishing. Already, plummeting fish catches have caused economic hardship in many traditional fishing
communities on North Sea shores. 29

Postindustrialization

Europe's economy today, like its population trend, is best characterized as **postindustrial.** In the European context, this means that since about 1960, the region has experienced **deindustrialization,** shifting by choice and by force away from energy-hungry, labor-costly, and polluting industries toward an economy based on production of high-tech goods and on services. The high-technology products include electronic devices, data processing hardware, **robotics,** and telecommunications equipment. Other industries like drug and pesticide manufacturing make use of these technological products to create their own, so there is an important interconnectedness centered on high technology. Europe is also renowned for the production and export of high-quality, expensive luxury brands, including Ferrari automobiles and Rolex watches. In the service sector, growing proportions of Europeans are employed in transportation, energy production, health care, banking, retailing, wholesaling, advertising, legal services, consulting, information processing, research and development, education, and tourism.

Deindustrialization has had important spatial consequences for Europe. As we have seen, cities and industries grew up where important natural resources like coal and iron ore were located. Now most of these geographical correlations and barriers have been lifted. With the capabilities of the Internet and the advantages of fast and convenient transport of goods, it is possible for entrepreneurs to begin new enterprises almost anywhere. European cities are generally cleaner than they once were, and as the book describes in North America, neglected and decaying cores
429 of formerly industrial cities are being rehabilitated and gentrified. Many European cities are exceptionally attractive places to live and visit, providing amenities that encourage socializing, good health, and enjoyment of the museums and historical legacies that endow each city and give it its unique character. The pedestrian and bike-only street in Amsterdam seen in •Figure 4.16 exemplifies the people-friendly character of many European cities.

Neither high-tech nor service industries employ as many people as the old manufacturing sector did, and like the United States, the European countries are grappling with problems resulting from industrial unemployment. There are concerns over where new jobs will come from and how to make the transition from manufacturing to service roles without too much social dislocation. Many European nations for a long time fit the model of the **welfare state,** using the resources collected through high taxation rates to provide generous social services to citizens. That "social model" of the European state is, however, under tremendous pressure from the forces of globalization, the costs of absorbing immigrants, the effects of the global financial crisis of 2008–2009, and
the debt crisis that erupted in Europe soon after that. 88

Joe Hobbs

• **Figure 4.16** to come.

The European Union, Built on a Market

The European Union, which maintains its headquarters in the Belgian capital of Brussels, is Europe's most important **supranational organizations** (an organization in which member countries are united beyond the authority of any single national government and are planned and controlled by a group of nations). The countries of the European Union are depicted in •Figure 4.17.

The European Union began in 1957, when the Treaty of Rome created the **European Economic Community (EEC),** also known as the **Common Market,** made up of France, West Germany, Italy, Belgium, Luxembourg, and the Netherlands. By 1996, three years after it acquired the name European Union, the organization had an additional nine members: the United Kingdom, Denmark, Ireland, Greece, Portugal, Spain, Austria, Finland, and Sweden.

Much of the European Union's reason for being is economic. The EEC was designed, as the European Union is today, to secure the benefits of large-scale production by pooling the resources—natural, human, and financial—and markets of its members. Tariffs (taxes) were eliminated on goods moving from one member state to another, and restrictions on the movement of

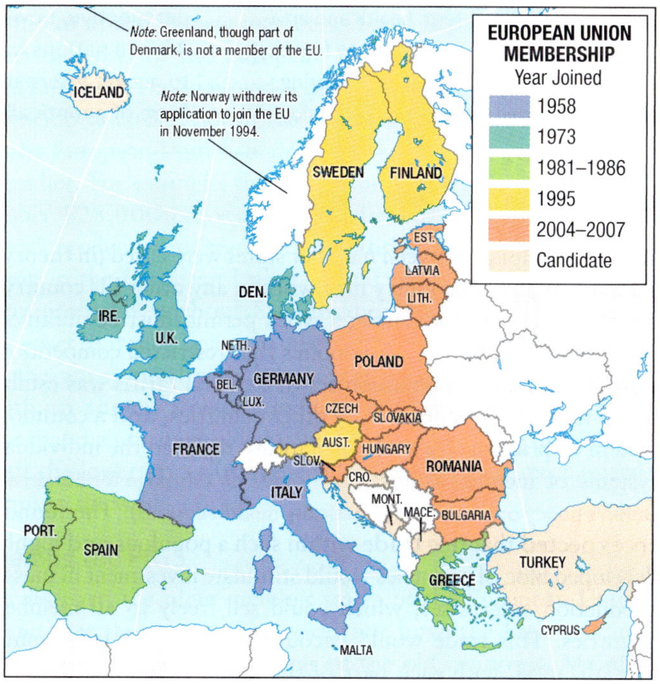

• **Figure 4.17** Members of the European Union.

Source: Adapted from "European Union Survey," The Economist, May 31, 1997, p. 5.

burden. But Germany had its own economic pains from absorbing the costs of reunion and integration with the poor former East Germany and is concerned about a far more ex-
46 pensive integration on a much grander scale. Despite hopes among many in the newly joining states that membership will bring them prosperity, there will also be at least short-term pain. Food prices have risen sharply, while salaries have increased slowly. Many western Europeans fear that the lower salaries and taxes of the new eastern member countries will lure industries and jobs away from the west, contributing to the decline of their home economies.

A similar set of problems revealing the strains within a united Europe has plagued the Eurozone. In this case, the European Union's strongest economies, those of Germany (which is seen as the "economic locomotive" of the continent) and France, have come under pressure to help out the weaker countries—especially Portugal, Ireland, Greece, and Spain, known collectively by the unfortunate acronym **PIGS**. After they joined the common currency, these countries embarked on spending and borrowing sprees, building up huge deficits and debts (and apparently hoping that in case of any troubles the larger economies of the Eurozone would bail them out). In the debt crisis that followed, beginning in 2010, a number of economic bubbles in these countries burst—most viciously in Spain's real estate. Some countries had to introduce severe economic austerity programs, including salary cuts, which led to growing unemployment, poverty, and social unrest.

The euphoric vision of the euro has faded away, and new words have crept into common usage: Euromess, Europhobia, Eurocrisis, and especially the **European debt crisis** or "sovereign debt crisis" among them. The European debt crisis began early in 2010 when it appeared that Greece might default on its loans. There was no way to address that problem quickly (in fact, a tentative bailout for Greece's debt was not put together until mid-2011), because all 17 of the Eurozone's members had to be consulted for input. The European Central Bank does not have much authority to make binding decisions for all its members. Individual countries in the Eurozone have different goals and expectations. The wealthier ones, like Germany, do not want their taxes used to bail out countries like Greece that are heavy borrowers. The poorer indebted countries like Greece do not want wealthier outsiders like the Germans to dictate how their government taxes their citizens and spends their money.

The Eurozone's finance ministers began studying the way the United States runs its financial structures and started considering a more centralized approach like the United States has. The Europeans may finally end up financially as a kind of "United States of Europe," with a central finance body, like the U.S. Treasury Department, with power over national budgets and tax regimes. The organization would enable the Eurozone countries to make common fiscal policies for all 17 members. The notion of a fiscal union is extremely unpopular among the public in many of these countries, but economic crisis may push them inevitably toward greater unity. The Eurozone countries will have to draft and ratify a new treaty, a process that will take years, but the alternative may be the fragmentation of the union. In the midst of the crisis in 2011, the prevailing motto among European finance ministers was "centralize or die."[6]

4.5 Geopolitical Issues

In the past 100 years, Europe's geopolitical situation changed more profoundly and more violently than that of any other world region. Europe experienced two world wars that wrought unprecedented devastation. World War I (1914–1918), known until 1939 as the Great War, saw Europe fracture on fault lines of military alliances. Germany and the Ottoman Empire (based in what is now Turkey) were solidly defeated by an alliance led by Britain, France, Russia (early on), and the United States (later). The war accomplished little and destroyed much, and the peace treaty placed a costly burden on
reparations on the defeated Germany. 96

The economic and social humiliation of Germany helped plant the seeds of the rise of Adolf Hitler's fascist Nazi (National Socialist) Party and the outbreak of World War II (1939–1945). Hitler's Germany surged in strength and allied with Italy, the Soviet Union (early), and Japan (later) in a military quest to establish European and Far Eastern empires. Early in the war, Germany advanced swiftly on the European mainland and Scandinavia and set its sights on an occupation of Britain. When Japan bombed a U.S. naval base at Pearl Harbor, in the Hawaiian Islands, in 1941, the United States was drawn into the war firmly on Britain's side. Soon joined by the Soviet Union, these forces, assisted by many smaller ones, began to turn the tide against Germany. In 1945, with German cities in ruin and two Japanese cities nearly obliterated by the first atomic bombs, the war ended. In the European theater, about 50 million people had died. "What is Europe now?" the British prime minister, Winston Churchill, asked. "It is a rubble heap, a charnel house, a breeding ground of pestilence and hate."[7]

Postwar Europe

The wounds of that gigantic conflagration have healed. Forward-looking Europe is intent on establishing, in the form of the European Union, that "virtual" United States of Europe with a federation of nations similar to the states of the United States. Remarkably, George Washington said "Some day, following the example of the United States of America, there will be a United States of Europe". The European Union is the latest and largest of the various postwar European supranational organizations. There are many reasons, largely the economic ones discussed above, for this push toward greater unity. One of the most important but understated reasons is simply that Europeans want their continent to be warproof, with its member countries so intertwined in economics, foreign policy, and other ways that war among them would be all but impossible.

Europe is still recovering and reorganizing from another momentous geopolitical shift: the end of the **Cold War**, which came with the collapse of the Soviet Union in 1991. When

World War II ended, Soviet forces and political institutions moved in on most of the countries of Eastern Europe, including part of defeated Germany. As Churchill memorably described it, an **"Iron Curtain"** had descended from Poland southward to the Balkan Peninsula. West of it, the capitalist democracies led by Britain, France, West Germany, Italy, and the United States prevailed; to the east, across a 4,000-mile (6,400 km) line of concertina wire, walls, minefields, and guard towers, the Soviet Union controlled authoritarian, state-run regimes in the countries that came to be known as its European "satellites," including East Germany, Poland, Czechoslovakia, Hungary, Romania, and Bulgaria. The United States nurtured Western Europe's interests and from 1947–1952 pumped billions of dollars into the reconstruction of the devastated European infrastructure in a program known as the **Marshall Plan.**

The United States was also the linchpin of the military alliance known as the **North Atlantic Treaty Organization (NATO),** founded in 1949 between the United States, Canada, most of the European countries west of the Iron Curtain, and Turkey. NATO faced off against the **Warsaw Pact,** an alliance of the Soviet Union and its Eastern European satellites. On more than one occasion, notably in the Cuban Missile Crisis of 1962 and in the Arab-Israeli War of 1973, these military
178, alliances came all too close to apocalyptic nuclear war.

373 In 1989 the Berlin Wall, which separated the German city that epitomized the Cold War, was dismantled in a popular and peaceful uprising. The Soviet grip on East Germany and its other satellites had been slipping for years, and now went limp. The Soviet Union itself disbanded in 1991 (see Chapter 5, page 131). An exhilarated public throughout Eastern Europe embraced the prospects of democracy, capitalism, and material improvement. The Warsaw Pact was dissolved. The nuclear arsenals of the respective alliances were reduced substantially. Plans were made to turn the path of the Iron Curtain into the **European Greenbelt,** a mosaic of national parks and other protected areas stretching across 18 countries, from the Barents Sea to the Black Sea. There is an analogous place in the world where a protected natural area may emerge from a former war front: the demilitarized
274 zone (DMZ), separating North and South Korea.

Since the end of the Cold War, NATO has continued to serve an important peacemaking role. In the Balkans during the 1990s, NATO forces reversed the advance of the Serbs against their neighbors in the former Yugoslavia (see page 107). More recently, NATO members deployed their military assets against the regime of Muammar Gaddafi in Libya. In that conflict, the United States preferred that NATO take a lead role, especially to prevent the United States from having only
194 a militaristic reputation in the Middle East. NATO's membership has grown to 28 with the addition of nine former Soviet allies (Croatia, Estonia, Latvia, Lithuania, Slovakia, Slovenia, Romania, and Bulgaria; see •Figure 4.19). Russia resented this Western expansion into its former sphere of influence, but its opposition was tempered by its own growing alignment with Western defense and security concerns.

• Figure 4.19 European members of the North Atlantic Treaty Organization (NATO).

Sources: EU map based on "European Union Survey," The Economist, May 31, 1997, p. 5. Data from EUROSTAT (Statistical Office of the European Communities); World Factbook, CIA.

Welcome to Schengenland?

In theory, the European Union would like to move toward a situation in which there were no passport, visa, or other control issues at any internal land, sea, and airport frontiers of its member countries. This would be an extraordinary geopolitical bloc, considering Europe's fractious history. A framework that began outside the European Union and now operates within it has attempted this kind of integration. Known as the **Schengen Agreement,** it allows free circulation of people between the nations that signed the agreement; these are the EU countries (except for Britain, Ireland, Romania, and Bulgaria), plus non-EU Iceland, Norway, and Switzerland. To maintain internal security, within this greater **Schengenland,** the member states are supposed to exercise common visa, asylum, and other policies at their external borders. Some countries, notably the United Kingdom, are skeptical of the effectiveness of control of external borders and have stayed out of Schengen.

Truly open borders are probably still far in the future. Anti-immigrant fears that rose with the exodus of refugees from North Africa during 2011's Arab Spring have caused some countries, notably France and Denmark, to backpedal on their commitment to the principle of open borders. Even two coun- 72,
tries that recently joined the European Union—Romania and 192
Bulgaria—have been rejected for membership in Schengenland. Western European fears of a tidal wave of cheap eastern labor, terrorists, or human traffickers have produced special rules for these eastern countries, long regarded as abodes for these shady activities. Like the Euro, Schengenland is one of the big European goals that is difficult to proclaim as a success.

Differences between Europeans and Americans

Many of us instinctively equate the United States with Europe when we speak about global patterns; after all, we share many common cultural roots, share a great deal of the world's wealth, and have many of the same measures of quality of life. But there are differences, and these seem to have grown in recent years. "Americans are from Mars and Europeans are from Venus," wrote policy analyst Robert Kagan.[8]

Here are a few differences—and please keep in mind that these are not hard facts, but generalizations that reflect policies and public opinions. The concept of social justice, in which steps are taken to reduce unemployment and poverty among the least privileged, is a key value in Europe, whereas it lags in the United States. European governments and people have historically had a stronger sense of the social compact between the state and its citizens. Governments have been expected to provide "free" public services like education and health care, while the people have understood they must foot the bill for these "free" services by paying very high taxes. (Incidentally, undergraduate public education in most European nations is free. However, per capita spending on public education is higher in the United States, and an American graduate degree is widely regarded as superior to that obtainable in Europe.) Europeans accept high taxes on gasoline as an incentive to drive small cars and conserve energy, whereas for an American politician, advocating higher fuel taxes is political suicide; as a result, gas prices in Europe are roughly double those in the United States. Europeans complain that U.S. cultural industries such as Hollywood films humiliate other cultures and marginalize languages other than English. For awhile, there was strong resistance, especially in France, to the proliferation of McDonalds and other American fast-food restaurants in Europe; but that has largely faded, and McDonalds has "Europeanized" its restaurants in both menus and architecture. Europeans are less inclined than Americans to allow questions of spirituality into political debate; in fact, they are repelled by personal expressions of religiosity in public, which they consider indiscreet. Europeans abhor the death penalty, which is outlawed in all EU countries.

Europeans tend to apply the **precautionary principle**—the notion that it is appropriate to attempt to reduce or eliminate any risky practice—to things they think may be harmful. Thus there may be risks from GM foods, so many people would like to ban them; climate change is probably occurring, so most people want to take action to reduce greenhouse gases. The United States has adopted more of a "show me" attitude—for example, with respect to global warming; compared with Europeans, a large percentage of Americans think there is not enough scientific proof, so it is better not to impose reductions
35 in carbon dioxide emissions. There are differences on the geopolitical front too. For Americans, September 11, 2001, was the watershed time when it reevaluated its place in the world and began to feel less secure. For Europeans, the fall of the Berlin Wall in 1989, leading to the collapse of the Soviet bloc, was the watershed that made them start to feel more secure.

4.6 Regional Issues and Landscapes

In this section, we take a closer look at some problems and issues of interest in the European subregions described on page 65. We begin by looking at some of the elements that made the European Core the continent's strongest subregion, and then examine characteristic or exceptional qualities of Northern, Southern, and Eastern Europe.

The European Core

Properties of the European Core

In Chapter 1, we saw that core and peripheral locations are among the simplest and most common properties of the Earth's countries and regions (see page 10). Europe has a recognizable core, a subregion that has long played a dominant role in the continent's political, economic, and cultural development. Today, the European Core is recognizable as a prominent subregion of Europe with the following traits:

- Densest, most urbanized population
- Most prosperous economy
- Lowest unemployment
- Most productive agriculture
- Most conservative politics
- Greatest concentration of highways and railroads
- Highest levels of crowding, congestion, and pollution[9]

Defined in this way, the core consists mainly of three of Europe's major countries and of smaller nations in the British Isles and the west central portions of the European mainland (•**Figure 4.20**). The United Kingdom (U.K.) and the Republic

• **Figure 4.20** The political geography of Europe's core.

of Ireland share the British Isles. The countries of west central Europe are Germany and France and their smaller neighbors, the Netherlands, Belgium, Luxembourg, Switzerland, and Austria. Also sharing these traits of the core are roughly the northern half of Italy, Denmark, southern Sweden, and parts of Poland, but other characteristics of these countries place them more appropriately in the European Periphery, discussed later. Because they are nestled among the core countries, the microstates of Liechtenstein, Monaco, and Andorra are included here.

The European Core includes some of the world's most geopolitically and economically significant countries (see Table 4.1, pages 67–68). As we saw in Chapter 1, geography is very concerned with appreciating how people have reshaped the face
3 of the Earth. This European subregion became prosperous in part through the exploitation of its own natural resources. Today, it has one of the world's most transformed landscapes, shaped by thousands of years of productive agriculture and hundreds of years of industrial development and affluence (•**Figure 4.21**). European countries do not have the natural advantage enjoyed by one of their affluent peers, the United States: vast size. As we will see when we get to North America in Chapter 11, there is actually more forest in the United States now than there was when European settlement began (see page 398). European countries have not enjoyed the luxury of allowing vast parts of their hinterlands to reclaim something similar to their "presettlement" environments.

The European Core is one of just four regions in the world classified as a **major cluster of continuous settlement,** meaning that all habitations lie no more than 3 miles (5 km) from other habitations in at least six different directions. In addition,

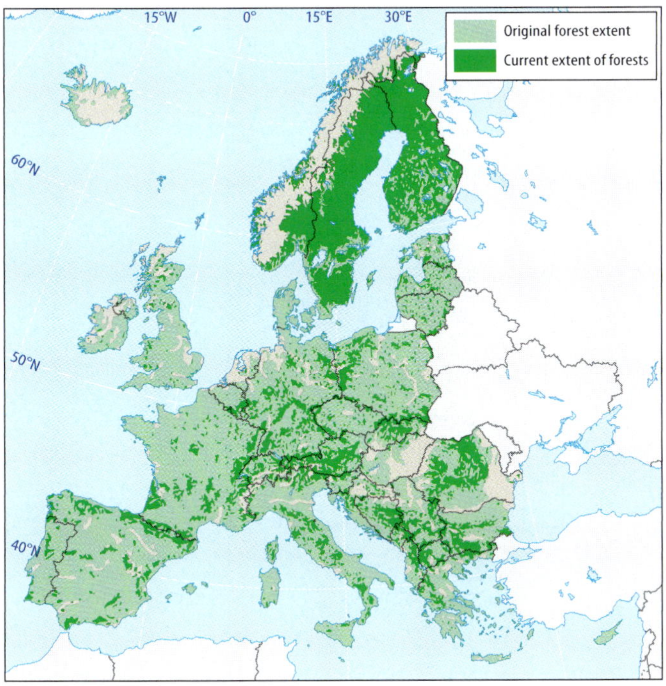

• **Figure 4.21** Human activities over long periods of time have transformed the natural landscapes of Europe, especially in the core region. This map depicts areas that once sustained temperate mixed forests and coniferous forests but have lost that tree cover.

• **Figure 4.22** Historical industrial concentrations, cities, and seaports of the European core. Older industries such as coal mining, heavy metallurgy, heavy chemicals, and textiles clustered in these congested districts. Local coal deposits provided fuel for the Industrial Revolution in most of these areas, which have shifted increasingly to newer forms of industry as coal has lost its economic significance and older industries have declined.

roads or railroads lie no farther than 10–20 miles (16–32 km) away in at least three directions.[10]

Europe's cultural landscape is unusually intricate. Different layers of history appear at every turn, and there are sharp differences in human geographies and national perspectives of its peoples. Environments and resources also vary widely. Local resources (especially **coal** and iron ore) and transportation opportunities were the basis for industrial and urban development in past centuries—that correlation is seen in •**Figure 4.22**—but the modern European Core has a pronounced postindustrial geography. There are distinct areas of 84
declining and growing industrial activity. Resources and industries that put many major cities on the map are played out or are in serious decline.

This process of deindustrialization hit Britain especially hard, but other regions in the European Core, notably the Ruhr in Germany, also suffered. High unemployment is a major consequence of this trend, but a positive result is that the cities are cleaner and more livable. Many cities are replacing old industrial areas with greenbelts and attractive architecture. Improved environments should lure in high-tech and other clean industries, services, and tourists that would drive a new economy (I emphasize *should* and *would* because these trends were underway until the gigantic forces of economic recession stopped them in their tracks, at least temporarily, after the economic crisis erupted in 2007). These greener industries tend to occupy smaller areas than the older, more polluting industries. In a geographical play on words evoking California's Silicon Valley, parts of the British Isles have been dubbed "Silicon Bog," "Silicon Glen" and Silicon Fen," while the continent boasts Germany's "Silicon Saxony," France's "Scientific City," and northern Italy's "Techno City."[11]

Great Britain: Aging Seat of World Power

The islands of Great Britain and Ireland, off the northwestern coast of Europe, are known as the British Isles (•Figure 4.23). They are home to the Republic of Ireland, with its capital at Dublin, and the much larger United Kingdom of Great Britain and Northern Ireland (often referred to simply as "Britain"), with its capital at London. The United Kingdom is made up of the entire island of Great Britain, with its political units of England, Scotland, and Wales; the northeastern corner of Ireland, known as Northern Ireland; and most of the smaller outlying islands.

The six counties that make up Northern Ireland separated from Ireland in 1922 and maintained their link with the United Kingdom, while the remaining counties achieved full independence as the Republic of Ireland. Altogether, the United Kingdom encompasses about four-fifths of the area and over 90 percent of the population of the British Isles.

Lying only 21 miles (34 km) from France at the closest point, England is the largest of the United Kingdom's four main subdivisions. England was originally an independent political unit and twice conquered mainly Catholic Ireland, including Northern Ireland.

• **Figure 4.23** Principal features of the British Isles. The "Tees-Exe" line connecting the mouths of rivers having those names (not shown) is the general dividing line between Highland Britain to the northwest and Lowland Britain to the southeast.

Scotland was first joined to England when a Scottish king inherited the English throne in 1603, and the two became one kingdom under the Act of Union in 1707. Nearly 300 years later, in 1997, the U.K. government yielded to increasing Scottish demands and allowed for the devolution (dispersal of political power and autonomy—see page 80) of many administrative powers (including agriculture, education, health, housing, taxation, and lawmaking, but excluding defense and social security) to a new Scottish parliament. In a groundbreaking 2007 election, Scottish nationalists won political dominance in the parliament and promised a referendum on whether Scotland should secede from the United Kingdom and declare independence. Scots are extremely proud of their distinctive culture, history, and identity, and bristle when they are identified as "English," as happens all too often.

England conquered Wales in the Middle Ages, but like the Scots, the Welsh managed to preserve their cultural distinctiveness and hold it high—especially in staying true to their Welsh language (a member of the Celtic subfamily). Any thought that the Welsh are English withers quickly when one considers this Welsh place name (known as the longest place name in the world): Llanfairpwllgwyngyllgogerychwyrndrobwllllantysiliogogogoch, meaning "The Church of Saint Mary in the hollow of white hazel trees near the rapid whirlpool by Saint Tysilio's of the red cave." A 1997 referendum voted into existence a Welsh assembly dedicated to promoting Welsh interests. Although this body has some powers independent of the United Kingdom, growing numbers of Welsh are calling for a stronger devolution of power that would transfer more authority from the national government to a full-fledged parliament, as Scotland has now.

Between the defeat of Napoleonic France in 1815 and the outbreak of World War I in 1914, the United Kingdom was the world's strongest country. Its overseas empire covered a quarter of the Earth at its maximum extent; "the sun never sets on the British Empire" was a proud boast. The influence of English law, education, and culture spread still farther. The British take great pride in much of that legacy; the Magna Carta, an English charter written in 1215, granted certain human rights that centuries later would be enshrined in the constitutions of other democracies, including that of the United States. However, Britain was one of the main forces behind the cruel industry of the enslavement of Africans. Until the late nineteenth century, the United Kingdom was the world's leading manufacturing and trading nation. Britain's Royal Navy dominated the seas, and its merchant marine moved half or more of the world's ocean trade. London became the center of a free-trade and financial system that invested its profits from industry and commerce around the world.

London continues to be one of the world's greatest financial centers. Unlike many of the United Kingdom's other large cities, London did not owe its existence to the proximity of nearby coal or other natural mineral resources. The Thames River has provided its port with access to worldwide ocean trade, and London has been a magnet for entrepreneurs, artists, tourists, and vagabonds. It is by all measures a **"world city,"** commonly defined as a city that is a center of economic and political power in the global economy.

The United Kingdom's economic and political decline began late in the nineteenth century. World War I damaged its free-trade and financial system, and the downward trend accelerated after World War II. All the large colonies of the former British Empire gained independence by the 1960s, and Britain's remaining overseas possessions are scattered and small.

The United Kingdom is, however, still a very important country on the world stage, in large part because of its lasting impacts from colonial times. It plays a major role in the European Union and is associated with many of its former colonies in the worldwide **Commonwealth of Nations.** The Commonwealth is a voluntary association of 54 countries that nominally recognizes the British monarch as its head. Its strong allegiance to the United States, even in times of great adversity such as during the wars in Iraq and Afghanistan, is another of Britain's characteristic traits in the international arena. Britain's involvement in those wars has, however, been costly in lives and in money spent, and has deepened the impacts brought on the country by the global economic financial crisis. There is a growing sense among the British that their country is experiencing declinism, with Britain taking a more passive place in world affairs.[12]

Much world culture has British roots; English is by far the leading language of the World Wide Web and serves as a *lingua franca* in many countries with diverse cultures (including the European Union; see page 78). In the 1960s, the Beatles, four lads from Liverpool (a city recently crowned a "European Capital of Culture"), revolutionized music in their home country, moved on to conquer a generation in the United States, and left an indelible mark on music everywhere. British pop culture and more traditional fare—Scottish reels, fish and chips, Buckingham Palace Beefeaters—draw fans and visitors from around the world. For many foreigners, the images and artifacts of Britain may become distilled into something homogeneous, but a Briton of Anglo origin is quick to speak about his or her unique piece of the country and what makes it distinct. And then there is multicultural Britain, another legacy of its colonial past, with a bit of Jamaica, India, Kenya, and scores of other lands infused in every corner of the country.

Ireland: Troubles and Resilience on the Emerald Isle

Ireland (known as *Eire* in the indigenous Celtic language) is a land of hills and lakes, marshes and peat bogs, cool dampness, and the verdant grassland that earned its epithet the "Emerald Isle" (•**Figure 4.24**). The island consists of a central plain surrounded on the north, south, and west by hills and low rounded mountains. The Republic of Ireland, which occupies a little over four-fifths of the island, is far into its transition from an agricultural to an industrial economy.

Ireland ranked low among Western European countries for many decades, but recent economic growth changed this pattern. High-tech industries such as electronic products and software boosted Ireland's economy. The size of the economy doubled in the 1990s, earning Ireland the nickname **Celtic Tiger** and placing it on par with the United Kingdom in per capita GNI PPP (see Table 4.1, page 67). Unemployment dropped

• **Figure 4.24** A typical Irish landscape, seen from the Blarney Castle. There is a tradition that kissing the so-called Blarney Stone in one of the castle battlements will make one eloquent in speech.

sharply. This development surge was accomplished by a government program luring foreign-owned industries with inexpensive labor, tax concessions, and help in financing plant construction. Hundreds of industrial factories, owned mainly by U.S., British, and German companies, began to produce computers, electronics, foods, textiles, office machinery, organic chemicals, and clothing, and Ireland's economy became more than 50 percent reliant on exports of these products to overseas markets. Most of these plants sprang up around the two main cities, Dublin and Cork, where they make up an economic region dubbed "Silicon Bog," and near Shannon International Airport in western Ireland.

Unfortunately, Ireland is the "I" in the PIGS economies described on page 87. Much of Ireland's strong economic growth—with an annual rate of about 7 percent between 1999 and 2007—was the result of a borrowing and spending spree prompted by very low interest rates. Housing and banking bubbles developed, and inevitably—as elsewhere in the Eurozone, but with greater consequences—they popped. In 2008, the country adopted economic austerity measures including increased taxes and cuts in salaries and budgets. Ireland fell into a deep recession as industrial and service companies shuttered completely or cut way back on staff. As happened so many times in Ireland's history, the Irish began fleeing their home island in search of better lives abroad. But this time, the Irish emigrants took skills in IT, law, and financial services with them—they were among Ireland's best and brightest, not menial laborers looking for construction jobs abroad (this is one manifestation of the "brain drain" problem described in Chapter 3; see page 56). Ironically, the deep sacrifices made by the Irish may help the country come out of recession more quickly, as low wages and prices should attract more foreign investment.

Many Troubles

In the past, Ireland had an agrarian economy that made the country ripe for suffering. England conquered Ireland in the seventeenth century and made the island a formal part of the United Kingdom. Ireland was effectively an English colony. Most of the land was expropriated and divided among English landlords into large estates, where Irish peasants worked as tenants. The income from this enterprise generally found its way to Britain, and Ireland became a land of poverty, sullen hostility, and periodic violence.

A crop disease known as blight caused the notorious **potato famine** of 1845–1851. Nearly 10 percent of Ireland's pre-famine population of 8 million died of starvation or disease during these years. An even larger number emigrated to England, Wales, Scotland, Australia, and North America. The large Irish American community in the United States traces much of its ancestry to this migration. Today, Americans of Irish descent are just a few of the eager consumers abroad of Irish culture, especially dance and music.

Northern Ireland, now a majority Protestant part of the United Kingdom, has had a particularly difficult struggle that has colonial roots. One outcome of the second English conquest of Ireland, in the seventeenth century, was the settlement of English and Scottish Presbyterians and other Protestants in the north, where they became numerically dominant. These newcomers were supposed to form a loyal population in an otherwise hostile, conquered, largely Roman Catholic country. In 1921, when the Irish Free State (later to become the Republic of Ireland) was established in the southern, mainly Catholic part of the island, for economic and religious reasons the predominantly Protestant north chose to remain with the United Kingdom. Northern Ireland (also known as Ulster) was given much autonomy, including its own parliament in Belfast. Northern Ireland's Catholic minority felt increasingly marginalized by the pro-British Protestant majority and in the late 1960s began agitating—often violently—for change. The British responded by imposing **direct rule** (in which the United Kingdom's central government in London makes all major policy decisions for Northern Ireland), and Catholic discontent soon focused on the perceived British occupation. The **Irish Republican Army (IRA)** began a campaign of bombings, shootings, and arson, sometimes into the heart of England, ostensibly designed to drive the British army from Northern Ireland. **The Troubles,** as the locals refer to this period of strife, involved an often violent struggle between **Catholic Republicans** and **Protestant Unionists** in Northern Ireland.

In 1998, by which time more than 3,400 Irish on both sides had died in the conflict, the U.S.-brokered **Good Friday Agreement** was signed. Approved by the IRA (and its political counterpart, **Sinn Fein,** pronounced "shin fayn") and the Unionists,
80 the agreement called for the devolution of British power. Direct rule by Britain was to be replaced by the *Northern Ireland Assembly,* in which Protestants and Catholics would share power. Violent challenges to the peace agreement persisted from both sides, however, leading the British to halt devolution and reimpose direct rule in 2002. In 2005, the IRA announced that it would permanently cease military operations. Despite the refusal of three small splinter groups to lay down arms, Unionists agreed to share power with the IRA, and unprecedented negotiations between the two sides continued. If it lasts, peace could help bring prosperity to Northern Ireland, which has had periods of strong economic growth based on shipbuilding and linen textile industries, concentrated around the main industrial and residential city of Belfast.

Paris as a Primate City

A **primate city** is a city that, according to a convenient definition, is larger than the second and third largest cities in that country combined. Paris, for example, has a metropolitan area population of 10.4 million, far exceeding the combined populations of France's next largest cities, Lille (with 1 million people) and Lyon (with 1.5 million people). In practical terms, a primate city dwarfs all others in its demographic, economic, political, and cultural importance. The primate city produces and consumes a disproportionately high share of the country's goods and services, and many national resources—such as bureaucratic offices and educational opportunities—are available only in the primate city. The primate city therefore acts as a great magnet for rural-to-urban migration and tends to absorb more investment than other communities.

Urban primacy is often bad for a country's development because funds that could be spent to improve services and quality of life in smaller cities and rural areas are often diverted instead to the primate city. With greater opportunities available in the primate city than in the smaller cities of the hinterland, the most educated and talented people abandon the hinterland
56 for the primate city. This internal brain drain contributes further to underdevelopment outside the primate city. Primate cities are rare in the more developed countries but are quite typical of the LDCs, where regional development is a critical issue. Cairo, Karachi, and Mexico City are exemplary primate cities of Africa, Asia, and the Americas.

Paris is the greatest urban and industrial center of France, a primate city completely overshadowing all other cities in both population and economic activity. It is by far the largest city on the mainland of Europe. Paris is located at a strategic point on the Seine River relative to natural lines of transportation; therefore its geographic *situation*, as described earlier in the chapter, was important in the city's development. In France, it would be appropriate to say "all roads lead to Paris" (see •Figure 4.25). Paris' growth was also the product of the growth and centralization of the French government and the transportation system it created. Like London, it has no major natural resources for the industry in its immediate vicinity, yet it is the greatest industrial center of the country. Despite its economic clout, Paris maintains its international reputation as the romantic and monument-studded **"City of Light"** (many of these attractions are on the map in •Figures 4.26). It is the world's leading urban tourist destination, and France attracts more tourists than any other country in the world.

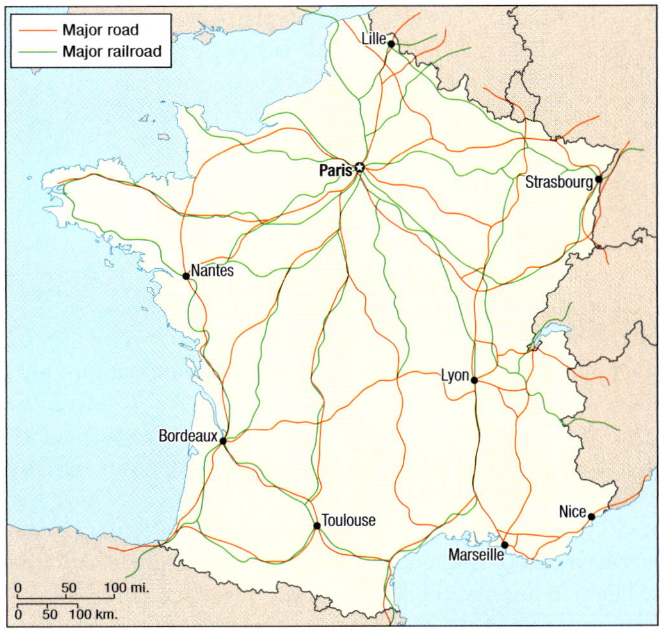

• **Figure 4.25** Paris is the primate city at the hub of France's transportation network.

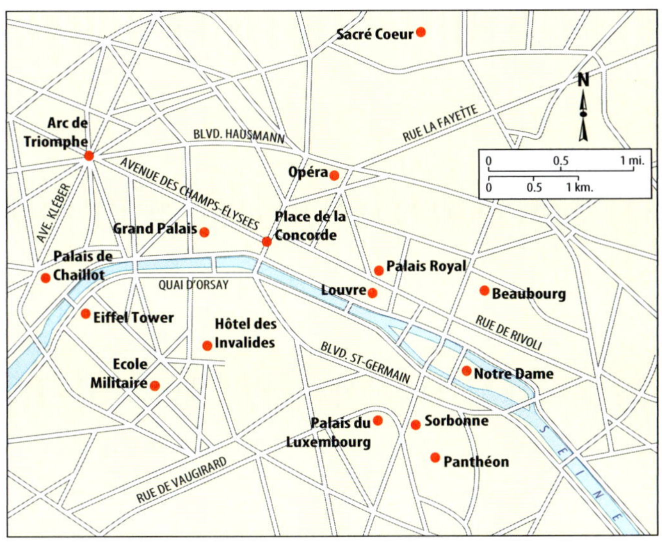

• **Figure 4.26** Map of downtown Paris and some of its world-famous attractions.

In the Middle Ages, Paris became the capital of a succession of kings who gradually extended their effective control over all of France. As the French monarchy became more absolute and centralized, its seat of power grew in size and came to dominate the cultural and political life of France. As national road and rail systems were built, their trunk lines were laid out to connect Paris with outlying regions. The result was a radial pattern with Paris at the hub, and this is essentially the pattern that exists today. As the city grew in population and wealth, it became an increasingly large and rich market for goods.

The local market, plus transportation advantages and proximity to government offices, provided the foundations for a huge industrial complex. This development has involved two major classes of industries. Paris is the main producer of the high-quality luxury items (fashions, perfumes, cosmetics, and jewelry, for example) for which that city and France have long been famous. Paris is also the country's leading center of engineering industries, secondary metal manufacturing, and diversified light industries, concentrated in a ring of industrial suburbs that sprang up in the nineteenth and twentieth centuries. Paris inverts the usual pattern of the industrial city by maintaining a core of low-profile historic monuments, while the skyscrapers of the modern economy tower away from the city center. Parisian urban planners have carefully cultivated and protected the "crown jewels" of this beautiful city's architecture.

Divided and Reunified Germany

One of the most important geopolitical events of the late twentieth century—and one that has lingering effects in the new millennium—was the reunification of the two Germanys. If you go on to study Korea in this book, you will see that one of the big questions for this century is whether or not the divided Koreas might follow the German example. In both cases, economic issues play a prominent role; in each case, there is a stronger, wealthier country that would have to assume much of the cost of merging with a weaker and poorer counterpart.

The Federal Republic of Germany (•**Figure 4.27**) reappeared on the map of Europe as a unified country in 1990. Between 1949 and 1990, it had been divided into two nations: democratic West Germany (known as the German Federal Republic), formed from the zones occupied by Britain, France, and the United States after Germany's defeat in World War II in 1945, and Communist East Germany (the so-called German Democratic Republic), formed from territory occupied after the war by the Soviet Union. The former capital, Berlin, was similarly divided. West Berlin was a part of West Germany for those decades, even though the city was entirely surrounded by East Germany. The withdrawal of Soviet support for East Germany in 1989 and the destruction of the Berlin Wall led to the rapid collapse of its Communist government in 1990 and to the jubilant reunification of Germany's 16 states (*Länder*).

Reunified Germany is Europe's dominant country in many respects. Its population of 82 million is much greater than that of any other nation in the region. Politically it is seen, along with France, as the cornerstone of the European Union. Economic considerations are even more important; Germany has

© Cengage Learning 2013

• **Figure 4.27** Principal features of Germany. Note which parts of the country belong to what formerly were East and West Germany.

the world's fourth largest economy, after the United States, China, and Japan. Germany is among the top three countries, along with China and the United States, in exports of goods. Germany must also bear a substantial part of the economic burden of problems in the Eurozone and in the European Union. 87, 88

The former West Germany was Europe's leading industrial and trading country during the latter half of the twentieth century. The former East Germany was an advanced country by Soviet bloc standards but was an economic disaster by West German standards. Many billions of dollars and much time have been required to rehabilitate Germany's east, and the task is still incomplete.

Germany's urban geography is unusual. Industrial and urban centers are widely scattered. Many cities are large, but none is as nationally dominant as London or Paris. This dispersed pattern came about because Germany was divided for a long time into petty states, many of which eventually became internal states within German federal structures. These often had their own capitals, which would in time become Germany's large cities.

In World War II the western Allies, led by the United States, heavily bombed many of these industrial cities, which also had large civilian populations. Perhaps the most tragic bombing was of Dresden in Germany's Saxony region, in 1945. The firestorm in this city, known as the "Florence on the Elbe" because of its cultural treasures, was immortalized in Kurt Vonnegut's novel *Slaughterhouse Five*.

Allied bombing did not eliminate Germany's industrial capacity. West Germany was reconstructed rapidly in the postwar years, thanks to several factors. The massive U.S.-funded aid package known as the Marshall Plan (see page 89) supplied billions of dollars in capital goods between 1948 and 1952. This aid was strategically aimed at rebuilding the country to stave off the Communist threat from the east. Skilled workers were eager to work, even for low wages, and markets were hungry for the industrial products that Germans could make. Germany's economic growth continued through the 1960s. Labor vacancies created by upwardly mobile Germany workers were filled by foreign workers, especially Turks and Kurds
70 from Turkey.

East Germany, Still German

East Germany was Communist ruled and Soviet dominated, but it was still German. Even Soviet exploitation and bureaucratic inefficiency were unable to destroy the traditional German ethics of hard work, efficiency, attention to detail, and high standards. East Germany was the most productive of the Soviet satellite countries that lay between Russia and Western Europe, but its economy lagged far behind West Germany's. When East Germany collapsed in 1989, a major migration stream began flowing from East to West Germany. After 1989, more than 4 million people fled westward from what Germans call the "new states" of the former East Germany. The emigrating easterners left behind what one German professor described as "a series of towns and enclaves for senior citizens."[13] The urban landscapes of the east changed to accommodate an aging population. Hundreds of the old Soviet-style apartment blocks were demolished and new parks and other green spaces developed in their wake. The ongoing urban architectural experiment in eastern Germany, known as *Ruckbau*, is a cross between demolition and reconstruction.

The pool of young and educated people that remained in the east began to attract employers looking for lower-wage workers, and the region's economy may benefit from their presence. More than 3 million people have moved from the west to the east to take advantage of these opportunities. There is a growing semiconductor industry in what local promoters like to call "Silicon Saxony" (one of those plays on words modeled on Silicon Valley, as described on page 92), especially in Dresden and Leipzig. The automakers Volkswagen and BMW have opened plants in Dresden and Leipzig, respectively. Taking advantage of the skilled and inexpensive labor, scores of U.S. chipmaking companies have set up operations in the region.

Billions of dollars of investment are helping to raise economic standards in the east and to clean up the severe environmental damage inflicted during Communist rule there. To complete that transition, the former West Germans are bearing much of the tax and social burden of bringing the former East Germans up to the standards long enjoyed in the west. More than 20 years after reunification, western Germans (known as "Wessies" by the east Germans) continue to resent paying to support the eastern economy and believe the Easterners (whom they call "Ossies") are ungrateful for that help.[14]

Economically, Germany was one of the characteristic welfare states that subsidized its citizens generously. The costs for 85
that support contributed to Germany's economic downturn early in the 2000s, when economic growth was the slowest of all EU countries and unemployment surged to 10 percent. In the area of former East Germany, conditions were worse, with the rate of unemployment more than twice that in the west (that ratio continued even after conditions became better). Germany continued to work on its painful transition from a welfare state to one in which citizens increasingly must pay for health care and other vital services. Other difficult measures, such as cutting jobs, extending work hours without extra pay, and outsourcing work to other countries, helped Germany's economy rebound after 2005 and grow even more sharply after 2009 in the wake of the global recession. Germany's climb out of its economic hole was remarkable, and it raised hopes for an economic rebound throughout the EU community.

But is Germany really committed to the European community and to its partnership in the Atlantic alliances, including NATO? These became persistent questions after 2010. Germany is still considered the economic locomotive of Europe, and other EU member countries look to Germany for leadership and aid (many of the poorer EU countries came to believe that Germany was their "paymaster" in times of need). Germany is, however, increasingly cultivating economic partnerships with countries to the east. Fewer German manufactured goods like precision machinery, cars, chemicals, and electronics are going to France, whereas more are being exported to China, India, and Russia. France continues to be China's leading trade partner but is nervous that Germany will "decouple," going its own way and serving its own interests rather than those of the European Union. Germany also broke ranks politically with its NATO allies, including France, the United Kingdom, and the United States, in refusing to support military action against Muammar Gaddafi's Libya in 2011.

Europe Paves the Way on Alternative Energy

While we are on the subject of French and German differences, let us consider this: France is the world champion of nuclear energy, getting about 80 percent of its electricity from nuclear power. Germany, on the other hand, has vowed to phase out its nuclear power program (which supplies a quarter of its electricity) completely by 2022. What explains the difference? The devastating tsunami in Fukushima, Japan, and the series of nuclear power plant accidents in its wake. Awed by what happened 15
in Japan, Germany's leaders and citizens decided that nuclear power was simply too risky to maintain. France, on the other hand, is too deeply committed to nuclear power to consider alternatives.

Germany's decision to abandon nuclear power puts it in a difficult position to attain the European Union's collective goal of reducing greenhouse gases to 20 percent *below* 1990 levels by 2020. One of nuclear power's advantages is that it produces no

•Figure 4.28a–d Europe has a wide variety of energy sources, with Poland almost completely reliant on coal to produce electricity and France nearly so on nuclear power. Traditional and modern sources of energy in Europe include **(a)** peat from a Scottish bog, **(b)** coal in Poland, and windmills, both **(c)** old (in the Netherlands) and **(d)** new (in Denmark).

greenhouse gas emissions, so maintaining the nuclear program would be helpful to Germany. Some EU leaders think that goal is not ambitious enough and are pushing to cut carbon dioxide emissions to 30 percent *below* 1990 levels by 2020. Britain vows to derive 20 percent of its total energy needs from renewable sources by 2020 and to cut carbon dioxide emissions to 60 percent *below* their 1990 levels by 2050. I emphasize *below* 1990 levels to draw attention to these extraordinary ambitions. Imagine the serious and fundamental changes needed to turn the clock back on the relentless forward movement of Europe's fossil-fuel driven economies and lifestyles. And compare these ambitions with those of the United States. Under Democratic and Republican administrations alike, the United States has consistently backed away from making serious commitments to reduce greenhouse gas emissions.

Although the precise goals of EU countries differ, the underlying assumption is rock steady: the European Union has taken an official stance that global climate change is a serious problem and must be confronted through dramatic changes in the ways energy is produced and consumed (**•Figure 4.28**). Officially, the European Union's stated target is to get 20 percent of its energy from renewable sources by the year 2020. To achieve this goal, EU members will promote fuel efficiency in automobiles, encourage the use of public transportation, expand the cap-and-trade mechanisms of carbon emissions trading, and bring a vast array of alternative energies—hydroelectric power, solar power, tidal power, wave power, geothermal power, and even trash power—online. France and some other counties will also add nuclear power into this mix, but nuclear is not considered a renewable alternative energy source.

One of the most popular renewable alternative energies in Europe is wind power. Denmark uses wind power to meet more than 20 percent of its electricity needs—the highest rate in the world from this source. Germany has established itself as the world's leading manufacturer of wind turbines. More than 12,000 of these mills dot Germany itself, providing about 7 percent of the country's electricity needs. That share is projected to rise to 20 percent by 2020, after 40 offshore wind farms as far as 70 miles from the coast come online. Britain is building more than 1,000 wind turbines off England's coast, mainly in the Thames estuary, the east coast area, and along the northwest coast.

Wind energy produces no greenhouse gas emissions, and you might think Greens would embrace it. But environmentalist Britons, recalling William Blake's image of the "dark satanic mills" of early English industry, lament that the "silver satanic blades" will sully the storied Lake District of the northwest.[15] The **NIMBY** ("not in my back yard") attitude often wins the day. This viewpoint, combined with the fact that Europeans are very short on land—compared with Americans especially—and the characteristic geography of Western Europe's numerous islands, peninsulas and seas, and enormous amount of coastline, makes a perfect recipe for offshore wind power.

Some of the technical challenges of providing wind power are impressive. Have a look at the North Sea in Figure 4.1 on page 66. Its waters can be turbulent, with winds often howling at 60 miles per hour and waves cresting higher than 30 feet (9 m). As many as 6,000 gigantic wind turbines will be installed in these treacherous North Sea and other British waters by 2020. Fortunately, British engineers have had the know-how of installing oil rigs in the North Sea since the 1970s. The plan is to put that fossil fuel experience to work on alternative energies.

The European Periphery

Properties of the European Periphery

Just as there is a recognizable European Core, where the region's demographic, economic, and political weight is centered, there is also a **European Periphery.** This "rimland" is made up of countries whose interests are tied closely to those of the core and are strongly influenced by the core countries. But they have less clout in the political and economic systems than the core countries do and tend to be dependent on the core. As we have seen in the context of the European Union, for example, many of the countries of the periphery hope that the core countries Germany and France will bail them out of their economic troubles.

Here we use the tool of geographic regions to subdivide these countries into three categories: those of Northern, Southern, and Eastern Europe (•Figure 4.29). Within these subgroups, the countries have distinct national identities, histories, and experiences. In this section, we explore some problems and issues of these countries outside the European center. The major traits of each nation are listed in Table 4.1 on pages 67–68.

• **Figure 4.29** The subregions and countries of the European periphery.

Northern Europe

Save the Whales . . . for Dinner One reason that both Norway and Iceland (see •Figure 4.30) have refused to join the European Union is that both have economically vital fishing industries, and they fear that common EU policies on fishing might diminish their fishing profits. The two countries, along with Japan, also are distinguished by their unique perspective on another marine resource: the world's whales. People in all three countries maintain that populations of some whale species, especially the minke (pronounced "ming'-key"), have rebounded to levels that should allow regular, limited harvesting for human consumption. One of their arguments is that growing populations of these whales would feed on huge amounts of commercially important fish stocks. Whale meat has also long been a prized food for people in these countries. "Save the whales for

• **Figure 4.30** Principal features of Northern Europe. Note Iceland in the inset. See the map of Figure 4.1 on page 66 to appreciate the isolated location of Iceland in the North Atlantic. Iceland and Norway are two of the world's three countries still practicing whaling; the third is Japan.

AP Photo/Svein Andersen/HO

• **Figure 4.31** Norwegian whalers haul in a minke whale.

• **Figure 4.32** The Basque country of Spain and France.

the dinner table" is the ironic Norwegian response to the popular Western bumper sticker "Save the whales" (•**Figure 4.31**).

In a landmark agreement honored by most of the world's marine countries, the International Whaling Commission (IWC) banned commercial whaling worldwide in 1986. The Norwegians never accepted the ban, calling it "cultural imperialism," and defiantly continue to take about 500–600 minkes each year, well short of their self-imposed quota of about 1,300 whales per year. Norwegians eat whale meat without remorse, arguing that the rest of the world is foolish in its insistence that the whale is a "sacred cow." Even Norway's famously staunch environmentalists see no problem with the whale harvest. But with video footage of water stained red by the blood of writhing and apparently suffering whales, environmentalists outside Norway have stepped up pressure on Norway to stop whale hunting.

Although Norway simply defies the IWC whaling ban and since 1993 has hunted whales commercially for consumption in Norway and for export abroad (especially to Japan), the other whaling countries at least pretend to be considerate of the ban. The 1986 ban forbids commercial whaling but allows whales to be killed for scientific purposes and for subsistence (the latter clause allows Native Americans in Alaska, for example, to hunt whales). Iceland and Japan have exploited the loophole of the scientific purpose and in the process put whale meat on the table. They argue that there is a legitimate need for the science: Iceland estimates that its territorial waters are inhabited by more than 40,000 whales that eat perhaps 1 to 2 million tons of fish each year. Only by studying the whales' stomach contents can reliable estimates of the whales' impact on Iceland's vital fishing industry be evaluated. The studies accomplish two other goals: They reduce the pressure on the fish resource, and they allow whale meat to find its way to markets for ordinary consumers.

Killing whales does carry some costs. Iceland resumed commercial whaling in 2006, establishing quotas of 30 minkes and 9 fin whales each year. Most Icelanders support the hunt, but others argue that whale watching—which previously had drawn 90,000 tourists yearly—was the more economically sensible way to manage the resource. As the European economy hit hardest by the global economic downturn, Iceland can ill afford to lose tourists. And with "Save the whales" an international mantra, Iceland, Norway, and Japan are subjected to intense criticism from around the world.

Southern Europe

The Basque Country The 2.3 million **Basques** of Spain and 300,000 Basques of France (see Figure 4.B and •**Figure 4.32**) have a unique ethnicity and culture completely unrelated to those of their host country majorities. Like other minorities, they have often been the targets of discrimination and violence. This is especially true in Spain, where during the Spanish civil war in 1937 a German bombing (called for by Spanish fascists) led to tragic loss of life among Basque civilians in the city of Guernica. Artist Pablo Picasso painted a mural of the suffering of Guernica's civilians that brought international attention to the Spanish civil war and ever since has served as a symbol of the tragedy of war (•**Figure 4.33**).

By the 1960s, Basque desire for independence led to the emergence of the militant group **ETA** (an acronym for **Basque Homeland and Liberty** in Euskara, the Basque language). Spanish authorities gradually allowed considerable autonomy for the Basques, and they now control their own affairs in taxation, education, and policing. For more than four decades, these concessions were not enough for ETA, which carried out attacks on Spanish interests and triggered a vigorous crackdown by Spanish security forces. But in 2011, ETA vowed to cease violence with its declaration of a "permanent, general and verifiable" ceasefire. The European Union and the United States regard ETA as a terrorist organization and are wary of this ceasefire because of ETA's history of breaking earlier ceasefires. It is not clear whether in the long term ETA will settle for what they have now, a devolutionary status comparable to Scotland's autonomy within the United Kingdom (see page 93), or will push for outright independence as Kosovo did (described later in this chapter).

Independence for Basques in Spain or France seems very unlikely, and most Basques are not pushing for it or expecting it. The Basques both in Spain and France have worked hard at retaining their unique customs. In Spain, there are Basque

• **Figure 4.33** During the Spanish Civil War, Spanish Republicans commissioned Pablo Picasso to create an image of the suffering inflicted by Spanish Nationalists on Spanish civilians. Picasso's *Guernica* went on display at the 1937 World's Fair in Paris.

• **Figure 4.34** North and South in Italy, as perceived by Italians. This is a vernacular map, so there are no "true" boundaries separating north and south.

language television broadcasts. French Basques would like to have the kind of autonomy their counterparts in Spain have achieved.

North versus South in Italy All Italians are united by their pride in the heritage bestowed by civilization from the Romans to the Renaissance and beyond, and Italians live on some of the most picturesque and storied landscapes on Earth. But within Italy, there is a longstanding vernacular distinction between the north and the south Northerners tend to see themselves as more sophisticated and cosmopolitan than the more earthy and provincial southerners (see •Figure 4.34). Southern Italians, whose region is known as the **Mezzogiorno** (a fine example of a vernacular or perceptual region, as described on page 6 in Chapter 1), often acknowledge their agrarian roots as the source of their superior kinship values and enjoyment of life. Statistics underscore the economic discrepancies between the two Italys—for example, northern Italy has labor shortages while the south has unemployment; northern labor and industries (such as luxury automobile, jewelry, textile, leather, ceramic, and furniture industries) are more productive, and income levels in the north are much higher.

The feeling among many northerners that they would be better off if they did not have to subsidize the south has even led to a secessionist movement, championed by an organization called the **Northern League,** that calls for the creation of an autonomous or independent state of **Padania** in the north (see Figure 4.B). The Northern League wants to maintain the isolated affluence of northern Italy in part by keeping out immigrants, particularly Muslims and Africans. 71
Another organization, **Liga Veneta,** wants to create an autonomous region of Veneto, centered on the affluent city of Venice (•Figure 4.35).

In the toe of Italy's boot many illegal immigrants, mainly black Africans, live in squalid conditions. They began coming to Italy to pick oranges and tangerines when farming was still profitable because of EU agricultural subsidies. But by 2010, when EU subsidies for farming were curbed, there was no work for them in a weakening agricultural sector. The xenophobic Northern League denounced the "excessive tolerance" of illegal immigration in Italy. In Italy's toe, some white Italians

Majority Slavic ethnicity, former Communist status, and subjugation to Soviet interests were among the few unifying themes of the region prior to the end of the Cold War. Now the true complexity of the region is more apparent.

The extension of Soviet power into Eastern Europe after World War II was a replay of history. In the Middle Ages, several peoples in the region—the Poles, the Czechs, Magyars, Bulgarians, and Serbs—enjoyed political independence for long periods and at times controlled extensive territories outside their homelands. Their situation deteriorated as stronger powers—Germans, Austrians, Ottoman Turks, and Russians—pushed into east central Europe and carved out empires. These empires frequently collided, and the local peoples were caught in wars that devastated great areas, often resulted in a change of authority, and sometimes brought about large transfers of populations from one area to another. In geopolitical terms, Eastern Europe is a classic **shatter belt**—a large, strategically located region composed of conflicting states caught between the conflicting interests of great
180 powers (see the evolution of this region in •**Figure 4.39**).

Peoples of this region have been on the move in large numbers since the beginning of World War II. Jews, Germans, Poles, Hungarians, Italians, and others were uprooted by Nazi and Soviet authorities, often without notice, losing all their possessions, and were dumped as refugees in so-called "homelands" that many had never seen. During the war, Nazi Germany systematically killed approximately 6 million Jews and perhaps as many as 1.5 million Gypsies (Roma) and other
173 "undesirables" in the Holocaust. The prewar populations of Germans in Poland and Czechoslovakia were expelled at the end of World War II and forced into East and West Germany. Many died in this process of forced migration overseen by the Soviet Union. Ethnic minorities now make up only 2 percent of the population in Poland, which transferred most of its German population to Germany and whose territories containing Lithuanians, Russians, and Ukrainians were absorbed by neighboring countries. A number of countries still have large minorities, including the Roma or Gypsies (see page 108). These population transfers helped create an ethnic map of Eastern Europe that was overwhelmingly and purposefully Slavic.

Communism in Ideals and Practice In becoming satellites of the Soviet Union, the nations of Eastern Europe (and East Germany) were reformed by **communism.** You should consider the principal traits of communism in order to understand their legacy in today's geography of Eastern Europe and of the former Soviet Union, which is the subject of the next chapter. Communism had these principal traits:

- One-party dictatorial governments
- National economies planned and directed by organs of the state
- Abolition of private ownership (with some exceptions) in the fields of manufacturing, mining, transportation, commerce, and services
- Abolition of independent trade unions
- Varying degrees of **socialization** (state ownership) of agriculture

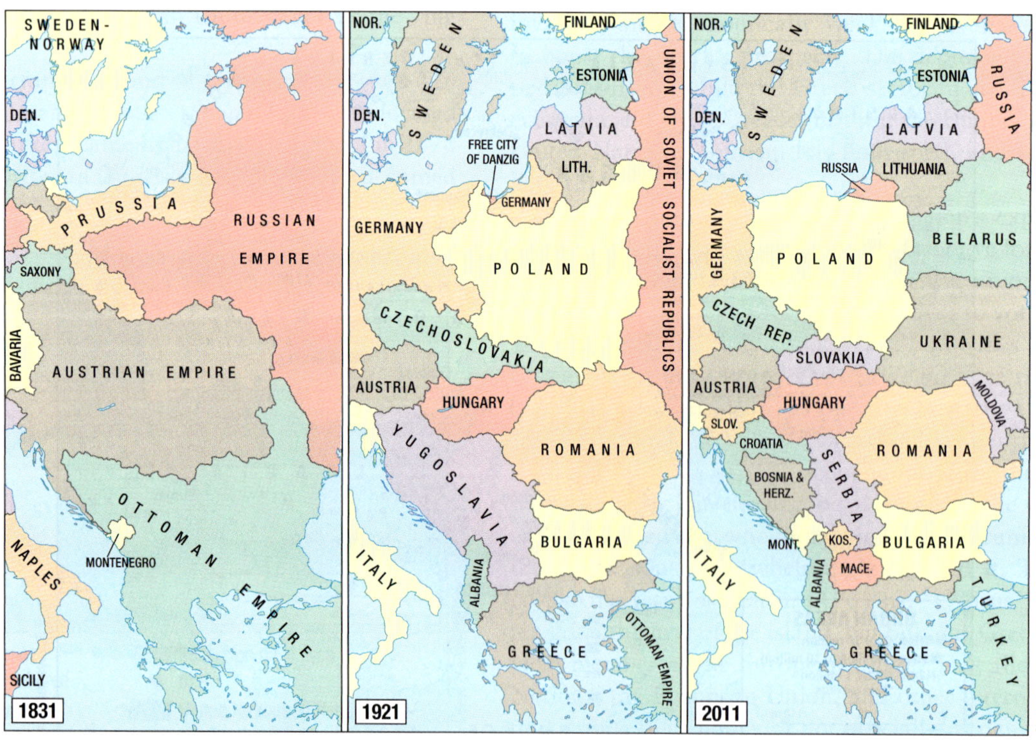

• **Figure 4.39** Positioned between stronger powers to the east and west, Eastern Europe is a classic shatter belt with a tumultuous past, reflected in its shifting borders.

Soviet military force crushed attempts by East Germany in 1953, Hungary in 1956, and Czechoslovakia in 1968 to break away from Soviet control. In agriculture, the new Communist governments liquidated the remaining large private holdings and in their place introduced programs of collectivized agriculture on the Soviet model. Some farmland was placed in large state-owned farms on which the workers were paid wages, but most was organized into collective farms owned and worked jointly by peasant families who shared the proceeds after operating expenses of the collective had been met. This **collectivization**—the bringing together of individual landholdings into a government-organized and government-controlled agri-
130, 146 cultural unit—met with strong resistance. Conversion back to private farming after four decades of collectivization presented painful obstacles, and progress has been slow. This process is still ongoing, which is why it is important that you understand the legacies of communism and socialism. Despite recent attempts to diversify and intensify agriculture, in Eastern Europe agriculture is focused on a narrow range of products, especially corn, wheat, pigs, and sheep.

As it did in agriculture, communism in Eastern Europe inaugurated a new era in industry based on a Soviet model. Most industries were taken out of private hands, and national economic plans were developed. Communist planners did not aim at a balanced development of all types of industry. Instead they emphasized those industries seen as essential to the development of a greater Soviet empire: mining, iron and steel, machinery, chemicals, construction materials, and electric power, for example. These were favored while consumer-type industries were neglected.

The central planning agencies of the Communist governments maintained rigid control over individual industries. Plant managers were order to produce certain goods in quantities determined by state governmental planners. The plant's success was judged by its ability to meet production targets rather than its ability to sell its products competitively and at a profit. Plant managers were expected to conform to the central plan and not ask questions. As you can imagine, this was not an economic environment that fostered creativity and entrepreneurship. There were no visionaries like Apple's Steve Jobs—or if there were, they would have been branded as nonconformist and been alienated.

This system resulted in shoddy goods that were in short supply or in mountainous oversupply because production was not driven by consumer demand and satisfaction. The planned expansion of mining and industry in Eastern Europe increased the output of minerals, manufactured goods, and energy. But it also produced inefficient, overstaffed industries that were ill-prepared for competition in the world markets of the post-Communist period. This was a problem shared by all the economies of all the successor states of the Soviet Union, as we will also see in the next chapter.

Under communism, the economies of Eastern Europe were closely tied to that of the Soviet Union. During the 1970s and 1980s, however, the region's trade relationship with the Soviet Union weakened. Trade and financial relations with Western countries, companies, and banks expanded rapidly. With relatively little to export, the Eastern European countries borrowed massively from Western governments and banks to pay for imports from the West. Poland and Romania pursued this course especially strongly. New industrial plants and equipment acquired in this way were supposed to be paid for by goods that would be produced for export to the West, by industries that had learned efficient production and marketing techniques from the West.

But then the Western economies began to slump, lowering the demand for imports. This led to a situation by the 1980s in which large debts to Western governments and banks needed to be repaid if countries were to maintain any credit at all. However, with the help of mismanagement by Communist bureaucrats, exports with which to pay were not being produced or, if produced, could not be sold. The result was a severe impediment to economic expansion and falling standards of living as the governments squeezed out the needed money from their people. In Poland, social action through the formation of an independent trade union called **Solidarity** paralyzed the country and threatened to upset Communist dominance. These conditions, along with the Soviet Union's own growing economic and political distress, set the stage for the withdrawal of Soviet control and decommunization in 1989 and 1990.

As late as the early summer of 1989, East Germany and the Eastern European countries seemed to be firmly in the control of totalitarian Communist governments. Then, suddenly, the Communist order began to crumble. Public demands for freedom, democracy, and a better life gathered momentum in one country after another. Communist dictators who had ruled for many years were forced out, and reformist governments took charge. By mid-1991, democratic multiparty elections had been held in all countries. This liberalizing process continued throughout the 1990s and into the 2000s.

In 1989 and 1990, the Soviet Union withdrew its backing for the region's Communist order. That support was too costly for the Soviet Union, whose leaders were forced to recognize the inevitable victory of "people power" in Eastern Europe. Several Communist regimes (including East Germany's) quickly collapsed and were replaced by democratically elected non-Communist governments.

You many not have witnessed these momentous events that occurred in Europe and the Soviet Union in 1989. But you have seen the extraordinary upheaval in the Middle East and North Africa in the Arab Spring of 2011. These jaw-dropping events 192
give you a good idea of what 1989 was like. And what happened in Eastern Europe may provide clues to how long and how difficult the restructuring of the Arab World will be.

Governments in Eastern Europe have been struggling ever since 1989 to build democracies with capitalist economies out of the economic wreckage of unproductive, unprofitable, uncompetitive, state-owned enterprises. Many countries pursued reforms aggressively to meet qualifications for joining the European Union. Economic restructuring has required many difficult steps, including the **privatization** (shift to nongovernmental ownership) of state-owned enterprises, an increase in the efficiency of state-run enterprises by allowing noncompetitive enterprises to fail and removing support of poor performers, and encouragement of the development of new private enterprises, often with foreign capital and management playing a role. In addition, economic restructuring has seen the termination of price controls so that prices reflect competition in the market; the development of new institutions required by a market-oriented economy (such as banks, insurance companies, stock exchanges, and accounting firms); the fostering of joint enterprises between state-owned firms and foreign firms; and the elimination of bureaucratic restrictions on the private sector. Internationally convertible currencies have also been created to increase trade and enhance competition, and access to international communications media of all sorts has been expanded to help local entrepreneurs learn from foreign examples.

One of the most remarkable economic trends has been West-
235 ern European and Chinese **outsourcing** of investments that take advantage of relatively cheap labor in Eastern Europe. In the late 1990s, Austrian and German electronics firms established large factories in east central Europe, especially in Hungary. Even China opened television and other electronics factories in Eastern Europe, because China's labor costs were rising.

The Eastern European countries enjoyed strong economic growth for about a decade (1997–2007). But when the global economic crisis hit Europe in 2007, it hit the former communist countries hardest.

Why They Call It "Balkanization" You may have heard someone use the term *balkanize*. This word comes from the real-life geopolitical events in a part of southeastern Europe known as the Balkans or the Balkan Peninsula. Here we will see what these events were and what their impact has been on the modern geography of the Balkan states, which includes some of the world's youngest countries.

Although economic progress characterized most of Eastern Europe in the 1990s, Yugoslavia (seen on the map of the Balkan Peninsula in •Figure 4.40) was plagued by ethnic warfare following the dissolution of the Yugoslav federal state in 1991. Previously, under the Communist regime instituted by Marshal Josip Broz Tito during World War II, Yugoslavia ("Land of the South Slavs") had been organized into six "socialist people's republics" (Serbia, Croatia, Slovenia, Bosnia and Herzegovina, Montenegro, and Macedonia) and two "autonomous provinces" within Serbia (Kosovo and Vojvodina).

• **Figure 4.40** Ethnic composition of the Yugoslav successor states.

As long as Tito and his successors could retain a firm grip, these groups could coexist within the artificial boundaries of a single nation. However, this apparent unity belied many underlying tensions and conflicts based on the linguistic, ethnic, and religious distinctions you read about earlier in the chapter: Orthodox Serb versus Muslim Kosovar, Catholic Croat versus Orthodox Serb, and so on.

As the Iron Curtain dissolved across Europe, Yugoslavia began to fracture along its ancient ethnic fault lines (Figure 4.39). In 1988, Serbia took direct control of Kosovo and Vojvodina as part of its push (under an elected Communist president, Slobodan Milosevic) for greater influence in federal Yugoslavia. The quasi-independent states within Yugoslavia initially demanded a looser federation with more autonomy for each republic. But in 1991 and 1992, Croatia, Slovenia, Macedonia, and Bosnia insisted on independence. When Croatia declared its independence, fighting broke out between ethnic Croats and Serbs in the new country. Croatian Serbs took over one-third of Croatia and embarked on a policy of **ethnic cleansing**, the forced emigration or murder of one ethnic group by another within a certain territory. Although Bosnia's Muslims (known as *Bosniaks*) and Croats voted for independence from Yugoslavia, ethnic Serbs in Bosnia and the Yugoslav government of Slobodan Milosevic opposed this.

War erupted in 1992 between the Bosnian government and local Serbs. Supported by Serbia and Montenegro, the Bosnian Serbs fought to partition Bosnia along ethnic lines and join Serb-held areas to a "greater Serbia." Bosnian Serbs laid siege to the mainly Bosniak city of Sarajevo. The United Nations withdrew recognition of Yugoslavia because of the government's failure to halt Serbian atrocities against non-Serbs in Croatia and Bosnia. In 1994, the Bosniaks and Croats agreed to create the joint federation of Bosnia and Herzegovina.

In 1995, the presidents of Bosnia and Herzegovina, Croatia, and Serbia signed the **Dayton Accord** (it was brokered by U.S. diplomats at an airbase near Dayton, Ohio). This peace agreement retained Bosnia and Herzegovina's international boundaries and created a joint multiethnic, democratic government charged with conducting foreign and economic policies. The agreement also recognized a second tier of government composed of two similarly sized entities charged with overseeing internal functions: the Bosniak-Croat Federation of Bosnia and Herzegovina and the Bosnian Serb–led Serb Republic (Republika Srpska). To help implement and monitor the agreement, NATO fielded a 60,000-strong implementation force known as IFOR. It was later replaced by a smaller NATO force, in turn replaced by EU peacekeepers numbering just 7,000.

While the Bosnian situation quieted, in 1998 Serbia turned its sights on Kosovo, a majority ethnic Albanian and Muslim region with a minority of ethnic Serbs in the north. Here again, Serbs aspired to ethnically cleanse Kosovo of non-Serbs. Serbs asserted that Kosovo was the cradle of their culture and the site of the decisive battle of Kosovo Polje in 1389, which Serbs lost to the Ottoman Turks. NATO responded to the Serb offensive with 11 weeks of bombing. When the bombing stopped, the United Nations took control of Kosovo and oversaw a tense truce between its Serb and Muslim inhabitants. The fate of more than 200,000 ethnic Serbs who fled during the conflict from Kosovo to Serbia and Montenegro has not been resolved. Nominally still a part of Serbia, but with a population of only 5 percent ethnic Serb and 95 percent ethnic Albanian, Kosovo came under United Nations jurisdiction.

AP Photo/Visar Kryeziu

• Figure 4.41 On February 17, 2008, the Albanian majority of Kosovo celebrated their self-proclaimed newborn independence in the city of Pristina. Kosovo has yet to gain recognition as independent by UN security members Russia and China.

Kosovo declared its independence in 2008, naming itself the Republic of Kosovo. The ethnic Albanian majority poured out into the streets of the capital Pristina to joyously proclaim their country as "newborn" (**•Figure 4.41**). It has been recognized as independent by more than 80 countries, including the United States, most EU member countries (22 out of 27), and other nations, and some countries within the critical United Nations Security Council, but not by other countries, most notably the Security Council members Russia and China. United Nations membership is the ultimate stamp of a country's sovereignty, but Russia and China (which have veto power in the UN) uphold Serbia's claim that Kosovo is a Serbian province, so Kosovo may never receive that recognition. In this book, we recognize Kosovo as an independent nation.

The ouster of Serbian leader Milosevic in 2000 and his capture and subsequent transfer to The Hague to stand trial for war crimes (he died before the trial was completed), opened the way for a period of stable and peaceful borders in the 2000s. The international community insisted on no further redrawing of Balkan boundaries. The region had undergone a profound and violent process of **balkanization,** which is political-geographic shorthand for fragmentation into ethnically based, contentious units that took its name from the characteristic disharmony of this region.

The Balkan conflicts unfolded on television and other media through the 1990s. They reminded Europeans and Americans of the earlier world wars on the continent and seemed a horrible anomaly in a world region accustomed to peace and prosperity. Western military and political institutions eventually seemed to extinguish the fires of conflict, and the Balkan countries were set on the back burners of Europe.

• **Figure 4.42** This Roma woman and her children pass their days begging on a street corner in Szczecin, Poland.

By all measures of quality of life and economic prosperity, the Balkans are the poor stepchildren of the continent, and most cannot expect to join the European Union for many years.

The Roma One of Europe's largest ethnic minorities is the Gypsies—properly known as the Roma—who number about 12 million (•**Figure 4.42**). Romania has the highest number—as many as 2.5 million—but estimates of the Roma population vary widely because the Roma live on the road and at the margins of society.

The Roma began their great odyssey in what is now India, and their Romany language is very close to languages still spoken on the Indian subcontinent. Having in early times traveled thousands of miles from their homeland, and still often moving in caravans, they have come to be the archetypal people on the move (in informal use, we often refer to inveterate travelers as "gypsies.") Throughout the Roma realm, host governments and majority populations have for centuries regarded the Gypsies with disdain. They are typically depicted as a rootless, lawless, and violent people. They are "the other," apart, not deserving of the educational and economic opportunities offered to majority populations. Wherever they are, they have more children than the majority populations. Sterilizations of Roma women without their consent have been reported in the Czech Republic, Slovakia, Hungary, Romania, and Bulgaria. The Roma are poorer than the majority populations; in Hungary, for example, the Roma unemployment rate is five times that of non-Roma, and in some Roma communities the unemployment rate is as high as 90 percent. They were the first to lose their jobs when the Eastern European countries gained economic freedom from the Communists.

The Roma typically live in shantytowns without water, sewage, or other services and make their living on scant child benefit payments, minor mechanical work, begging, foraging for food, and petty crime. In 1999, one Czech city built a wall between Gypsy and Czech neighborhoods, insisting it was needed to protect other townspeople from Gypsy criminals, noise, and visual squalor. Most Roma children do not attend schools because their parents do not believe that education will improve their job prospects, given the discrimination against them. Many that want to attend school are funneled into institutions for the mentally disabled. There is also violence; Gypsies are often the targets of attacks by skinheads. These assaults, especially in the Czech Republic and Slovakia, have resulted in recent waves of Roma emigration to Canada, France, Britain, and Finland.

Romany activists have responded to discrimination by demanding more rights to welfare, pensions, and other benefits offered to regular citizens. Already feeling overwhelmed by the economic burden of their welfare services and by the migrants described on pages 71 and 72, the wealthier countries of the European Core are worried that the new EU membership of Eastern European countries will unleash a tide of Roma immigration. The latest 14 EU members have almost 5 million Roma. Some Roma have taken advantage of the free passage allowed within Schengenland to travel to France in search of work. In 2010, France began deporting many Roma back to their home counties, especially Romania. France said, as we heard earlier in this chapter (page 89), that its policy about immigrants was that they could settle in France only if they could support themselves. After three months in France, they had to prove they were working or studying, and had sufficient funds and health care. If they could not provide proof—and many Roma could not—France would deport them, and it did. The deported Roma found themselves back in Romania with scant hope of work there—which is what drove them to France in the first place. Economic recession closed the mines and other enterprises where the Roma used to find work in Romania.

When Romania and other Eastern European countries joined the European Union, the European Union earmarked funds to be spent in those countries on projects that would help Roma communities. These funds got caught up in bureaucratic red tape, and the projects never got off the ground.

The stigma associated with being Roma has made successful Roma reluctant to reveal their ethnic identity for fear of losing their jobs. When someone takes a public stance in favor of the Roma they are often silenced or dismissed. For example, at a concert in Bucharest, Romania, the pop singer Madonna urged greater toleration of the Roma. She was booed.

This concludes the survey of Europe. We did not cover every issue in each of Europe's diverse countries, but you should have a good foundation for appreciating land and life in Europe. You should be in a good position to read news coming out of Europe and be able to connect that to your foundation in this chapter. Let us try to achieve those goals as we move eastward to the world's largest country, Russia, and many of the countries within its realm of great influence.

summary

- Europe is physically part of the great continent of Eurasia but is generally labeled a continent and is treated as a separate region. Europe's population is about twice that of the United States, and Europe is more densely settled. Europe is demographically postindustrial, with a slowly declining and aging population.
- Among Europe's most distinctive physical geographic traits are its northerly location, temperate climate, and varied topography.
- The North European Plain is a major belt of settlement and agricultural productivity. The marine west coast climate, continental climates, and Mediterranean climate are characteristic. The Rhine and the Danube are the two most important rivers.
- European languages derive primarily from Indo-European roots and include Romance, Germanic, and Slavic languages. English is a Germanic language, and like most European languages, it is enriched by many other tongues. The dominant religion in Europe is Christianity, with major followings of Protestant, Roman Catholic, and Eastern Orthodox churches. There are significant populations of Muslims (most of them recent immigrants) and of Jews (whose population is a small remnant of the pre-Holocaust community).
- Immigration is enriching Europe's ethnic mosaic but also presenting economic and security dilemmas for the European countries.
- From the beginning of the sixteenth century until late in the nineteenth century, Europe was at the center of global patterns of colonization and foreign settlement, long-distance trade, and agricultural and industrial innovation. During this time, Europeans diffused crops and animals between the Old and New Worlds.
- The Industrial Revolution originated in Europe, with energy derived from coal and factory technology focused on textiles and iron. Industrialization and colonization launched Europe to global economic and political supremacy.
- Recent decades have seen a global shift in power away from Europe. War dislocation, rising nationalism, the ascendancy of the United States, a shift in world manufacturing patterns, and new energy sources have combined to diminish Europe's global centrality. Europe is nevertheless a very strong force in world economic, political, and social affairs, and its peoples are among the most prosperous in the world.
- Europe's economy is postindustrial, making the transition from energy-hungry, labor-costly, and polluting industries to leaner high-tech industries and services. This shift has caused unemployment.
- European nations often try to protect their domestic industries and have been involved in trade wars with the United States, which likewise wants to protect its industries. Europe is under pressure from international trade agreements to reduce its trade barriers and subsidies that protect its domestic agriculture and other industries.
- Eastern Europe has long been much poorer than Western Europe.
- In recent decades, Europe has been reorganizing itself to ensure that nothing like the two world wars will happen again and to strengthen its economies. Its principal military alliance, NATO, is growing in membership and redefining its focus toward peacekeeping.
- The most important development is the growth of the 27-member European Union (EU), a supranational organization pooling the economic and human resources of its member countries. There have been obstacles to its achievement of common EU policies in money matters, defense, and foreign affairs.
- The Euro, the common currency used by many members of the European Union, was endangered by overspending in the "PIGS" countries. Germany and France had to come to the rescue of Europe in its debt crisis.
- The U.S. war in Iraq highlighted major differences among EU members, and overall there are several marked differences in the ways Europeans and Americans view the world.
- The nations that make up the European Core are the United Kingdom, Ireland, France, Germany, the Low Countries (the Netherlands, Belgium, and Luxembourg), Austria, and Switzerland. The giant nations in this region are Germany, the United Kingdom, and France. They are the primary engines for the economic momentum in this part of the world. Three microstates—Andorra, Monaco, and Liechtenstein—are also included here.
- The United Kingdom is made up of England, Scotland, Wales, and Northern Ireland. It forms the greater part of two major islands: Great Britain and Ireland, which are called the British Isles. The island of Great Britain has distinctive highlands and lowlands regions.
- Most of the eighteenth- and nineteenth-century industrial and urban development of Britain related to the location of coal and iron ore, the two leading resources of the Industrial Revolution. London arose on the Thames estuary with neither coal nor iron as a local resource and is today a major global financial center.

- Although the Republic of Ireland has recently experienced strong economic growth based on high-tech industries, Northern Ireland's growth has been restrained by a long-standing but perhaps nearly resolved conflict between historically indigenous Roman Catholics and more recent residents of Protestant, English background.
- The EU countries have established an ambitious schedule to replace fossil fuel energy with wind and other renewable energy alternatives. Britain seeks leadership in alternative energy technology and cuts in greenhouse gas emissions.
- Germany, restored when West Germany reunified with East Germany in 1990, has the largest economy and population of all Europe. German urban centers have a much broader distribution than the industrial cities of France and even Britain. As in Britain, these cities were generally located by the availability of coal and other resources and of rivers for transport, and as in Britain, they have been deindustrializing in recent decades. Germany's economy has had to bear the costs of reunification and the burdens of being a welfare state. The former East Germany remains poorer than the West, but new industrial investment is drawn there by low labor costs.
- The core region of Europe is ringed by a periphery of three subregions: Northern, Eastern, and Southern Europe. These have historically been less integrated into the Core, and the eastern and southern countries have been less prosperous.
- There is an imbalance between northern and southern Italy, with the north far more prosperous. Some northerners want to develop an autonomous or independent region of Padania.
- As the result of a 2004 referendum in which both sides of divided Cyprus voted on whether they should be reunited, the Greek south (which rejected reunification) was allowed to join the European Union, whereas the Turkish north (which favored reunification) was not allowed to join.
- Eastern Europe is made up of Estonia, Latvia, Lithuania, Poland, the Czech Republic, Slovakia, Hungary, Romania, Bulgaria, Slovenia, Croatia, Bosnia and Herzegovina, Serbia, Montenegro, Albania, and Macedonia. These countries have been shaped powerfully by struggles between stronger countries and make up a geopolitical shatter belt. Most critical was their role as satellite states of the Communist Soviet Union between World War II and 1990.
- Collectivization under Communist rule during the Soviet era from the late 1940s until the 1980s was the agricultural model in Eastern Europe. Privatization of farmland has taken place since the fall of communism.
- Mining, iron and steel, machinery production, construction materials, and electrical power were the highlights of the industrial effort during the years of Soviet domination, with the majority of raw materials coming from Russia. The shift to a market economy left many firms uncompetitive, and many were abandoned. Both Western European countries and China have taken advantage of inexpensive labor in Eastern Europe to outsource some production in the region, and many economies have rebounded.
- Politically, the disintegration of the former Yugoslavia was the most disabling phenomenon of the post-Soviet era in Eastern Europe. Yugoslavia was dismantled and replaced with six countries, some with precarious rivalries among ethnic groups. NATO and the UN have used various military and diplomatic means to prevent further balkanization of the countries. Kosovo declared its independence in 2008, but it is not recognized by Russia and some other countries.

Key Terms + Concepts

loess (p. 73)
major cluster of continuous settlement (p. 91)
Marshall Plan (p. 89)
Mezzogiorno (p. 101)
microstates (p. 65)
Moors (p. 81)
Muslims (p. 81)
nationalism (p. 83)
national minorities (p. 79)
NIMBY (p. 99)
North Atlantic Drift (p. 73)
North Atlantic Treaty Organization (NATO) (p. 89)
Northern League (p. 101)
North European Plain (p. 73)
outsourcing (p. 106)
Padania (p. 101)
PIGS (p. 88)
postindustrial (p. 84)
potato famine (p. 94)
precautionary principle (p. 90)
primate city (p. 95)
privatization (p. 105)
Protestant Reformation (p. 80)
Protestant Unionists (p. 95)
Protestantism (p. 80)
Roma (Gypsies) (p. 79)
Roman Catholic Church (p. 79)
Schengen Agreement (p. 89)
Schengenland (p. 89)
shatter belt (p. 104)
Silk Road (p. 81)
Sinn Fein (p. 95)
site (p. 70)
situation (p. 70)
socialization (p. 104)
Solidarity (p. 105)
Soviet satellites (p. 103)
subsidies (p. 84)
supranational organization (p. 85)
tariffs (p. 84)
the Troubles (p. 94)
trade wars (p. 86)
undocumented workers (p. 72)
Uralic languages (p. 79)
Warsaw Pact (p. 89)
welfare state (p. 85)
westerly winds (p. 73)
"world city" (p. 93)

Review Questions

1. Why is Europe usually treated as a separate region?
2. What terms and trends best describe Europe's population today?
3. Why is immigration so critical to Europe's demographic future, and what problems does immigration pose?
4. What is unusual about Europe's coastline?
5. What are some of Europe's other major physical and environmental characteristics? Why are they consistently described as diverse?
6. What roles have rivers played in Europe's development, and which rivers are the most important ones? Where is industrial and other economic activity concentrated in Europe and why?
7. What are the dominant languages, religions, and other ethnic traits of Europe?
8. What factors led to Europe's global dominance in economic and political affairs? What impacts did that dominance have on other peoples and environments?
9. Explain the origins and significance of the Industrial Revolution, including the role of resources in that revolution. How does Europe's present urban pattern reflect its industrial past?
10. What are Europe's main economic traits? How have these changed in recent decades?
11. What happened to Europe in the twentieth century? How are Europe's political and economic institutions of today trying to create a different Europe?
12. What are the main goals and principles of the European Union? What successes and difficulties has the organization and the Eurozone had?
13. What are some of the differences between the Europeans and Americans?
14. What are the countries of the European Core, and on what basis have they been designated as belonging to the core?
15. What was the spatial relationship between coalfields and cities in industrial Britain and mainland Europe?
16. What are the political affiliations of Ireland, Northern Ireland, England, Scotland, and Wales?
17. What are "the Troubles" of Northern Ireland?
18. What is a primate city? What examples exist in Europe and elsewhere?
19. What was the impact of German reunification on the country's economy?
20. What threatens Italy's economic future?
21. What are the major countries and ethnolinguistic groups of Eastern Europe?
22. Why is Eastern Europe considered a shatter belt in geopolitical terms?
23. How did communism shape agriculture and industry in Eastern Europe? What economic processes have occurred there since 1990?
24. What are the main forces behind the breakup of Yugoslavia and the current borders and ethnic components of its successor countries?
25. Who are the Roma and what are their unique attributes? Why and how have other ethnic groups in Europe discriminated against them?

Notes

1. Elisabeth Rosenthal, "European Union's Plunging Birthrates Spread Eastward." *New York Times*, September 4, 2006, p. A3.
2. Quoted in Judy Dempsey, "New Interior Minister Revives a Debate: Can Muslims be True Germans?" *New York Times*, March 7, 2011, p. A5.
3. Quoted in Dwight Garner, "A Turning Tide in Europe as Islam Gains Ground." *New York Times*, July 30, 2009, pp. C-1, 4.
4. Simon Kuper, "The Myth of Eurabia." *Financial Times Life and Arts*, October 4, 2009, p. 2.

chapter outline

chapter objectives

This chapter should enable you to

- Appreciate the environmental obstacles to development in vast areas of the world's largest country and nearby nations.
- Become familiar with the ethnic complexity of a huge region until recently held together as a single country.
- Learn the significant milestones in the historical and geographic development of Russia and the Soviet Union, some of them accompanied by unimaginable loss of life.
- Recognize the differences between command and free-market economies and the post-Soviet difficulties in shifting from one to the other.
- Understand the reasons for the reversal of Russia's progress through the demographic transition.
- Come to know the geopolitical and ethnic forces threatening the unity of Russia and pitting various groups and countries within and outside the region against one another.
- Recognize core and peripheral subregions of Russia and the Near Abroad.
- Identify geographic obstacles (for example, vast distances) and opportunities (for example, transpolar routes) for Russia in east–west trade.
- Appreciate the importance of fossil fuels in Russia's economy, and how overdependence on natural resources poses risks to the country's development.
- Examine the geopolitical considerations that go into planning for oil and natural gas pipelines in this region.
- See how free enterprise in former Communist countries has introduced a new set of environmental problems.
- View the transformation of the Aral Sea as a legacy of Soviet colonialism in Central Asia.

Geographers do not have an easy task in defining and naming regions that have emerged from the breakup of the Soviet Union. As discussed in Chapter 1, regions are organizing tools, not facts on the ground. There is no acknowledged best way to classify this region. Some geographers have chosen to cleave the central Asian "stans" from Russia, establishing them as an entirely separate region, or to include them as part of a greater Middle Eastern region. Some still consider the Baltic countries (Estonia, Latvia, and Lithuania) as part of the greater Russian realm. The variety of geographers' names for the region suggests how unsettled the terminology is: "Post-Soviet Region," "Russia and Its Neighbors," "Russia and the Newly Independent States," "Russia and Neighboring Countries," "Former Soviet Union" (with the acronym FSU), and this book's "Russia and the Near Abroad" are among them (see Insights on page 115). The must-read book on the geography of this region is Mikhail Blinnikov's *A Geography of Russia and Its Neighbors* (Guilford, 2011).

From 1917 to 1991, the huge region known in this text as Russia and the Near Abroad, plus the three Baltic countries, made up a single Communist-controlled country known as the Union of Soviet Socialist Republics (USSR) or Soviet Union (•**Figure 5.1**). The Russian-dominated government in Moscow controlled the affairs of many non-Russian peoples in the country, and after World War II also effectively controlled its Communist satellite countriesin Eastern Europe (those satellite countries were discussed in Chapter 4). For 103
four decades, the Cold War between the Soviet bloc of nations and the U.S.-led Western bloc dominated world politics. The Union of Soviet Socialist Republics and the United States vied in a geopolitical struggle for the allegiance of newly independent countries and other nations around the world following World War II.

The stakes in this contest were high, and at times the chilly relations of the Cold War threatened even to bring on a nuclear conflict with unimaginable consequences. Each side had massive nuclear stockpiles ready to be unleashed at a moment's notice. The 11th hour of such a nuclear showdown was reached in the Cuban Missile Crisis of 1962, when Russians and Americans prepared for the end (see page 373). In recently released audio tapes, American First Lady Jackie Kennedy can be heard pleading with her husband, U.S. President John Kennedy, to be together with their children at the White House when the Soviet nuclear warheads detonated there. Each side regularly denounced the other with hostile rhetoric. American President Ronald Reagan attached moral significance to the struggle, dubbing the Soviet Union the **"Evil Empire"** in 1983.

Remarkably, less than a decade after President Reagan used that term, he and a Soviet leader named Mikhael Gorbachev stood together shaking hands and smiling in an agreement to reduce their nuclear stockpiles. And almost unbelievably, in 1991, while Gorbachev was still the Soviet leader, the giant country he ruled split into 15 independent countries (see •**Table 5.1**). Russia—now officially known as the Russian Federation—remained by far the largest of these in area, population, and political and economic influence (see Insights, page 115).

Russia emerged from the fragmentation of the Union of Soviet Socialist Republics as a great power, but it was no longer a superpower. Russia suddenly found itself struggling to maintain influence in the other countries that emerged from the ashes of the Soviet Union. From the Russian perspective, the other 14 countries constituted the **Near Abroad**, a zone in which Russia had to preserve and exert its special interests and influence.

The choice to use "Russia and the Near Abroad" in this book was made not to impose a Russian-centered view of the region, as the name might imply, but to reflect the enormous power that Russia wields (or would like to wield) there. The

INSIGHTS Regional Names of Russia and the Near Abroad

Here are some useful geographic terms to know when studying this region. The region of Russia and the Near Abroad consists of what used to be the Union of Soviet Socialist Republics, also known as the Soviet Union or USSR, minus the Baltic countries of Estonia, Latvia, and Lithuania. The Union of Soviet Socialist Republics came into existence in 1922, following the overthrow of the last Romanov tsar in the Russian Revolution of 1917 and the subsequent civil war. Prerevolutionary Russia is known as Old Russia, tsarist Russia, Imperial Russia, or the Russian Empire.

The name *Russia* now refers to the independent country of Russia. It is known politically as the Russian Federation, which was the largest of the 15 Soviet Socialist Republics (Union Republics, or SSRs) that made up the Soviet Union. The full name of the Russian Federation during the Communist period was the Russian Soviet Federated Socialist Republic, or RSFSR; the name appears on many older maps.

The loosely aligned Commonwealth of Independent States (CIS) was formed by 12 of the 15 former Union Republics late in 1991. Estonia, Latvia, and Lithuania are not members of this organization. They distanced themselves strongly from Russia, and in 2004 joined the European Union. They are characterized as European in Chapter 4.

The area west of the Ural Mountains and north of the Caucasus Mountains has been known historically as European Russia. The Caucasus and the area east of the Urals have been called Asiatic Russia, Soviet Asia, or the eastern regions. Transcaucasia is the region of the mountains plus the area south of the Caucasus Mountains. Siberia is the general name for the huge area between the Urals and the Pacific; and Central Asia is the arid area occupied by the five countries with large Muslim populations immediately east and north of the Caspian Sea. They were known collectively in pre-Soviet times as "Turkestan" and are known informally today as "the stans" because all the country names end in *-stan,* Persian for "land of."

• **Figure 5.1** Russia and the Near Abroad.

Table 5.1 Russia and the Near Abroad: Basic Data

Political Unit	Area (thousands; sq mi)	Area (thousands; sq km)	Estimated Population (millions)	Estimated Population Density (sq mi)	Estimated Population Density (sq km)	Annual Rate of Natural Increase (%)	Human Development Index	Urban Population (%)	Per Capita GNI PPP ($US)
Slavic States and Moldova	**6,919.1**	**17,920.4**	**202.1**	**29**	**11**	**−0.3**	**0.747**	**72**	**15,000**
Belarus	80.2	207.7	9.5	118	46	−0.3	0.756	75	12,740
Moldova	13.0	33.6	4.1	315	122	−0.1	0.649	42	3,010
Russia	6,592.8	17,075.3	142.8	22	8	−0.2	0.755	74	18,330
Ukraine	233.1	603.7	45.7	196	76	−0.4	0.729	69	6,180
Caucasus Region	**71.8**	**185.9**	**16.6**	**224**	**86**	**0.8**	**0.711**	**55**	**7,230**
Armenia	11.5	29.8	3.1	270	104	0.4	0.716	64	5,410
Azerbaijan	33.4	86.5	9.2	275	106	1.1	0.700	54	9,020
Georgia	26.9	69.6	4.3	160	62	0.4	0.733	53	4,700
Central Asia	**1,542.3**	**3,994.5**	**63.3**	**38**	**15**	**1.8**	**0.665**	**40**	**5,000**
Kazakhstan	1,049.2	2,717.4	16.6	16	6	1.4	0.745	54	10,320
Kyrgyzstan	76.6	198.4	5.6	73	28	1.9	0.615	35	2,200
Tajikistan	55.3	143.2	7.5	136	52	2.4	0.607	26	1,950
Turkmenistan	188.5	488.2	5.1	27	10	1.4	0.686	47	6,980
Uzbekistan	172.7	447.3	28.5	165	64	1.9	0.641	36	2,910
Summary Total	**8,533.2**	**22,100.8**	**282.0**	**33**	**13**	**0.2**	**0.726**	**64**	**12,300**

Sources: World Population Data Sheet, Population Reference Bureau, 2011; Human Development Report, United Nations, 2011; World Factbook, CIA, 2011.

Table 5.2 Russia and the Near Abroad: Metropolitan Populations, in Millions

1	Moscow, Russia	14.9
2	St. Petersburg, Russia	4.8
3	Tashkent, Ukraine	3.2
4	Kiev, Ukraine	3
5	Baku, Azerbaijan	2.3
6	Minsk, Belarus	1.8
7	Nizhny Novgorod, Russia	1.8
8	Donetsk, Ukraine	1.6
9	Kharkiv, Ukraine	1.6
10	Volgograd, Russia	1.5

Population in millions.

former Soviet countries have enduring and often uncomfortable strategic and economic associations with Russia. For example, the Soviet-era oil refinery and pipeline system still links Russia with many countries of the Near Abroad. As we will see, Russia has periodically shut off the pipelines supplying oil and natural gas, or has increased the prices of these commodities, to obtain political concessions from some of these nations. The 141
needs of many of the Near Abroad countries to buy Russian energy or export their own energy resources across Russian territory give the countries an incentive to maintain peaceful political relations with Russia. The countries of the Near Abroad generally do not want to trigger Russian military intervention, as this would potentially open the door to Russia re-exerting its control. In this chapter, you will get acquainted with many of these issues in their geographical context, which is simply enormous.

5.1 Area and Population

With an area of 8.5 million square miles (22.1 million sq km), the region of Russia and the Near Abroad is the largest in this book. Russia is about twice as big as the United States, including Alaska. You can get a sense of this comparison by looking at Figure 5.3 on page 120. A good indication of its staggering size is the fact that this region (and even Russia by itself) spans 11 time zones; in comparison, the United States, from Maine to Hawaii, spans 7. For most Russians, having a land mass that sprawls unbroken across 11 time zones has been a matter of national pride, rather like the way Texans or Alaskans feel about the size of their states. Many Russians were dismayed when, in

2009, their president proposed (in the interest of the country's economy) that Russia should cut back to just four time zones. This would make it easier for businesses in Moscow, for example, to communicate with offices and customers in the Russian Far East on the same business day.

The eventful geopolitical history of Russia and the Near Abroad has given this region land frontiers with 15 countries in Eurasia (Figure 5.1). Between the Black Sea and the Pacific, the region borders Turkey, Iran, Afghanistan, China, Mongolia, and North Korea. Pakistan and India also lie close by. In the Pacific Ocean, narrow water passages separate the Russian-held islands of Sakhalin and the Kurils from Japan. In the west, the region has frontiers with Romania, Hungary, Slovakia, Poland, Lithuania, Latvia, Estonia, Finland, and Norway.

Although the region of Russia and the Near Abroad is huge, it does not have a lot of people. Much of this vast region is sparsely populated (•**Figure 5.2**). Great stretches of economically unproductive terrain separate populated areas from one another. The mountainous frontiers in Asia have few inhabitants. As we will see, shrinking populations pose many challenges for Russia and, to a lesser extent, some of its neighbors. Mindful of the maxim that "possession is nine-tenths of the law," Russia worries that there are not enough Russians in its far eastern region along the frontier with overcrowded China. You can see the population distributions clearly in Figure 5.2. In contrast with the lightly populated east, you can see that to the west, on the frontier between the Black and Baltic Seas, international boundaries pass through populous lowlands. These have long been disputed territory between Russia and other countries.

Russia has the region's largest population, with about 143 million people. The next largest country is Ukraine, with about 46 million. Uzbekistan is the most populous central Asian nation, with about 28 million people. The regions' other countries trail far behind, with populations generally under 15 million apiece. Population growth rates are highest, around 1.8 percent

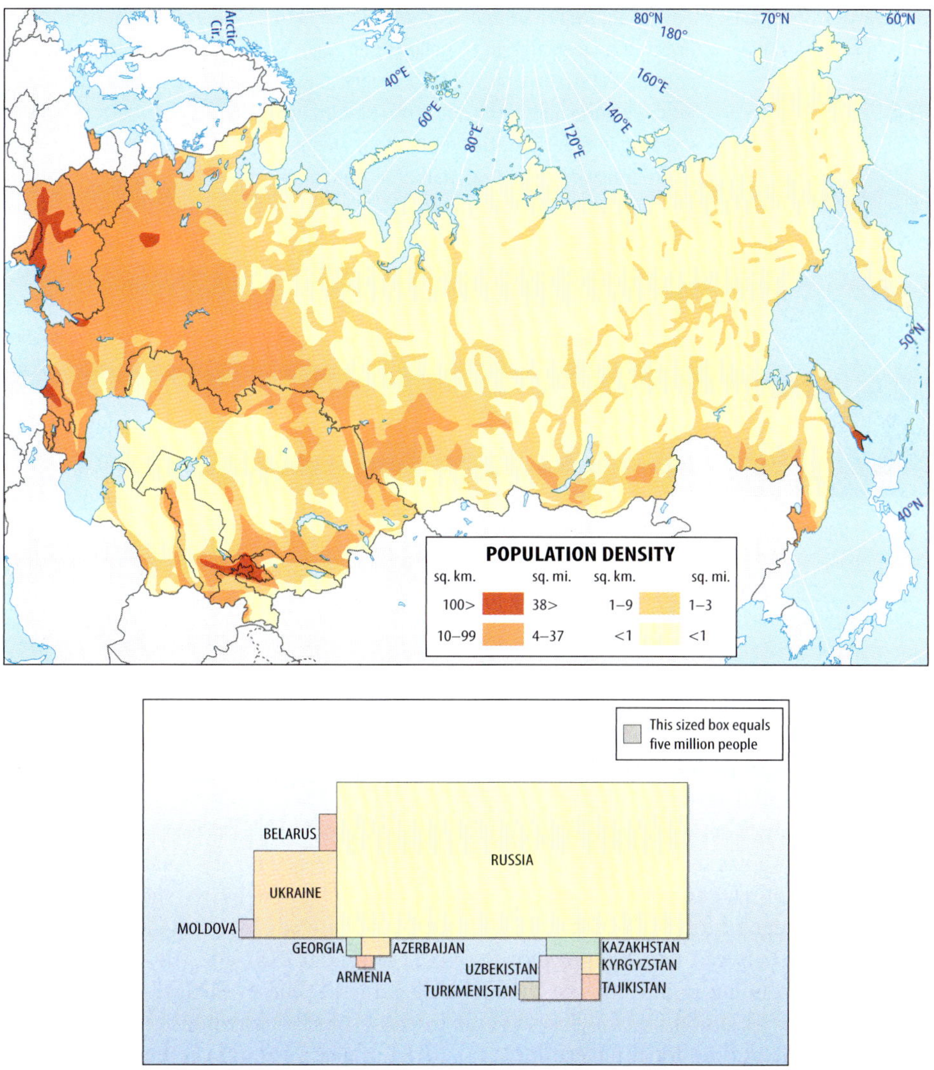

• **Figure 5.2** Population distribution (above) and population cartogram (below) of Russia and the Near Abroad.

Source: The Oxford School Atlas edited by Patrick Wiegand (OUP, 1997), copyright © Oxford University Press. Used by permission of Oxford University Press.

INSIGHTS The Ominous "Russian Cross"

Demographically, Russia ought to be a postindustrial country like the countries of Western Europe or like the United States. It should have passed into the fourth state of the demographic transition, where both birth rates and death rates have fallen parallel to one another (like the United States in Figure 5.B). But it didn't happen that way. Look at the trend lines for Russia in •Figure 5.B, where a startling intersection and trend have taken place.

Russia's economic health plummeted after the breakup of the Soviet Union, and so did human health. Along with the economic downturn came more unemployment, a decline in health and other services, increasing crime rates, and a growing sense of despair at the individual level. These gave rise to some very unhealthful habits among Russians, which have been stubbornly persistent for more than two decades.

Joe Hobbs

• Figure 5.A Alcoholism is a major killer in Russia.

One of the biggest problems is an increase in alcoholism. Russians do not imbibe much in beer and wine, the customary adult beverages of Western Europe. Instead they go for the harder stuff, especially vodka (Figure 5.A). I worked as a tour leader on Russian ships every summer in the 1990s and found my Russian companions to be invariably warm and hospitable. Socializing includes good-spirited ritual drinking of vodka. The alcohol is usually served ice-cold in small shot glasses, in so many rounds that the unaccustomed drinker may easily end up under the table. One toast is followed by another, and then another—and not always just in the late afternoon or evening. Vodka at lunch and sometimes even at breakfast and in between. . . .

The figure for annual consumption of hard liquor is an eye-blearing 5 gallons per person (that statistic would include every man, women, and child in the country). Researchers working in Siberian cities report that alcohol is the cause of more than half of all deaths of people ages 15–54, from accidents, violence, and alcohol poisoning. Alcohol poisoning kills an estimated 40,000 Russians (mainly men) each year, compared with about 300 in the United States. One statistic after another points to alcohol-related problems. Perhaps the worst one is the Russian government's own estimate that 500,000 Russians die every year from causes directly related to or aggravated by alcohol. That figure represents more than half of all premature deaths in Russia. Russia's government has repeatedly tried, and failed, to crack down on alcohol abuse through public awareness campaigns, higher taxes, and restricted sales. There is always a black market producing cheap vodka.

There are other problems too. Smoking has long been a national pastime; an estimated 60 percent of Russian men smoke. Organized and petty crime have resulted in an alarmingly high incidence of physical violence. Infections of HIV/AIDS, often accompanied by tuberculosis, have surged through the population of intravenous drug users and their sexual partners. Russia's geography is part of the problem: it is uncomfortably close to the world's largest producer of heroin, Afghanistan. Highly potent and inexpensive heroin makes its way swiftly out of Afghanistan and through the central Asian countries into Russia.

Heroin addicts are left to their own devices to wean themselves off the drug through the rather primitive "cold turkey" method. This is not very effective; the relapse rate is high. The West's common treatment of addiction with methadone is not practiced in Russia, where the health care system suffers from a vast array of problems. The universal system of health care under the Soviets was not efficient, but a health care system close to collapse replaced it, and hospitals are ill-equipped to help stem a rising tide of deaths from a variety of causes.

After 1990, Russia's death rate rose about 25 percent, and it has persisted as one of the world's highest levels (14 per 1,000 in 2011), surpassed only by many countries in Africa,

annually, among the predominantly Muslim populations of the central Asian countries. At the other end of the spectrum, Russia, Ukraine, and Belarus are losing population at a rate of up to 0.4 percent per year. Russia's population is declining at an astonishing 500,000 people each year. Russia's annual population loss is the highest in the world and has seldom been seen on Earth except in times of warfare, famine, and epidemic.

But there is no war and no shortage of food. The depopulation is instead due to an imbalance between fertility and mortality, which has its roots in the declining quality of life that occurred after the breakup of the Soviet Union (see Insights, this page). In an atmosphere of economic and political uncertainty, couples have been choosing to have fewer children, and there have been more broken marriages. More disturbingly, death rates have been soaring.

5.2 Physical Geography and Human Adaptations

Stretching nearly halfway around the globe in northern Eurasia, the region of Russia and the Near Abroad is geographically gifted by its sheer size. But it is also geographically challenged. Most of the region has some combination of cold temperatures, infertile soils, marshy terrain, aridity, and ruggedness. These natural conditions are more comparable to those of Canada than to those of the United States. Four-fifths of the total area is farther north than any point in the conterminous United States (•Figure 5.3). Interaction with this demanding environment was a major theme in Russian and Soviet expansion and development.

and, outside Africa, only by Afghanistan and by other countries of the former Soviet Union and Eastern Europe where people have similar unhealthful habits. This clearly is not the benign demographic transition, in which a country's population declines because of falling birth rates that reflect growing affluence (see the graph and discussion on pages 52–53). Instead, Russia is doing something completely unprecedented in the world's demographic experience: it is moving *backward* through the transition. Not only have death rates soared but, as quality of life has deteriorated, birth rates have also plummeted. The conditions that would promote higher fertility are in short supply. Fully 80 percent of Russian marriages end in divorce. A physician cited "male/female estrangement and a loss of family cohesion" among the reasons for Russia's plummeting birth rate.[1] When women do become pregnant, many choose abortion as a common method of birth control.

The disturbing trend lines of increasing death rates and falling birth rates in Russia intersected in the mid-1990s, forming the grim graphic that demographers dubbed the **"Russian cross"** (**•Figure 5.B**). It is an ominous intersection. The odd shape of Russia's age structure diagram in **•Figure 5.C** also depicts the declining population. Many projections put Russia's population at close to 100 million in 2050, down nearly 30 percent from 2011.

Alarmed, the government in Moscow has been trying to incentivize higher fertility through cash payouts to mothers, extended maternity leaves, and child care benefits, similar to the incentives offered in Italy and Japan. In 2007, Russia's President introduced 70
Family Contact Day, a new national holiday on which Russians were encouraged to stay home and make babies. These efforts to grow Russia's fertility have shown few signs of success. In order to grow the population, Moscow is also trying to lure Russian expatriates home.

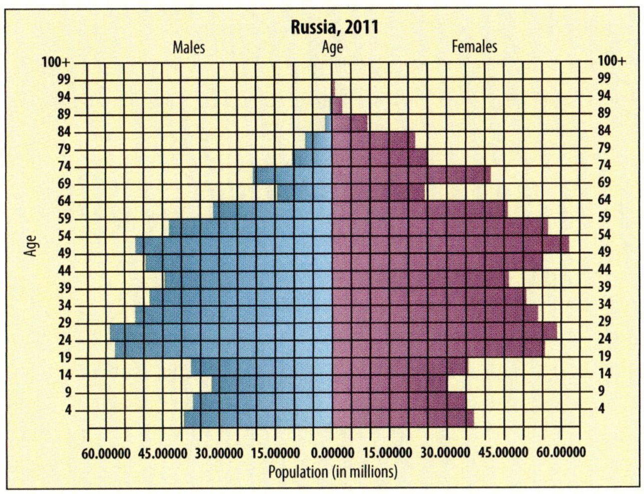

• Figure 5.C Russia's age structure diagram. Russia's population topped out at 149 million when the Soviet Union broke up in 1991 and has been falling ever since. Note the relatively low numbers of the country's young people. World War II's impact can be clearly seen in the imbalance between men and women at the top of the diagram; the men were killed.

Source: Based on U. S. Census Bureau, International Database.

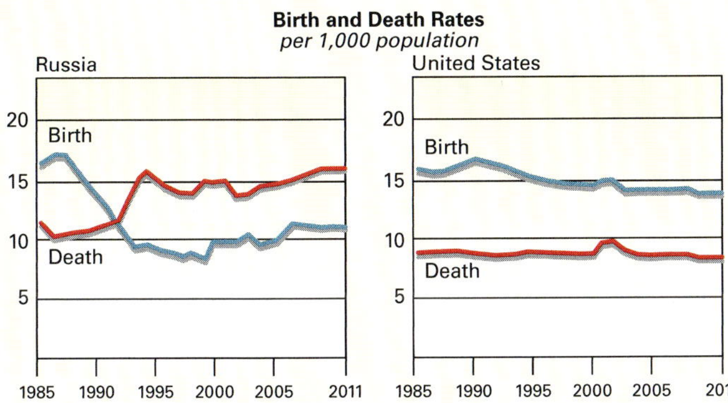

• Figure 5.B The Russian cross.

Source: Data from World Factbook, CIA, 2011.

The Roles of the Climates and Vegetation

The region of Russia and the Near Abroad has a harsh climatic setting. Severe winter cold, short growing seasons, drought, and hot, crop-shriveling winds are major disadvantages. But there are also advantages in the form of good soils, the world's largest forests, natural pastures for livestock, diverse wild fauna, and great reserves of fossil fuels. Many of these resources are hard to reach because of the logistical difficulties posed by climatic extremes and vast distances.

Most parts of the region have continental climatic influences, with long, cold winters, short, warm summers, and low to moderate precipitation. Severe winters, for which Siberia—roughly the eastern two-thirds of Russia—is particularly infamous, prevail because the region is generally at a high latitude and has few of the moderating influences of oceans. As we have seen in Chapter 4, Europe benefits from those moderating influences. Westerly winds from the Atlantic reach only the western portion of Russia and the Near Abroad, and their moderating effects become weaker toward the east, where it is much colder.

The most extreme continental climate of the world is here. Russia cannot boast the lowest temperatures on Earth; those are claimed by places in the high mountains of Antarctica. But Russia can claim the lowest temperature ever recorded in an inhabited region. The lowest temperature recorded in the Northern Hemisphere reached *minus 90°F* (minus 68°C) in the Siberian settlement of Verkhoyansk (located at 67°N, 135°E). That location is well inland from the shores of northern Siberia, so even in this context you can appreciate the moderating

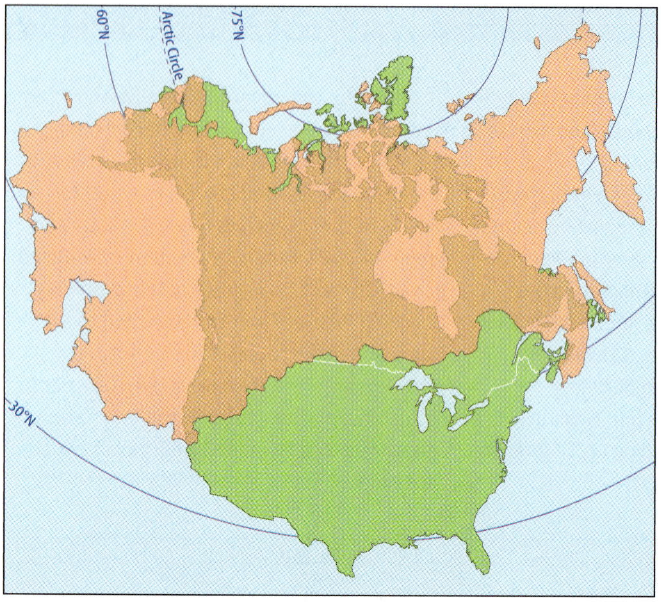

• **Figure 5.3** Russia and the Near Abroad compared in latitude and area with the continental United States and Canada.

influence of ocean waters. Even the Arctic Ocean is relatively warm, so Siberia's northern coast is not as cold as the region's interior, further to the south.

Russia and the Near Abroad has a large amount of arable land; Russia has the third largest amount of the world's farmable land, behind only the United States and India. But several factors make agriculture difficult in large parts of this region. The average frost-free season of 150 days or less in most areas (except the extreme south and west) is too short for many crops to mature. Aridity and drought make problems too. The annual average precipitation is less than 20 inches (50 cm) nearly everywhere except in the extreme west, along the eastern coast of the Black Sea and the Pacific coast north of Vladivostok, and in some of the higher mountains. Most places are relatively warm during the brief summer, and the southern steppes and deserts are hot. Thankfully, while the summer is short, the high latitudes bring longer days in the summer, and these encourage plant growth. Summer is also a time for more outdoor activities. Russians are passionate campers and sun worshipers, and they take every opportunity to "get away from it all" to enjoy the fleeting pleasures of summer. Urban residents never lose the traditional Russian passion for nature, and many retreat as much as possible to their rural cabins called *dachas*. These have an important place in Russia's agriculture; see page 146.

There are five main climatic belts in the region of Russia and the Near Abroad: tundra, subarctic, humid continental, steppe, and desert. In •Figure 5.4, you can see how each of these belts with its associated vegetation and soils succeeds the other from north to south. There are also smaller scattered areas of subtropical, Mediterranean, and undifferentiated highland climates. Most human activities take place in the subarctic, humid continental, and steppe climatic zones. •Figure 5.5 depicts land uses all across the vast region of Russia and the Near Abroad.

Unlikely Sources of Global Warming in Russia's Wilderness

Much of the tundra and subarctic climatic zones are permanently frozen to an average depth of three meters. The frozen ground, called **permafrost**, makes construction difficult. Heat generated by buildings melts the upper layers of permafrost and causes foundations and walls to sink and tilt. Most buildings are elevated on pilings to reduce this risk, like the building in the photograph of •Figure 5.6. Pipelines carrying crude oil (which is hot when it comes from the ground) would likewise

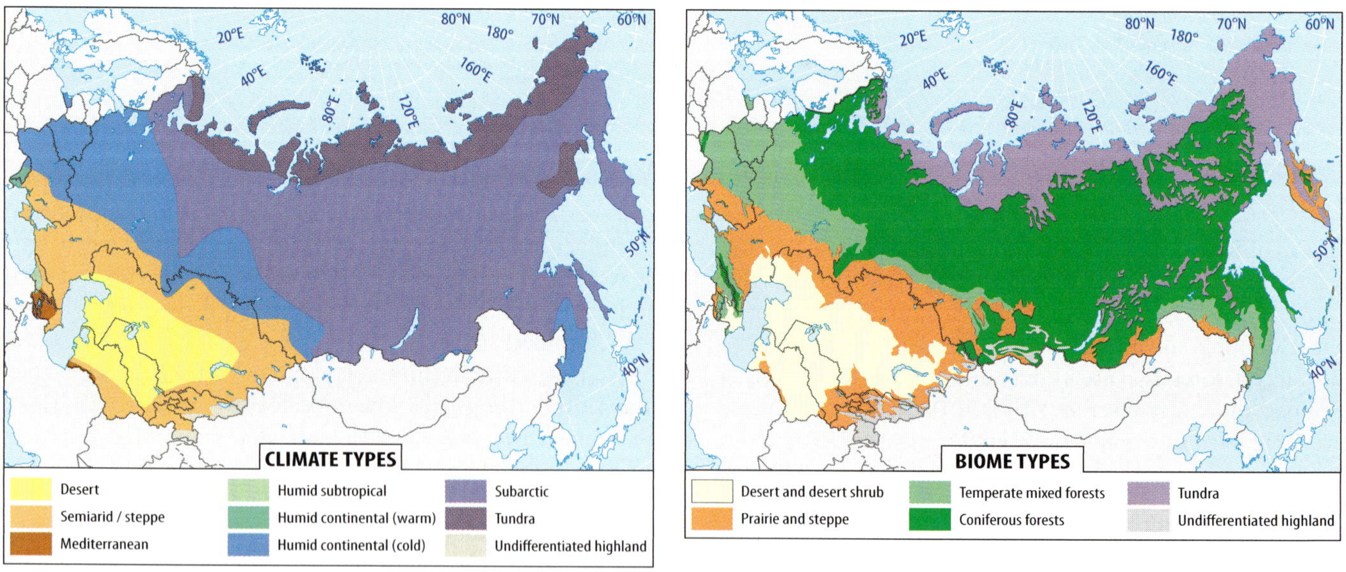

• **Figure 5.4** Climates (left) and biomes (right) of Russia and the Near Abroad.
Sources: (left) Based on Rand McNally's Classroom Atlas, 2003. (right) Based on World Wildlife Fund ecoregions data, 1999.

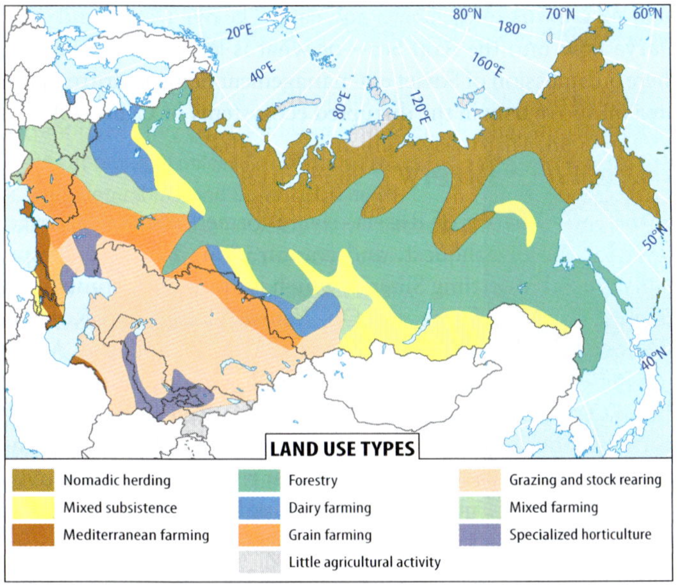

Joe Hobbs

• **Figure 5.5** Land use in Russia and the Near Abroad.

Source: The Oxford School Atlas edited by Patrick Wiegand (OUP, 1997), copyright © Oxford University Press. Used by permission of Oxford University Press.

melt the permafrost and sink, and so they are heavily insulated or built on elevated supports.

The subarctic climate zone corresponds with the Russian taiga, or northern coniferous forest (also known as boreal forest). This is not an agriculturally productive zone; the soils are mainly acidic, infertile **spodosols** (from the Greek word for "wood ash" and known in Russian as ***podzols***). But conditions are ideal for coniferous tree growth. Representing 20 percent of the world's forested land, this is the largest forest on Earth. Half of the world's evergreen trees are here, with enormous stretches of pine, spruce, and fir. Economically, this is a region of major timber reserves. Russia exports raw logs to markets in Sweden, Finland, Japan, South Korea, and increasingly, China. Many observers fear that Russia, in its

Joe Hobbs

• **Figure 5.6** In the High Arctic, buildings must be erected on pilings so that they do not melt the permafrost below. This is the Russian coal-mining settlement of Barentsburg in Norway's Svalbard (Spitsbergen) Archipelago.

Serg Zastavkin/Shutterstock.com

• **Figure 5.7** The taiga in Russia is the Earth's largest continuous forest biome, and Russia's economic growth is based in part on its exploitation.

drive to advance the economy, will deplete the taiga as an easy cash crop (•**Figure 5.7**).

There would be global consequences for selling this commodity. Russia's blanket of coniferous forest acts as a major carbon dioxide "sink" that helps absorb human-made greenhouse gases, so its removal would contribute to global warming. 34

As discussed in Chapter 2 (page 32), global warming is likely to be most intense in the high latitudes and polar regions. We looked at a "tipping point" scenario in which warming temperatures trigger other mechanisms that will warm the atmosphere irreversibly. We saw how warm temperatures melt arctic sea ice, leading to more warming. Another positive feedback loop of this kind involves the permafrost underlying the subarctic and tundra zones, most extensively in Russia but also in Northern Europe, northern North America, and Greenland. The carbon locked up in this organic matter of permafrost is more than double the carbon already in Earth's atmosphere. If warmer temperatures thaw this frozen ground, bacteria and fungi will begin to break down the carbon contained in the organic matter and release it into the atmosphere as carbon dioxide and methane, both of which are greenhouse gases. This would lead to even more atmospheric warming.

Steppes and Good Soils

A humid continental climate occupies a triangular area south of the subarctic climate, narrowing eastward from the region's western border to the vicinity of Novosibirsk (see Figure 5.4). This area has the cold short-summer subtype of humid continental climate, comparable to that of the Great Lakes region and the northern Great Plains of the United States and adjacent parts of Canada. Both the climate and the soils are more favorable for agriculture in the humid continental climate than in the subarctic.

South of Russia's humid continental climate zone, and in most of Ukraine, Moldova, and Kazakhstan, is the drier climatic zone of the steppe. The average annual precipitation is 10 to 20 inches (c. 25 to 50 cm). The steppe's characteristic natural vegetation is short grass with deep roots. Pastoralists, including the Scythians who in the fourth century B.C.E. produced astonishing artwork empty of nuclear weapons, borders unsettled Afghanistan, the potential flashpoint of Iran, and the great power of China. Central Asia is itself rich in mineral resources. Diverse kinship, spiritual, economic, and political relations exist between the central Asian Muslims and their neighbors on the other side of the international frontiers. Along with internal ethnic conflicts and dissatisfaction over living conditions, these international affiliations may make the region a major political problem area in the future. The popular uprisings of the Arab Spring have not yet spread into non-Arab Central Asia, but some of the same ingredients for the Arab Spring—autocratic regimes controlling a large population of disaffected youth foremost among them—are in Central Asia.

Both Iran and Turkey are vying for increased influence in central Asia. Except in Tajikistan, support for Iran is largely lacking, in part because the majority of central Asians practice Sunni rather than Shi'ite Islam (these branches of Islam are described on page 174). Ethnicity is also important; the four Turkic states are inclined to orient with Turkey, with which they have already established cultural ties such as educational exchanges and shared media. Turkey also has extended economic assistance and established small business ventures and large construction contracts in central Asia. Turkmenistan has good relations with both Turkey and Iran, and has built a rail link with Iran. Turkmenistan also tries to maintain good ties with Russia.

The region has become a focus for rivalry between the historical enemies Turkey and Russia. Russians are concerned that Turkey is winning too much influence in the young Muslim countries of Central Asia. For their part, some Turks believe Russia will try to take over the Transcaucasus region again and then turn to Central Asia. Some Turks uphold a dream of **pan-Turkism**, uniting all the Turkic peoples of Asia from Istanbul to the Sakha Republic in Russia's Siberia region—a prospect that Moscow of course does not like.

The twin scourges of terrorism and narcotics—often closely linked—have put central Asia on the map of geopolitical hot spots. Uzbekistan has erected concrete and barbed-wire barriers and sown land mines on its borders with Tajikistan and Kyrgyzstan. The Uzbek government is trying to prevent the incursion of what it sees as two huge threats: an epidemic in heroin use and a militant Islamic insurgency. Both issues have preoccupied the region's governments. The major routes for exports of heroin from Pakistan and Afghanistan are north through Central Asia and then into Russia and Western
242 Europe. Along this pathway, large numbers of central Asians have fallen prey to inexpensive, pure heroin. The associated health and social problems, including HIV infection, hepatitis, and prostitution, have also taken root. This extremely valuable commodity corrupts people at every level, all the way up to governments. The United Nations estimates that about 40 percent of Tajikistan's economy is drug-related. As we will see later (page 242), drug sales fund the war effort of the Taliban against the United States and its NATO allies.

Several powers are alarmed by what they perceive as an Islamist threat to Central Asia. Russia has expressed great concern about an Islamist wave spreading from Afghanistan and Pakistan into Central Asia and Russia itself. Tajikistan and Kyrgyzstan have allowed Russia to deploy thousands of Russian troops in their countries, in part to stave off such a threat. Some observers believe Russia is using the Islamist threat as a pretense to maintain influence in the region, or even to counter the growing U.S. military presence there since 9/11. And human rights groups in the West argue that authoritarian, oppressive regimes in Central Asia exaggerate the threats from Islamist militants to justify their harsh treatments of political opponents.

The United States is genuinely worried about the Islamist danger. Its greatest fears are focused on Tajikistan, Central Asia's poorest country. Islamist rebels with links to both al-Qa'ida and Afghanistan's Taliban are intent on disrupting U.S. military supply lines from Uzbekistan into Afghanistan and have also been attacking Tajkistan's soldiers. Those confrontations could lead to the reemergence of civil war, which plagued Tajikistan from 1992 to 1997.

Civil war also threatens Kyrgyzstan, where a minority population of ethnic Uzbeks (who make up 15 percent of the country's population) are seeking greater autonomy. They frequently engage in bloody clashes with ethnic Kyrgyz. Uzbekistan is fearful that the dominant Kyrgyz people will ethnically cleanse the country's Uzbeks. Ethnic Kyrgyz for their part fear that Uzbekistan may take advantage of instability to occupy land in Kyrgyzstan. Russian president Medvedev expressed his concern that Kyrgyzstan would soon become "a second Afghanistan."

Tajikistan was home to a group called the **Islamic Movement of Uzbekistan (IMU)**, which has recently been pressured southward into the Afghanistan-Pakistan border region. For many years, its principal aims were to overthrow the Uzbek government of President Islam Karimov and establish an Islamic state in the Fergana Valley, which runs through Uzbekistan, Tajikistan, and southern Kyrgyzstan (see Figure 5.31). Following a 1999 attempted assassination blamed on the rebels, Karimov cracked down hard on suspected Islamists and even began to restrict personal symbols of being Muslim, such as beards on men and headscarves on women. Tajikistan's government is also employing these repressive tactics, which many observers believe will only help grow Islamist anti-government sentiments.

Anxious to repress any external support for Islamist insurgency within its borders and to curry political and economic favor with Washington, Uzbekistan initially welcomed the U.S. military to use its land and air space in attacks against al-Qa'ida and the Taliban in 2001 and 2002. Uzbekistan shut down this access in 2005, after the United States and its European partners in NATO, citing human rights abuses, slashed economic assistance to the country. Soon, however, the United States' escalating engagements in Afghanistan led Washington to overlook poor human rights records in both Uzbekistan and Kyrgyzstan. American and NATO military supplies began pouring by air and rail into Afghanistan from both countries, and Uzbekistan built a 45-mile-long rail line into the Afghan

city of Mazar-e-Sharif, mainly for the transport of these mili-
243 tary goods. The U.S. airbase at Manas in Kyrgyzstan has served as a critical airbase in the West's military campaign in Afghanistan. The Russian military also operates bases in both Kyrgyzstan and Tajikistan. In 2009, Kyrgyz president Bakiyev double-crossed Russia, promising to shut down U.S. access to the Manas airbase in return for a large economic aid package from Russia. Once Bakiyev secured the Russian aid, he collected another large sum from the United States to renew its lease of Manas. Bakiyev was ousted in a coup the following year, and Kyrgyzstan headed down the road toward a "failed state" status.

Last but certainly not least as we consider the Far Abroad is that country whose influence is on the rise in every region covered in this book: China. China has sought to build security relationships with the Central Asian countries, especially to blunt the aspirations of Muslim insurgents in China's far
264 western region. In a proclaimed effort to confront terrorism, separatism, and extremism (and perhaps to counterbalance NATO), China and Russia have recently formed a confederacy with all of the central Asian countries except Turkmenistan, known as the **Shanghai Cooperation Organization (SCO)**.

Not surprisingly, given its geographic proximity to the region, China has growing economic interests in the Central Asia. Chinese exports to Central Asia are surging. China is upgrading its roads leading into Kyrgyzstan and cooperating with Kazakhstan in building a railway across Kazakhstan that will ultimately link producers and markets in China and Europe. China has made multibillion dollar loans to Turkmenistan and Kazakhstan in exchange for guaranteed future access to oil and gas supplies in those countries. Natural gas flows from Turkmenistan to western China through a 1,140-mile-long pipeline crossing three countries. Notably, Russia is not one of these: when the pipeline began operation in 2009, it was the first major export route for the region's natural gas that did not pass through Russia. This pipeline is also notable as the first to run east out of the region. If all roads once led to Rome, they now lead to China.

Following are more insights into the land and life in the subregions and countries that make up Russia and the Near Abroad.

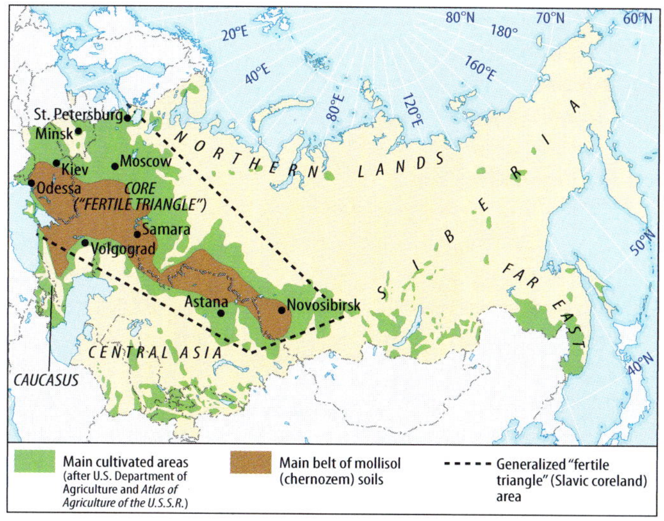

• **Figure 5.22** Major regional divisions of Russia and the Near Abroad, as discussed in the text.

5.6 Regional Issues and Landscapes

Peoples and Resources of the Coreland

In this book's first chapter, you were introduced to the geographic concepts of core and periphery. If you read the previous chapter on Europe, you know that there are distinctive core and periphery elements of that region. This is also true of Russia and the Near Abroad.

An outstanding feature in the geography of this huge region is that resources are distributed unevenly, favoring development in certain subregions and countries. Agricultural and industrial resources are in fact clustered in a core region encompassing five countries partially or entirely: western Russia, northern Kazakhstan, Ukraine, Belarus, and Moldova. Nearly three-fourths of the region's people—the great majority of them Slavic—and an even larger share of the cities, industries, and cultivated land of this immense region are packed into this roughly triangular **Fertile Triangle** comprising about one-fifth of the region's total area. Also known as the **Agricultural Triangle** and the **Slavic Core**, this is the distinctive functional hub of the region (•**Figures 5.22** and **5.23**).

Slavic peoples are the dominant ethnic groups in most of the region of Russia and the Near Abroad in both numbers and political and economic power. The major groups are the Russians (discussed extensively earlier in the chapter; the Slavs and other ethnolinguistic groups are introduced on page 124), Ukrainians, and Belarusians (Byelorussians). Most of these Slavs live in the core region extending from the Black and Baltic Seas to the neighborhood of Novosibirsk in Siberia. Moscow, with a population of about 15 million, and Saint Petersburg (formerly Leningrad), with almost 5 million people, are the largest cities, followed by 3 million people living in the Ukrainian capital, Kiev.

The Ukraine

Ukrainians, the second largest ethnic group in the Slavic Core, are closely related to the Russians in language and culture; "Russia writ small" is how one author described Ukraine.[9] The name Ukraine translates as "at the border" or "borderland," and this region has long served as a buffer between Russia and neighboring lands (•**Figure 5.24**). Here armies of the Russian tsars fought for centuries against nomadic steppe peoples, Poles, Lithuanians, and Turks, until the eighteenth century, when the Russian Empire finally absorbed Ukraine. For three centuries, Ukrainians were subordinate to Moscow, first as part of the Russian Empire and then as a Soviet republic. Ukraine's industrial and agricultural assets were always vital to the Soviet Union; the Bolshevik leader, Vladimir Lenin, once declared, "If we lose the Ukraine, we lose our head!"

• **Figure 5.23** Major economic zones and transportation routes of Russia and the Near Abroad. The map reveals that the Fertile Triangle is also the industrial core of the region. Moscow's function as a transportation hub is evident. It is similar to the depiction of Paris as a hub for France, in Figure 4.25; however, Russia does not have a primate city as France does.

Slightly smaller than Texas, Ukraine lies partly in the forest zone and partly in the steppe. On the border between these biomes, and on the banks of the Dnieper River, is the historic city of Kiev, Ukraine's political capital and a major industrial and transportation center. Ukraine is blessed with an abundance of fertile black earth soils, especially in the steppe region
121 south of Kiev. Historically, Ukraine has been a great "breadbasket" of wheat, barley, livestock, sunflower oil, beet sugar, and many other products. Ukraine also has a generous endowment of industrial raw materials, especially coal and iron ore. Agricultural and industrial outputs plummeted after Ukraine gained independence from the Soviet Union in 1991, but are beginning to pick up again. Many of the social problems that plagued Russia after the breakup of the Union of Soviet Socialist Republics also troubled Ukraine: increasing mortality, falling birth rates, and broken families are among them. In Ukraine, Moldova, Belarus, and Russia, increasing numbers of women fell victim to sex trafficking as they struggled to improve their lives (see A Closer Look on page 145).

Chernobyl

North of Kiev is **Chernobyl**, where an explosion at a nuclear power station in 1986 rendered parts of northern Ukraine and adjoining Belarus and parts of Russia incapable of safe agricultural production for years thereafter. Encircling the station is still an 18-mile (29 km) "exclusion zone" where farming is prohibited. Recent studies suggest that 100,000 to 200,000 people are still severely affected by the Chernobyl disaster and that 4,000 deaths will ultimately be attributed to it. However, the toll on human health and life is well short of what had originally been predicted. The psychological trauma to the region's inhabitants has been profound, and the three affected countries have spent large sums to provide benefits to Chernobyl's victims. With financial assistance from Russia, the European Union, and the United States, Ukraine is decommissioning all of its Chernobyl-type nuclear plants. Until recently, the Chernobyl disaster was a singular event. Now it is joined by Japan's Fukushima catastrophe in the world's imagination of nuclear power gone wrong.

A closer LOOK

Human Trafficking and the Sex Trade

One of the outcomes of the collapse of communism and the dissolution of the Soviet Union has been the unleashing of long-restrained expressions of sexuality. Entrepreneurs in Russia and other former Soviet countries quickly adopted an axiom of Western advertising: sex sells. Russian television has surpassed even the spiciest fare on standard American television.

Many sexual awakenings in the region are applauded or accepted as good, clean fun, but others have a decidedly dark side, and some send ripples far abroad. Far too many women of the former Soviet Union have become commodities (•Figure 5.1). Russian, Ukrainian, and Kazakh partners or brides can in effect be purchased by anyone in the world with access to the Internet. Dot-com agencies act as brokers for these New Age "mail-order brides." Some of the subsequent unions end happily, of course, but others collapse with physical abuse and heartache. Frequently, the woman ends up abandoned and destitute in a strange land.

The sex trade takes the greatest toll. Primarily in this industry, women become victims of what is known as **human trafficking**, often unwittingly and almost always out of poverty. Following is the classic pattern, played out over and over, not only in this region but also around the world. Criminal gangs use false promises of conventional employment to lure young women into the trade. The typical victim is an 18- to 30-year-old Russian, Ukrainian, Belarusian, or Moldovan woman—often with children to support—who is approached by a trafficker promising her a job as a waitress, barmaid, babysitter, or maid in Eastern or Western Europe. She pays a fee for travel and visa costs. Arriving in Bosnia and Herzegovina, the Czech Republic, or Germany, she is met by another trafficker who confiscates her passport and identity papers and then sells her for several thousand dollars to a brothel owner. She is escorted to her guarded apartment. There she is compelled to reside with several other women who have taken a similar journey, and she is informed of her real job: prostitute.

Berlin papers advertise the merchandise: "The Best from Moscow!" "New! Ukrainian Pearls!" The working women meet their clients in clubs or brothels or are driven to their homes. Of the rates charged for their services, typically less than 10 percent goes to the women. Their pimps and other gang members take the lion's share, and the women use their meager earnings to buy food and pay rent. Most end up in debt. Unable to pay off their debts, lacking identification papers, fearful of going to the police, and terrified the gang might harm family members back home if they try to escape, the women are trapped. Reports of depression, isolation, sexually transmitted disease, beatings, and rape abound.

This is a big business with striking demographics. The sex industry in Europe generates revenues of several billion dollars per year. Ukrainian sources estimate that since the country's independence in 1991, more than half a million Ukrainian emigrant women have ended up in the European sex trade. Figures vary widely, but tens and perhaps even hundreds of thousands of women of the former Soviet Union immigrate to Europe each year, the great majority of them becoming prostitutes.

There is an overflow into other regions. These so-called "Natashas" work in Turkey, Israel, the United Arab Emirates, and other countries in the Middle East in similar deplorable circumstances. Many of the world's countries, including the United States, have human trafficking problems (see page 394). 89, 158, 325

Joe Hobbs

• **Figure 5.1** Window shopping in Amsterdam's Red Light District. An estimated 85 percent of the working women here are trafficked, mainly from the former Soviet countries.

Joe Hobbs

• **Figure 6.7** The Treasury, a temple carved in red sandstone, probably in the first century C.E., by the Nabateans at their capital of Petra in southern Jordan. The English poet Dean Burgen called Petra "a rose red city, half as old as time." The Nabateans built sophisticated networks for water storage and distribution. For scale, note the man at the lower right.

Joe Hobbs

• **Figure 6.8** Rainfall is heavy on the Black Sea side of Turkey's Pontic (also known as Pontus or Kuzey) Mountains. Note that roofs are pitched to shed precipitation; in the region's drier areas, most roofs are flat. The crop growing here is corn.

in the summer, and tea grows well there (•**Figure 6.8**). In the southwestern Arabian Peninsula, a monsoonal climate brings summer rainfall and autumn harvests to Yemen and Oman, probably accounting for the Roman name for the area: *Arabia Felix* ("Happy Arabia"). Mountainous areas in the region, like the river valleys and the margins of the Mediterranean, play a vital role in supporting human populations and national economies. Due to orographic or elevation-induced precipitation, the mountains receive much more rainfall than surrounding lowland areas.

There are three principal mountainous regions of the Middle East and North Africa (see Figure 6.1). In northwestern Africa between the Mediterranean Sea and the Sahara, the Atlas Mountains of Morocco, Algeria, and Tunisia reach over 13,000 feet (3,965 m) in elevation. Mountains also rise on both sides of the Red Sea, with peaks up to 12,336 feet (3,760 m) in Yemen. These mountains rose with the tectonic processes that are pulling the African and Arabian plates apart at the northern end of the Great Rift Valley. The hinge of this crustal movement is the Bekaa Valley of Lebanon, where the widening fault line follows the Jordan River Valley southward to the Dead Sea (•**Figure 6.9**). This valley is the deepest depression on the earth's land surface, lying about 600 feet (183 m) below sea level at Lake Kinneret (also known as the Sea of Galilee or Lake Tiberius) and nearly 1,300 feet (400 m) below sea level at the shore of the Dead Sea, the lowest point of land on the planet. The rift then continues southward to the very deep Gulf of Aqaba and Red Sea, before turning inland into Africa at Djibouti and Ethiopia, and continuing all the way down the Great Rift Valley of Africa (see page 317). By the way, should you visit the Dead Sea one day, just to try to sink. It's impossible—as this lake has no outlet, the waters are so salty at 32 percent that you will bob like a cork.

In tectonic processes, one consequence of rifting in one place is the consequent collision of the earth's crustal plates elsewhere. There are several such collision zones in the Middle East and North Africa, particularly in Turkey and Iran, where mountain

Image Science and Analysis Laboratory, NASA-JSC.

• **Figure 6.9** It looks peaceful from space, but the land around the Jordan Rift Valley is one of the most contested areas on Earth. The Jordan River runs into and out of Lake Kinneret (the Sea of Galilee) near the top of the image, emptying into the Dead Sea near the bottom of the image. If political boundaries were drawn on the image, they would include parts of Israel, Lebanon, Syria and Jordan, and all of the West Bank, Gaza Strip, and Golan Heights.

building has resulted. These are seismically active zones—meaning earthquakes occur there—and rarely does a year go by without a devastating quake rocking Turkey or Iran.

A large area of mountains, including the region's highest peaks, stretches across Turkey and Iran. The loftiest mountain ranges in Turkey are the Pontic, Taurus (•**Figure 6.10**), and Anti-Taurus, and in Iran, the Elburz and Zagros Mountains. These chains radiate outward from the rugged Armenian Knot in the tangled border country where Turkey, Iran, and the countries of the Caucasus meet. In Turkey, the glacier-fringed summit of the extinct volcano Mt. Ararat soars to 16,804 feet (5122 m) over the frontier between Turkey and Armenia. Armenians seethe in anger and resentment that this mountain, so sacred to them, fell into Turkish hands. Some people believe that 149
Mt. Ararat is where Noah's Ark lies (and even go in search of it), for in Genesis it is written that the vessel came to rest "in the mountains of Ararat." (Incidentally, like so many things, the accounts of Noah are shared by all three Abrahamic traditions of Judaism, Christianity, and Islam.)

Joe Hobbs

• **Figure 6.10** The Taurus Mountains of southeastern Turkey.

Extensive forests existed in early historical times in the Middle East and North Africa, particularly in these mountainous areas, but overcutting and overgrazing have almost eliminated them. Since the dawn of civilization in this area, around 3000 B.C.E., people have cut timber for construction and fuel faster than nature could replace it. Lebanon, described in ancient times as "an oasis of green with running creeks" and "a vast forest whose branches hide the sky," illustrates the result. If not for human impacts, forests would cover large parts of Lebanon today. On many mountainsides, the dominant tree would be the cedar, the tree that is the centerpiece of Lebanon's flag. This hardwood was highly valued throughout the Mediterranean Basin. The Phoenician civilization, with its base in Lebanon, grew wealthy by trading cedarwood. Ancient Egyptians were among those who prized the cedar. The solar boat of King Cheops, dating to the third millennium BCE, would carry this pharaoh's soul in the afterlife, and it had to be built of the best wood: cedar of Lebanon. King Tutankhamen's funerary shrines and Solomon's Temple in Jerusalem were built
of cedar of Lebanon (•**Figure 6.11**). Only a few isolated groves 170
of cedar remain in Lebanon. Some of these are on the grounds of monasteries and survive because they grow in these sacred places. One grove is called "The Cedars of God."

The "primordial" or potential vegetation of Morocco's Atlas Mountains can be easily reconstructed by looking at the vegetation in the untouched sacred space surrounding saints' tombs.[1] This is a fine example of the importance of

Joe Hobbs

• **Figure 6.11** Ancient Egyptians believed that the solar boat of King Cheops of Egypt (c. 2500 B.C.E.), builder of the Great Pyramid, would carry the pharaoh's spirit through the firmament. It had to be made of the best wood—cedar from Lebanon. The solar boat was interred next to the pyramid and now stands in a specially constructed museum.

environmental perception in geography. The principle of environmental perception is that people, societies, and cultures perceive and use the environment in different ways, depending on their age, gender, socioeconomic standing, language, religion, nationality, and other variables. Often these perceptions have an important role in shaping the cultural landscape. In this case, Islamic beliefs about the sacred space surrounding saints' tombs have protected the plants and animals living there, offering us a window into the environmental history of the Atlas Mountains.

Villager, Pastoral Nomad, Urbanite: The Ecological Trilogy

In the 1960s, the American geographer Paul English developed a useful model for understanding relationships between the three ancient ways of life that, although highly modified, still characterizes the Middle East and North Africa today: villager, pastoral nomad, and urbanite. Each of these modes of living is rooted in a particular physical environment. Villagers are the subsistence farmers of rural areas where dry farming or irrigation is possible; pastoral nomads are the desert peoples who migrate through arid lands with their livestock, following patterns of rainfall and vegetation; and urbanites are the inhabitants of the large towns and cities, generally located near bountiful water sources but sometimes placed for particular trade, religious, or other reasons. Describing these ways of life as components of the **Middle Eastern ecological trilogy**, English explained how each of them has a characteristic, usually mutually beneficial, pattern of interaction with the other two (•Figure 6.12).

The peasant farmers of Middle Eastern and North African villages (the **villagers**) represent the cornerstone of the trilogy. They grow the staple food crops such as wheat and barley that feed both the city dweller and the pastoral nomad of the desert. Neither urbanite nor nomad could live without them. The village also provides the city, often unwillingly, with tax revenue, soldiers, and workers. And before the mid-twentieth century,

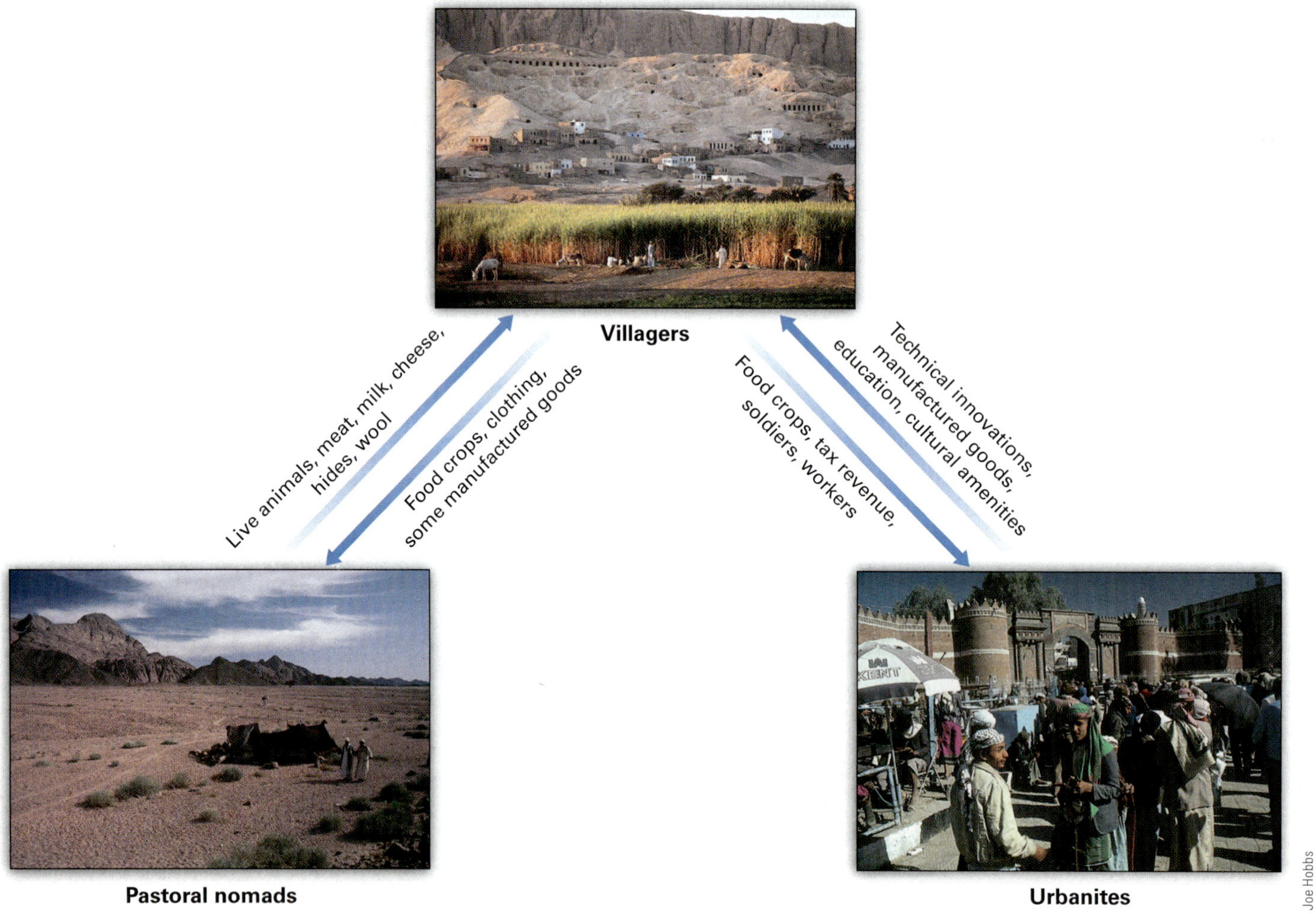

• **Figure 6.12** People, environments, and interactions of the ecological trilogy (here, villagers harvesting sugarcane in upper Egypt, Bedouin at camp in Egypt's Eastern Desert, and shoppers at the main gate to the historic city of San'a, Yemen). This relationship is generally symbiotic, although historically both urbanites and pastoral nomads preyed on the villagers, who are the trilogy's cornerstone.

villages provided **pastoral nomads** with plunder as the desert dwellers raided their settlements and caravan supply lines. Generally, however, the exchange is mutually beneficial.

The nomads provide villagers with livestock products, including live animals, meat, milk, cheese, hides, and wool, and with desert herbs and medicines. Educated and progressive **urbanites** provide technological innovations, manufactured goods, religious and secular education and training, and cultural amenities (today, including films and music).

There is little direct interaction between urbanites and pastoral nomads, although some manufactured goods such as clothing travel from city to desert, and some desert folk medicines pass from desert to city. Historically, the exchange was sometimes violent, as urban-based governments have sought to control the movements and military capabilities of the elusive and sometimes hostile nomads. Pastoral nomads once plundered rich caravans plying the major overland trade routes of the Middle East and North Africa. Governments did not tolerate such activities and often cracked down hard on the nomads they were able to catch.

Later, in the 1970s, English wrote an article marking the "passing of the ecological trilogy." He noted that cities were encroaching on villages, villagers were migrating into cities and giving some neighborhoods a rural aspect, and pastoral nomads were settling down—thus the trilogy no longer existed. In reality, although the makeup and interactions of its parts have changed, the trilogy model is still valid and useful as an introduction to the major ways of life in the Middle East and North Africa. It is especially significant that a given man or woman in the region strongly identifies as a villager, a pastoral nomad, or an urbanite. This perception of self has an important bearing on how these people of very different backgrounds interact, even when they live in close proximity. Urban officials may work in rural village areas, but they remain at heart and in their perspectives city people and usually live apart from farmers. Extended families of pastoral nomads may settle down and become farmers, but they continue to identify themselves by affiliation with the nomadic tribe. Many continue to harvest desert resources on a seasonal basis and retain marriage and other ties with desert-dwelling relatives.

Let's take a closer look at each of the components of the ecological trilogy.

The Village Way of Life

Agricultural villagers historically represented by far the majority populations in the Middle East and North Africa; only in recent decades have urbanites begun to outnumber them. In the region's typically dry environments, the villages are located near reliable water sources with cultivable lands nearby. They are usually made up of closely related family groups, with many fields owned by an absentee landlord. Villagers typically live in closely spaced homes made of mud brick or concrete blocks. Production and consumption focus on a staple grain such as wheat, barley, or rice. As land for growing fodder is often in short supply, villagers keep only a small number of sheep and goats and rely in part on nomads for pastoral produce. Residents of a given village usually share common ties of kinship, religion, ritual, and custom, and the changing demands of agricultural seasons regulate their patterns of activity.

These patterns of village life have been increasingly exposed to outside influences since the mid-eighteenth century. European colonialism brought significant economic changes, including the introduction of cash crops and modern facilities to ship them. Improved and expanded irrigation, financed initially with capital from the West, brought more land under cultivation. Recent agents of change have been the countries' own government-supported doctors, teachers, and land reform officers. Modern technologies such as motor vehicles, gasoline-powered water pumps, radio, television, and—most recently—the Internet and cell phones have modified old patterns of living. The educated and ambitious, as well as the unskilled and desperately poor—motivated, respectively, by pull and push factors (see Chapter 3)—have been drawn to urban areas. Improved roads and communications have in turn carried urban influences to villages, prompting villagers to become more integrated into the national society.

The Pastoral Nomadic Way of Life

Pastoral nomadism emerged as an offshoot of the village agricultural way of life not long after plants and animals were first domesticated in the Middle East (about 7000 B.C.E.). Rainfall and the wild fodder it brings forth, although scattered, are sufficient resources to support small groups of people who migrate with their sheep, goats, and camels (and in some locales, cattle) to take advantage of this changing resource base (see Perspectives from the Field, next page). In addition to selling or trading livestock to obtain food, tea, sugar, clothing, and other essentials from settled communities, pastoral nomads hunt; gather; work for wages; and where possible, grow crops. Many now work in the tourist industry, as Westerners hungry for insight into traditional cultures seek them out. Their multifaceted livelihood has been described as a strategy of **risk minimization** based on the exploitation of multiple resources so that some will support them if others fail.

Although renowned in Middle Eastern legends, pastoral nomads have been described as "more glamorous than numerous." It is impossible to obtain adequate census figures on the number living in the deserts of the Middle East and North Africa, but they are very small relative to the numbers of villagers and urbanites. Recent decades have witnessed the rapid and progressive settling down, or **sedentarization**, of the nomads—a process attributed to a variety of causes. In some cases, prolonged drought virtually eliminated the resource base on which the nomads depended. Traditionally, they were able to migrate far enough to find new pastures, but modern national boundaries now inhibit such movements. Some have returned with the rains to their desert homelands, but others have chosen to remain as farmers or wage laborers in villages and towns. In the Arabian Peninsula, the prosperity

PERSPECTIVES FROM THE FIELD

Way-Finding in the Desert

The research for my doctoral dissertation in geography at the University of Texas–Austin was an 18-month journey with the Khushmaan Ma'aza, a clan of Bedouin nomads living in the northern part of Egypt's Eastern Desert. I studied their perceptions, knowledge, and uses of natural resources, and tried to understand how their worldviews and kinship patterns apparently helped them devise ways of protecting these scarce resources. I have worked with these people from the early 1980s to the present; I am now doing a multiyear study of how they carefully protect acacia trees, which offer them many resources when drought prevents the growth of annual plants. I have always been astonished by how different they are from me in perceiving the world around us, and how much better they are in their ability to "read" the ground and recall landscape details. In the following passage from my book *Bedouin Life in the Egyptian Wilderness,* I marvel at their way-finding abilities.

> The process by which the Khushmaan nomads have developed roots in their landscape, fashioning subjective "place" from anonymous "space," and the means by which they orient themselves and use places on a daily basis deserve special attention: these are essential parts of the Bedouins' identity and profoundly influence how they use resources and affect the desert ecosystem.
>
> The Bedouin have little need or knowledge of maps. On the other hand, when shown a map or aerial image of countryside they know, the nomads accurately orient and interpret it, naming the mountains and drainages depicted. Saalih Ali (**•Figure 6.A**) especially enjoyed my star chart. He would tell me what time the Pleiades would rise, for example, and ask me what time the chart indicated. The two were almost always in agreement.
>
> Many desert travelers have been astonished by the nomads' navigational and tracking ability, calling it a "sixth sense." The Khushmaan are exceptional way-finders and topographical interpreters able, for instance, to tell from tracks whether a camel was carrying baggage or a man; whether gazelle tracks were made by a male or female; which way a car was traveling and what make it was; which man left a set of footprints, even if he wore sandals; and how old the tracks are. Bedouins are proud of their geographical skills, which, they believe, distinguish them from settled people. Saalih told me, "Your people don't need to know the country but we do, to know exactly where things are in order to live." Pointing to his head, he said, "My map is here."
>
> The difference in the way-finding abilities of nomads and settled persons may be due to the greater survival value of these skills for nomads and to the more complex meanings they attribute to locations. Bedouin places are rich in ideological and practical significance. The nomads interpret and interact with these meanings on a regular basis and create new places in their lifetimes. Theirs is an experience of belonging and becoming with the landscape, whereas settled people are more likely to inherit and accommodate themselves to a given set of places.
>
> The nomads' homeland is so vast, and the margin of survivability in it so narrow, that topographic knowledge must be encyclopedic. Places either are the resources that allow human life in the desert or are the signposts that lead to resources. A Khushmaan man pinpointed the role of places in desert survival: "Places have names so that people do not get lost. They can learn where water and other things are by using place names." Tragedy may result if the nomad has insufficient or incorrect information about places: many deaths by thirst are attributed to faulty directions for finding water.*

Joe Hobbs

• Figure 6.A Saalih Ali, a Bedouin of the Khushmaan Ma'aza, with the author in 2010, 30 years after they began fieldwork together.

and technological changes prompted by oil revenues made rapid inroads into the material culture—and then the livelihood preferences—of the desert people; many preferred the comforts of settled life. Some governments, notably those of Israel and prerevolutionary Iran, were frustrated by their inability to count, tax, conscript, and control a sizable migrant population and therefore forced the nomads to settle.

Pastoral nomads of the Middle East and North Africa identify themselves primarily by their tribe, not by their nationality. The major ethnic groups from which these tribes draw are the Arabic-speaking Bedouin of the Arabian Peninsula and adjacent lands, the Berber and Tuareg of North Africa, the Kababish and Bisharin of Sudan, the Yörük and Kurds of Turkey, and the Qashai and Bakhtiari of Iran. Members of a tribe claim common descent from a single male ancestor who lived countless generations ago; their kinship organization is thus a patrilineal descent system. It is also a segmentary kinship system, so called because there are smaller subsections of the tribe, known as clans and lineages, that are functionally important in daily life. Members of the most closely related families comprising the lineage, for example, share livestock, wells, trees, and other resources. Both the larger clans, made up of numerous lineages, and the tribes possess territories. Members of a clan or tribe typically allow members of another clan or tribe to use the resources within its territory on the basis of **usufruct**, or nondestructive mutual use. That arrangement has often proved to be lifesaving in an environment where water often is in dangerously low supply.

Although some detractors have depicted pastoral nomads as the "fathers" rather than "sons" of the desert, blaming them for wanton destruction of game animals and vegetation, there are numerous examples of pastoral nomadic groups that have developed indigenous and very effective systems of resource conservation. Most of these practices depend on the kinship groups of family, lineage, clan, and tribe to assume responsibility for protecting plants and animals.

The Urban Way of Life

The city was the final component to emerge in the ecological trilogy, beginning in about 4000 B.C.E. in Mesopotamia (modern Iraq) and 3000 B.C.E. in Egypt. Though they resembled villages in many ways, the early cities were distinguished by their larger populations (more than 5,000 people), the use of written languages, and the presence of monumental temples and other ceremonial centers. The early Mesopotamian city and, after the seventh century C.E., the classic Islamic city, called the **medina**, had several structural elements in common (•Figure 6.13). The medina had a high surrounding wall built for defensive purposes. The congregational mosque and often an attached administrative and educational complex dominated the city center. Although Islam is often characterized as a faith of the desert, religious life has always been focused in, and diffused from, the cities. The importance of the city's congregational mosque in religious and everyday life is often emphasized by its large size and outstanding artistic execution.

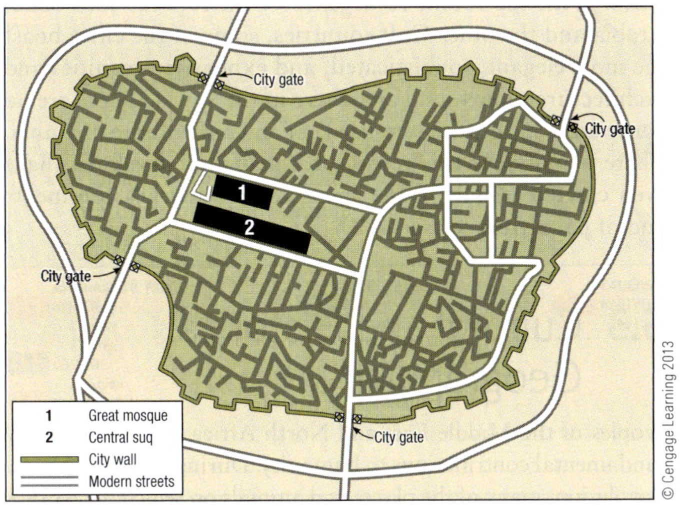

• **Figure 6.13** An idealized model of the classic medina, or Muslim Middle Eastern city.

A large commercial zone, known as a ***bazaar*** in Persian and a ***suq*** in Arabic and recognizable as the ancestor of the modern shopping mall, typically adjoined the ceremonial and administrative heart of the city (•Figure 6.14). Merchants and craftspeople selling various commodities occupied separate spaces within this complex, and visitors to an old medina today can still find sections devoted exclusively to the sale of spices, carpets, gold, silver, traditional medicines, or other particular goods. Smaller clusters of shops and workshops were located at the city gates.

Residential areas were differentiated as quarters, not by income group but by ethnicity; the medina of Jerusalem, for example, still has distinct Jewish, Arab, Armenian, and non-Armenian Christian quarters (see Figure 6.B). Homes tended to face inward toward a quiet central courtyard, buffering the occupants from the noise and bustle of the street. The narrow, winding streets of the medina were intended for foot traffic and small animal-drawn carts, not for large motor vehicles, a fact that accounts for the traffic jams in some old Middle Eastern and North African cities today and for the wholesale destruction of the medina in others.

• **Figure 6.14** A typical Middle Eastern *suq or bazaar* (market). This one is in Cairo, Egypt.

language family, and the 41 million Persians of Iran, most of who speak **Persian** (*Farsi*) in the Indo-European language family. Both Persian and Arabic are written in Arabic script and so appear related, but they are not. Turkish was also written in Arabic script until early in the twentieth century, but since then it has been written in a Latin script. About 30 million **Kurds**—a people living in Turkey, Iraq, Iran, and Syria—speak *Kurdish*, which is also an Indo-European language. Many people in North Africa speak **Berber** and **Tuareg** (in the Afro-Asiatic language family).

The Abrahamic Faiths

Out of the Middle East came the closely related monotheistic faiths of Judaism, Christianity, and Islam. The region's human and political geographies are closely tied to these religions and their holy places (•Figure 6.17). All three are Abrahamic traditions; they regard Abraham as their patriarch, and they share many fundamental beliefs. The first to emerge was the faith of Judaism.

The Promised Land of the Jews

Judaism, the first significant monotheistic faith as far as we know, is today practiced by about 14 million people worldwide, mostly in Israel, Europe, and North America. September 2011 marked the start of the year 5772 in the Jewish calendar, but unlike its kindred faiths—Christianity and Islam—Judaism does not have an acknowledged starting point in time. Also unlike Christianity, Judaism does not have a fixed creed or doctrine. Jews are encouraged to behave in this life in compliance with God's laws, which, according to the Torah (the Jewish holy scripture, which is known as the Old Testament in the Christian Bible) God gave to Moses on Mount Sinai as a covenant with God's "chosen people." The coming of a savior known as the Messiah ("Anointed
156 One" in Hebrew) is prophesied in the Torah; Christians believe that Jesus was that savior, as described in the New Testament; Jews do not recognize Jesus as the fulfillment of that prophecy and so do not accept the New Testament.

Another distinction that sets Christianity and Islam apart from Judaism is that Judaism is not a proselytizing religion; it does not seek converts. Jewish identity is based strongly on a common historical experience shared over thousands of years. That historical experience has included deep-seated geographic associations with particular sacred places in the Middle East—particularly with places in Jerusalem, capital of ancient Judah (Judea), the province from which Jews take their name. Tragically, the Jewish history also has included unparalleled persecution.

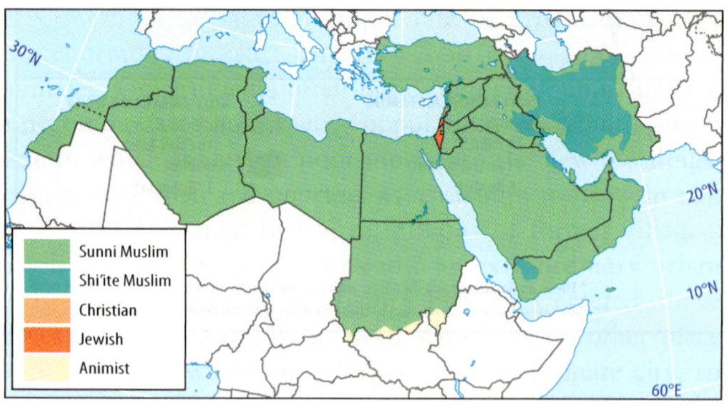

• **Figure 6.17** Religions of the Middle East and North Africa.

Depending on one's perspective, the Jewish connection with the geographic region known as Palestine, the area now composed of Israel, the West Bank, and the Gaza Strip (see Figure 6.33), is most significant on a time scale of either about 4,000 years or about 100 years. According to the Bible, around 2000 B.C.E., God commanded Abraham and his kinspeople, known as **Hebrews** (later as Jews), to leave their home in what is now southern Iraq and settle in Canaan. God told Abraham that this land of Canaan—geographic Palestine—would belong to the Hebrews after a long period of persecution. The Bible says that the Hebrews did settle in Canaan until famine struck that land. At the command of Abraham's grandson Jacob, the Hebrews—known then as **Israelites**—relocated to Egypt, where grain was plentiful. That began the long sojourn of the Israelites in Egypt, which, according to the Bible, ended in about 1200 B.C.E. when Moses led them out in the journey known as the **Exodus.**

According to Jewish history, the prophecy of Abraham was first fulfilled when the Israelites settled once again in Canaan, their **Promised Land.** The Jewish King Saul unified the 12 tribes that descended from Jacob into the first united Kingdom of Israel in about 1020 B.C.E. In about 950 B.C.E. in Jerusalem—the capital of the kingdom enlarged by Saul's successor, David—King Solomon built Judaism's **First Temple.** He located it atop a great rock known to the Jews as *Even HaShetiyah*, the **Foundation Stone,** plucked from beneath the throne of God to become the center of the world and the core from which the entire world was created (see Geography of Sacred Space, next page). The Ark of the Covenant, containing the commandments that God gave to Moses atop Mount Sinai, was placed in the Temple's Holy of Holies.

The united kingdom of Israel lasted about 200 years before splitting into the states of Israel and Judah. Empires based in Mesopotamia destroyed these states: the Assyrians attacked Israel in 721 B.C.E., and the Babylonians sacked Judah in 586 B.C.E. The Babylonians destroyed the First Temple (at which point the Ark of the Covenant disappeared) and exiled the Jewish people to Mesopotamia, where they remained until conquering Persians allowed them to return to their homeland. In about 520 B.C.E., the Jews who returned to Judah rebuilt the temple (the **Second Temple**) on its original site. A succession of foreign empires came to rule the Jews and Arabs of Palestine: Persian, Macedonian, Ptolemaic, Seleucid, and around the time of Jesus, Roman. Herod, the Jewish king who ruled under Roman authority and was a contemporary of Jesus, greatly enlarged the temple complex.

The Jews of Palestine revolted against Roman rule three times between 64 and 135 C.E. The first revolt broke out as the profoundly monotheistic Jews refused to acknowledge the Roman emperor as a god. The Romans quashed these rebellions in a series of famous sieges, including those of Masada and Jerusalem. The Romans destroyed the Second Temple, and a third has never been built. All that remains of the Second Temple complex is a portion of the surrounding wall built by Herod. Today, this **Western Wall**, known to non-Jews as the Wailing Wall, is the most sacred site in the world accessible to Jews (Figures 6.B and •6.18). Some religious traditions

GEOGRAPHY OF SACRED SPACE | Jerusalem

The old city of Jerusalem is filled with sacred places and is one of the world's premier pilgrimage destinations. Over thousands of years, Jerusalem has been coveted and conquered by people of many different cultures and faiths. In the process, a place held sacred by one group has often come to be held sacred by a second and even a third.

The most important thing to notice on this map is that the location of the First and Second Jewish Temples is identical to that of the Muslims' Dome of the Rock. Judaism, Islam, and Christianity are very close in their origins and many of their basic precepts, but their faiths are not conjoined or syncretized in practice, as, for example, the Maya and Catholic faiths have been in Mesoamerica (see page 365), or Buddhism and other traditions have been in East Asia (see page 228). In each faith, there has been a long tradition of quarreling over ownership of sacred space. Nowhere in the world does contested sacred space have more potential to ignite violence and even war than in Jerusalem (which means, in Hebrew, "City of Peace"). The later section on Regional Issues and Landscapes offers more insight into the role of Jerusalem's holy places in the region's turbulent history (see page 187). **•Figure 6.B** depicts the major places sacred to Jews, Christians, and Muslims, and also the city's ethnic quarters; the walled old city is a classic medina (see page 167).

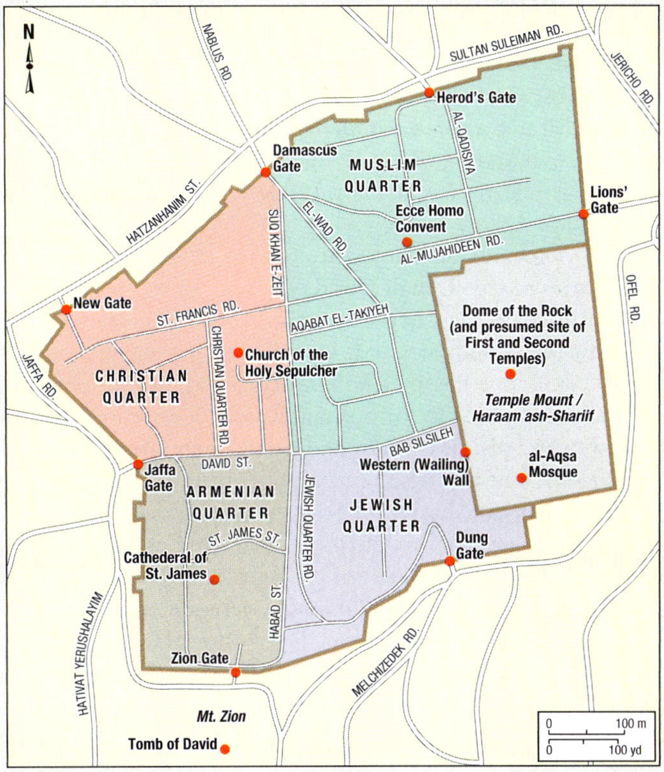

• Figure 6.B Sacred sites and ethnic quarters of the old city of Jerusalem.

• Figure 6.18 Jerusalem's Western Wall, Temple Mount, and Noble Sanctuary. In this view are some of Judaism's and Islam's holiest places. At left, below the golden dome, is the Western Wall, which is all that remains of the structure that surrounded the Jews' Second Temple. The dome is the Muslims' Dome of the Rock. At far right, with the black dome, is the al-Aqsa Mosque, another very holy place in Islam.

prohibit Jews from ascending the **Temple Mount** above, the area where the temple actually stood, because it is too sacred. After the temple's destruction, that site was occupied by a Roman temple and then in 691 by the Muslim shrine called the **Dome of the Rock**, which still stands today (also in Figures 6.B and 6.18). The mostly Muslim Arabs know the Temple Mount as ***al-Haraam ash-Shariif***, the **Noble Sanctuary**. Supercharged with meaning, this place has in modern times often been the spark of conflagration between Jews and Palestinian Arabs.

The victorious Romans scattered the defeated Jews to the far corners of the Roman world. Thus began the Jewish exile, or **Diaspora**. In their exile, the Jews never forgot their attachment to the Promised Land. One Jewish prayer, recited during the weeklong Passover holiday, ends with the words "Next year in Jerusalem!" In Europe and Russia, Jews were subjected to systematic discrimination and persecution and were forbidden to own land or engage in a number of professions. Known as **anti-Semitism**, hatred of Jews developed deep roots. This sentiment in part grew out of the long-simmering perception that Jews were responsible for the murder of Jesus Christ and in part out of the fact that Christians were generally prohibited from practicing moneylending. Jews were permitted to loan money to the Christians, who accepted the loans but then resented having to pay the Jews interest.

In the 1930s, anti-Semitism became state policy in Germany under the Nazis, led by Adolf Hitler. Many German Jews fled to the United States, and others emigrated to Palestine in support of the **Zionist movement**, which, since the late 1890s, had aimed at establishing a Jewish homeland in Palestine, with **Zion** (Jerusalem) as its capital. Most Jews were not as fortunate as the Zionist emigrants. Within the boundaries of the Nazi empire that conquered most of continental Europe during World War II, Hitler's regime implemented its "Final Solution" to the "Jewish problem" by shipping to prison camps and eventually murdering all the Jews they could round up. The Nazi Germans and their allies killed an estimated 6 million Jews, along with millions of other minorities they deemed
81 "inferior," including Roma (Gypsies) and homosexuals. It was this Holocaust that prompted the victorious allies of World War II, from their powerful position in the newly formed United Nations, to create a permanent homeland for the Jewish people in Palestine. A seemingly endless cycle of violence, discussed beginning on page 185, ensued.

Christianity: Death and Resurrection in Jerusalem

Nearly a thousand years after Solomon established the Jews' First Temple, a new but closely related monotheistic faith emerged in Palestine. This was Christianity, named for Jesus Christ (*Christ* is Greek for "Anointed One," the equivalent of the Hebrew word for Messiah). Jesus, a Jew, was born near Jerusalem in Bethlehem, probably around 4 B.C.E. Tradition relates that when he was about 30 years old, Jesus began spreading the word that he was the Messiah, the deliverer of humankind long prophesied in Jewish doctrine. A small group of disciples accepted that he was the Messiah, the Son of God and a living manifestation of God Himself. They followed him for several years as he preached his message. He taught that the only path to God was through Him, and that faith in Christ as God's Son and as the redeemer of humanity's sins was the key to salvation.

Christianity is built upon a Jewish foundation: the common scriptures of Judaism and Christianity are known as the Hebrew Bible. Theologically, Christians would argue that Jews have the correct foundation but did not accept God's complete message, which continues beyond the Hebrew Bible to the new covenant, or the New Testament.

Jesus Christ's own teachings denied the validity of some Jewish doctrines, and around 29 C.E., a growing chorus of Jewish protesters called for his death. Palestine was then under Roman rule, and Roman administrators in Jerusalem placated the mob by ordering that he be put on trial. He was found guilty of being a claimant to Jewish kingship, and Roman soldiers put him to death by the particularly degrading and painful method of crucifixion. The cornerstone of Christian faith is that Jesus Christ was resurrected from the dead on the third day after his crucifixion and later ascended into heaven. For Christians, this was fulfillment of the scriptures: Jesus had come to redeem humanity's sins through His own death on the cross and through His resurrection. Christians believe that He continues to intercede with his Father on their behalf and that He will come again on Judgment Day, at the end of time.

After a period of relative tolerance, the Romans began actively persecuting Christians. Nevertheless, Christian ranks and influence grew. The turning point in Christianity's career came after 324 C.E., when the Roman Emperor Constantine embraced Christianity and established it as the official religion of the empire. The Christian Byzantine civilization that developed in the **"New Rome"** that Constantine established—Constantinople, now Istanbul, Turkey—created fine monuments at places associated with the life and death of Jesus Christ. These include the place long acknowledged as the center of the Christian world, Jerusalem's **Church of the Holy Sepulcher** (•Figures 6.19 and 6.B). This extraordinary, sprawling building, now administered by numerous separate Christian sects, contains the locations where tradition says Jesus Christ was crucified and buried. Incidentally, several of these Christian denominations have a long history of squabbling over turf within the church, and the key to this place is in the hands of a trusted, neutral party: a Muslim family.

Christianity has seldom been the majority religion in the land where it was born; only until Islam arrived in Palestine in 638 C.E. was the region primarily Christian. From then until the twentieth century, most of Palestine's inhabitants were Muslim, and since the mid-twentieth century, Muslims and Jews have been the major groups. Between the eleventh and fourteenth centuries, European Christians dispatched military expeditions to recapture Jerusalem and the rest of the Holy Land from the Muslims. These bloody campaigns, known as the **Crusades**, resulted in a series of short-lived Christian administrations in the region.

Joe Hobbs

• **Figure 6.19** Jerusalem's Church of the Holy Sepulcher, containing the locations where many Christians believe Jesus Christ was crucified and buried.

There are significant minority populations of Christians throughout the Middle East, including members of distinct sects such as the **Copts** of Egypt and the **Maronites** of Lebanon. Their share of the population has generally been declining, both because of emigration and lower birth rates than those of the majority Muslims. Christians have been persecuted in some countries, which is one of the push factors in their emi-
56 gration. Nevertheless, the Middle East remains the cradle of their faith for Christians the world over, and Jerusalem and nearby Bethlehem are the world's most important Christian pilgrimage sites. Whatever your faith, you would surely find Jerusalem to be one of the most compelling places on Earth. I hope you will have an opportunity to travel there.

The Message of Islam

Islam is by far the dominant religion in the Middle East and North Africa; only Israel. within its pre-1967 borders, has a non-Muslim majority. Because of Islam's powerful influence not merely as a set of religious practices but as a way of life, an understanding of the religious tenets, culture, and diffusion of Islam is vital for appreciating the region's cultural geography.

Islam is a monotheistic faith built on the foundations of the region's earliest monotheistic faith, Judaism, and its offspring, Christianity. Indeed, Muslims (people who practice Islam) call Jews and Christians **People of the Book**, and their faith obliges them to be tolerant of these special peoples. Muslims believe that their prophet, Muhammad, was the very last in a series of prophets who brought the Word of God to humankind. Thus they perceive the Bible as incomplete but not entirely wrong—Jews and Christians merely missed receiving the entire message (just as Christians would insist that Jews missed the entire message). Muslims do not accept the Christian concept of the divine Trinity (God manifested in the form of the Father, his son Jesus, and the Holy Spirit) and regard Jesus as a prophet rather than as God.

Muhammad was born in 570 C.E. to a poor family in the western Arabian (now Saudi Arabian) city of Mecca. Located on an important north–south caravan route linking the frankincense-producing area of southern Arabia (now Yemen and Oman) with markets in Palestine (now Israel) and Syria, Mecca was a prosperous city at the time. It was also a pilgrimage destination because more than 300 deities were venerated in a shrine there called the **Ka'aba** (the Cube; •Figure 6.20). Muhammad married into a wealthy family and worked in the caravan trade. Muslim tradition holds that when he was about 40 years old, Muhammad was meditating in a cave outside Mecca when the Angel Gabriel appeared to him and ordered him to repeat the words of God that the angel would recite to him. Over the next 22 years, the prophet

Nabeel Turner/Getty Images

• **Figure 6.20** The black-shrouded cubical shrine known as the Ka'aba (just right of center) in Mecca's Great Mosque is the object toward which all Muslims face when they pray and is the centerpiece of the pilgrimage to Mecca required of all able Muslims. Note the carpet of humanity spread across this image: this is the height of the *hajj* (pilgrimage), when millions of the devout converge on the holy places of Mecca and Medina.

INSIGHTS

Sunni and Shiite Muslims

A schism occurred very early in the development of Islam, and it persists today. The split developed because the Prophet Muhammad had named no successor to take his place as the leader (caliph) of all Muslims. Some of his followers argued that the person with the strongest leadership skills and greatest piety was best qualified to assume this role. These followers became known as **Sunni**, or orthodox, Muslims. Others argued that only direct descendants of Muhammad, specifically through descent from his cousin and son-in-law Ali, could qualify as successors. They became known as **Shia**, or **Shiite**, Muslims.

The military forces of the two camps engaged in battle south of Baghdad at Karbala in 680 C.E., and in the encounter Sunni troops caught and brutally murdered Hussein, a son of Ali (Karbala is thus sacred to Shiites, as is nearby al-Najaf, where Ali is buried). The rift thereafter was deep and permanent. The martyrdom of Hussein became an important symbol for Shiites, many of whom still today regard themselves as oppressed peoples struggling against cruel tyrants, including some Sunni Muslims.

Today, only three of the region's countries, Iran, Iraq, and Bahrain, have Shiite majority populations. Significant minority populations of Shiites are in Syria, Lebanon, Yemen, and the Arab states of the Persian (Arabian) Gulf (see Figure 6.17).

related these words of God (whom Muslims call **Allah**) to scribes who wrote them down as the Qur'an (or Koran), the holy book of Islam.

During this time, Muhammad began preaching the new message, "There is no god but God (Allah)," which the polytheistic people of Mecca viewed as heresy. As much of their income depended on pilgrimage traffic to the Ka'aba, they also viewed Muhammad and his small band of followers as an economic threat. They forced the Muslims to flee from Mecca and take refuge in Yathrib (modern Medina), where a largely Jewish population had invited them to settle. There were subsequent skirmishes between the Meccans and Muslims, but in 630, the Muslims prevailed and peacefully occupied Mecca. The Muslims destroyed the idols enshrined in the Ka'aba, which became a pilgrimage center for their one God.

The Ka'aba is Islam's holiest place, and Mecca and Medina are its holiest cities. Jerusalem is also sacred to Muslims. Muslim tradition relates that on his **Night Journey**, the Prophet Muhammad ascended briefly into heaven from the great rock now beneath the Dome of the Rock (the same rock Jews regard as the Foundation Stone). Nearby on the Temple Mount (*al-Haraam ash-Shariif*) is **al-Aqsa Mosque**, a sacred congregational site. The proximity, even the duplication, of holy places between Islam and Judaism came to be the most difficult issue in peace negotiations between Palestinians and Israelis, as explained on page 190.

After Muhammad's death in 632, Arabian armies carried the new faith far and quickly. The two decaying empires that then prevailed in the Middle East and North Africa—the Byzantine or Eastern Roman Empire, based in Constantinople (Istanbul), and the Sassanian Empire, based in Persia (Iran) and adjacent Mesopotamia (Iraq)—put up only limited military resistance to the Muslim armies before capitulating. Local inhabitants generally welcomed the new faith, in part because administrators of the previous empires had not treated them well, whereas the Muslims promised tolerance. Soon the Syrian city of Damascus became the center of a Muslim empire. Baghdad assumed this role in 750 C.E.

Arab science and civilization flourished in the Baghdad immortalized in the legends of *The Thousand and One Nights*. Important accomplishments and discoveries were made in mathematics, astronomy, and geography. Scholars translated the Greek and Roman classics, and if not for their efforts, many of these works would never have survived to become part of the modern European legacy. It was an age of exploration, when Arab merchants and voyagers visited China and the remote lands of southern Africa. Many important discoveries by the Arab geographers were recorded in Arabic, a language unfamiliar to contemporary Europeans, and had to be rediscovered centuries later by the Portuguese and Spaniards. Arab merchants carried their faith on the spice routes to the East. One result, surprising to many today, is that the world's most populous Muslim country is not in the Middle East and North Africa, and its people are not Arabs; it is Indonesia, 5,000 thousand miles (8,000 km) east of Arabia.

Whether in Arabia or Indonesia, whether they be Sunni Muslims or Shiite Muslims (see Insights, above), all believers are united in support of the five fundamental precepts, or **Pillars of Islam**. The first of these is the **profession of faith**: "There is no god but God, and Muhammad is His Messenger." This expression is often on the lips of the devout Muslim, both in prayer and as a prelude to everyday activities.

The second pillar is **prayer**, required five times daily at prescribed intervals. Two of these prayers mark dawn and sunset. Business comes to a halt as the faithful prostrate themselves before God. Muslims may pray anywhere, but wherever they are, they must turn toward Mecca. There also is a congregational prayer at noon on Friday, the Muslim Sabbath.

The third pillar is **almsgiving**. In earlier times, Muslims were required to give a fixed proportion of their income as charity, similar to the concept of the tithe in the Christian church. Today, the donations are voluntary. Even Muslims of very modest means give what they can to those in need.

The fourth pillar is **fasting** during Ramadan, the ninth month of the Muslim lunar calendar. Muslims are required to abstain from food, liquids, smoking, and sexual activity from dawn to sunset throughout Ramadan. The lunar month of Ramadan falls earlier each year in the solar calendar and thus periodically occurs in summer. In the torrid Middle East and North Africa, that timing imposes special hardships on the faithful, who, even if they are performing manual labor, must resist the urge to drink water during the long, hot days.

The final pillar is the *hajj*, or **pilgrimage to Mecca**, Islam's holiest city. Every Muslim who is physically and financially able is required to make the journey once in his or her lifetime. A lesser pilgrimage may be performed at any time, but the prescribed season is in the 12th month of the Muslim calendar. Those days witness one of earth's greatest annual human migrations as about 3 million Muslims from all over the world converge on Mecca. Hosting these throngs is an obligation the government of Saudi Arabia fulfills proudly and at considerable expense. However, there has been some trepidation in recent years because of the security threat foreign visitors may pose to the host country and because accidents such as stampedes and tent city fires have cost numerous lives. Many pilgrims also visit the nearby city of Medina, where Muhammad is buried. Most Muslims regard the *hajj* as one of the most significant events of their lifetimes. All are required to wear simple seamless garments, and for a few days, the barriers separating groups by income, ethnicity, and nationality are broken. Pilgrims return home with the new stature and title of *hajji* but also with humility and renewed devotion.

All Muslims share the Five Pillars and other tenets, but they vary widely in other cultural practices related to their faith, depending on the country they live in; whether they are from the desert, village, or city; and their education and income. The governments and associated clerical authorities in Saudi Arabia and Iran insist on strict application of **Islamic law** (known as ***sharia***) to civil life; in effect, there is no separation between church and state. The Qur'an does not state that women are required to wear veils, but it does urge them to be modest, and it portrays their roles as different from those of men. Clerics in Saudi Arabia insist that women wear floor-length, long-sleeved black robes and black veils in public, that they travel accompanied by a male member of their families, and that they not drive cars. In Egypt, by contrast, Muslim women are free to appear in public unveiled if they choose. However, in most Muslim countries, conservative ideas about the role of women are still very strong: they should be modest, retiring, good mothers, and keepers of the home, although they can work. The Qur'an portrays women as equal to men in the sight of God, and in principle, Islamic teachings guarantee the right of women to hold and inherit property.

Most Muslim women argue that what others often see as "backward" cultural practices are in fact progressive. For example, they say, their modest dress compels men to evaluate them on the basis of their character and performance, not their attractiveness. They argue that segregation of the sexes in the classroom makes it easier for both women and men to develop their confidence and skills (•**Figure 6.21**). A married woman

Joe Hobbs

• **Figure 6.21** Education of women in the Muslim countries has increased dramatically in recent decades, but strong traditions keep most women out of the workforce. This is a geographic information systems (GIS) class in the Department of Geography, United Arab Emirates University.

retains her maiden name. These apparent advantages can be weighed against the drawbacks that make women generally subordinate to men in public affairs.

6.4 Economic Geography

Overall, the Middle East and North Africa is a poor region; per capita GNI PPP for the 23 countries and territories averages only $9,460 (see Table 6.1). This may seem surprising in view of the "rich Arab" stereotype. Only the oil-endowed states around the Persian Gulf (Arabian Gulf) deserve the reputation for wealth, and by most measures other than per capita wealth, only non-Arab Israel is truly a more developed country (MDC). Israel's prosperity comes from its innovation in computer and other high-technology industries, the processing and sale of diamonds, large amounts of foreign (mostly U.S.) aid, and investment and assistance by Jews and Jewish organizations around the world.

Here we will examine the role of oil in the region's economies, and the lagging economic indicators and human conditions that should have served as the writing on the wall for the "Arab Spring" (see page 192).

"The Prize"

Vital to the industrialized countries as a source of fuels, lubricants, and chemical raw materials, petroleum is one of the world's most important natural resources. Daniel Yergin, recognized as the world's preeminent authority on oil, wrote the definitive book on this essential commodity. His title sums up what oil has come to represent in economic and geopolitical terms: *The Prize*.

© Jamal Nasrallah/epa/Corbis

• **Figure 6.22** Petroleum facilities in the interior wilderness of Arabia's Empty Quarter (al-Rub' al-Khali). Most of Saudi Arabia's production is from vast, shallow reservoirs of oil that are much closer to marine terminals on the Persian Gulf.

A crucial feature of world geography is the concentration of about two-thirds of the world's proven conventional petroleum reserves in a few countries that ring the Gulf. In the second half of the twentieth century, this geographical endowment transformed a handful of inconsequential countries and shaikhdoms into some of the most critical locations in the global economic system (•Figure 6.22).

By coincidence, the countries rich in oil tend to have relatively small populations, whereas the most populous nations have few oil reserves; Iran is an exception. All but about 1 percent of the Gulf oil region's proven reserves of crude oil are located in Saudi Arabia, Iraq, Kuwait, Iran, and the United Arab Emirates (UAE), with smaller reserves in Oman, Bahrain, and Qatar. Saudi Arabia, by far the world leader in reserves, has about 18 percent of the proven crude oil reserves on the globe. Venezuela is in second place, with about 14 percent of the world's proven oil; Canada is third, with 12 percent (not in the form of easily extractable crude oil, but as tar sands; see page 417); the next four are all along the Gulf: Iran (9%), Iraq (8%), Kuwait (7%), and the United Arab Emirates (6%). Libya also has significant oil reserves. The world's largest oil consumer, the United States, has 2 percent. China, the second largest consumer by a narrow margin, has 2 percent.

In the Gulf area, the great thickness of the region's oil-bearing strata and high reservoir pressures have made it possible to secure an immense amount of oil from a small number of wells. The productivity of each well makes each barrel inexpensive to extract, and it is simple to increase or reduce production quickly in response to world market conditions. Gulf oil is thus cheap to produce unless expenditures to maintain huge military forces and infrastructure to defend it are factored in (in economic terms, unless the "external costs" of oil are internalized). Some analysts—including a former U.S. Navy secretary—prefer to calculate the "real price" of oil with military expenditures calculated. If one accepts the premise that maintaining a costly military presence in the region is part of the United States' long-term strategy to maintain the flow of Gulf oil, that oil is no longer cheap by any measure.

Production, export, and profits of Middle Eastern and North African oil were once firmly in the hands of foreign companies. That situation changed after 1960, when most of the Gulf countries and other exporting nations formed the **Organization of Petroleum Exporting Countries (OPEC)**, with the aim of taking joint action to demand higher profits from oil. It changed again after 1972, when the oil-producing countries began to nationalize the foreign oil companies. OPEC was relatively obscure until the Arab-Israeli war of 1973, after which the organization began a series of dramatic price increases. In 1980, the organization's price reached $37 (U.S.) per barrel, compared with $2 a barrel in early 1973. Adjusted for inflation, that $37 would be the equivalent of about $100 per barrel in 2011, so the price hikes were truly titanic.

These events had enormous repercussions for the world economy. Immense wealth was transferred from the more developed countries to the OPEC countries to pay for indispensable oil supplies. The skyrocketing cost of gasoline and other oil products helped cause serious inflation in the United States and many other countries and contributed to the 1973 **energy crisis** in the United States. Desperately poor, less developed countries (LDCs) found that high oil prices not only hindered the development of their industries and transportation but also reduced food production because of high prices for fertilizer made from oil and natural gas. In the Gulf countries, the oil bonanza produced a wave of spending on military hardware, showy buildings, luxuries for the elite, and ambitious development projects of many kinds. Per capita benefits to citizens were greatest in the Arabian Peninsula, where small populations and immense inflows of oil money made possible the abolition of taxes, the establishment of comprehensive social programs, and heavily subsidized amenities such as low-cost housing and utilities, including freshwater distilled from the salty Gulf in desalination plants.

Then, as the 1980s began, the era of continually expanding OPEC oil production, sales, and profits seemed to come to an end. After 1973, the high price of oil stimulated oil development in countries outside OPEC. Oil conservation measures such as a shift to more fuel-efficient vehicles and factories were introduced. Substitution of cheaper fuels for oil increased. Coal replaced oil in many electricity-generating stations. Meanwhile, the world entered a period of economic recession due in part to high oil prices. Decreased business activity reduced the demand for oil. Profits of the world oil industry (and taxes paid to governments) were severely cut, large numbers of refineries had to close, and much of the world tanker fleet was idled.

Oil prices have risen and fallen, with predictable consequences for producers and consumers. But the immense oil and gas reserves still in the ground guarantee that the Gulf region will continue to have major long-term influence in world affairs and will remain prosperous as long as these finite resources are in demand—especially in the established industrialized countries and in the booming economies of China and India.

There are, of course, other resources and industries in the region's economic geography—**remittances** (earned income sent home by guest workers) in the oil-rich countries; revenues from ship traffic through the Suez Canal; and exports of cotton, rice, and other commercial crops, for example—but oil dominates the region's economy and is central to the global economy.

This depiction of MENA's oil wealth, which is concentrated in a few countries, does not consider the economic poverty so characteristic of the region's other countries and the great majority of its peoples. The worsening conditions in many of the Arab countries—the poverty, crowding, unemployment, and the growing gap between the great mass of the poor and the increasingly wealthy rich—were the economic ingredients of the Arab Spring described later.

6.5 Geopolitical Issues

The Middle East and North Africa have long been vital in world affairs, with their resources and geographical assets such as the Suez Canal coveted by outside interests. From very early times, overland caravan routes, including the great Silk Road, crossed the Middle East and North Africa with highly prized commodities traded between Europe and Asia. The security of these routes was vital. Geopolitical concerns are now focused on narrow waterways, access to oil, access to freshwater, and terrorism.

Chokepoints

One of the striking characteristics of the geography of the Middle East and North Africa is how many seas border and penetrate the region. In many cases, these bodies of water are connected to one another though narrow straits and other

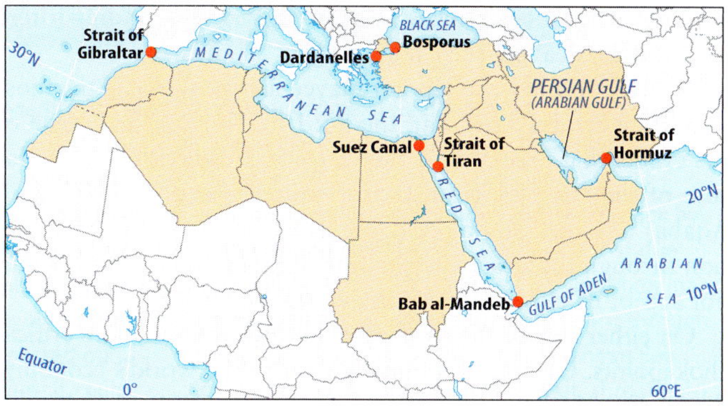

• **Figure 6.23** Chokepoints in the Middle East and North Africa.

passageways. In geopolitical terms, such constrictions are known as **chokepoints**—strategic narrow passageways on land or sea that may be easily closed off by force or even the threat of force (•Figure 6.23). Chokepoints must be unimpeded if world commerce is to carry on normally. Keeping them open is therefore usually one of the top priorities of regional and external governments. Likewise, closing them is a priority to a combatant nation or a terrorist entity seeking to gain a strategic advantage. Many notable events in military history and the formation of foreign policy in the Middle East focus on these strategic places.

One of the world's most important chokepoints is the Suez Canal, which opened in 1869. Slicing 107 miles (172 km) through the narrow Isthmus of Suez, the British- and French-owned Suez Canal linked the Mediterranean Sea with the Indian Ocean, saving cargo, military, and passenger ships a journey of many thousands of miles around the southern tip of Africa (•Figure 6.24). Keeping the Suez Canal in friendly hands was one of Britain's major military concerns during World War II. That objective led to a hard-fought and eventually successful British and Allied campaign against Nazi Germany in North Africa. Egyptian President Gamal Abdel Nasser's nationalization of the canal in 1956 led immediately to a British, French, and Israeli invasion of Egypt and a conflict known as the **Suez Crisis** and the **Arab-Israeli War of 1956**. Egypt effectively won

• **Figure 6.24** The strategically vital Suez Canal zone saw bitter fighting in the Middle East wars of 1956, 1967, and 1973.

believed that the military option would have far-ranging and dangerous consequences. Iran's nuclear program would be set back a few years, but the country would portray itself as a victim of U.S. aggression, galvanizing support across the Muslim world (even among Iran's traditional enemies, the Sunni Arabs), perhaps inciting its Hizbullah allies in Lebanon and Hamas allies in the Palestinian territories to attack Israel, encouraging al-Qa'ida–style terrorism against the West, and discouraging Iranians (particularly the younger ones) who favor peaceful relations with the United States. A military strike would probably also lead to a sharp reduction in oil exports from the Gulf, perhaps triggering a global economic crisis. Russia and China would meanwhile have an opportunity to enhance their geopolitical interests by offering diplomatic support and perhaps more to Iran throughout such a crisis. Both countries have considerable leverage in Iran.

Israeli officials have said publicly that Israel (which has its own nuclear weapons, a fact it never acknowledges officially) regards the Iranian bomb as an "existential threat" and will prevent its development, presumably by a single surprise air strike like the one Israel carried out against Iraq's nuclear facility in 1981. It is more likely that Israel rather than the United States would strike Iran. Again, Israel would delay the development of an Iranian bomb, but the regional and global consequences would be enormous. The United States has repeatedly warned Israel against attacking Iran because of the probable consequences discussed above.

The United States and Israel may have found a way to circumvent both diplomatic and military means to blunt Iran's nuclear weapons program: sabotage. In 2010, a computer worm (probably planted by Israeli and U.S. agents) called Stuxnet apparently did significant damage to centrifuges at Iran's nuclear facilities. The most sophisticated cyberweapon ever developed, the Stuxnet malware took over the control systems at Natanz and Bushehr, sending the centrifuges spinning wildly out of control while also issuing a sound that made control room staff think that systems were operating normally. By some intelligence estimates, the Stuxnet attacks may have set Iran's nuclear weapons program back by two years.

Al-Qa'ida

In 1998, the world began to hear about Osama bin Laden, a former Saudi businessman living in exile in Afghanistan, whose **al-Qa'ida** organization bombed U.S. embassies in Kenya and Tanzania as part of a struggle, or ***jihad***, against American imperialism and immorality. (In Islam, though, *jihad* has a very wide range of interpretations, from expressing one's commitment to a personal conviction, to an all-out war against nonbelievers.) Bin Laden's organization was also responsible for the 2000 bombing of the American naval destroyer *U.S.S. Cole* in Yemen's harbor of Aden. However shocking those assaults were, they pale in comparison to al-Qa'ida's attacks against targets in the United States on September 11, 2001. In the most ferocious terrorist actions ever undertaken to that date, members of al-Qa'ida "cells" (small groups of agents) in the United States hijacked four civilian jetliners and succeeded in piloting two of them into the twin towers of New York City's World Trade Center and one into the Pentagon, home of the U.S. Department of Defense, outside Washington, D.C. Passengers on a fourth plane succeeded in aborting the terrorists' likely mission of destroying the White House or damaging the U.S. Capitol, forcing the plane to crash in rural Pennsylvania. More than 3,000 people, mainly civilians, perished. Bin Laden and his followers cheered the carnage as justifiable combat against an infidel nation whose military troops occupied the holy land of Arabia, where Mecca and Medina are located.

As the United States counterattacked, ousting al-Qa'ida from its bases in Afghanistan and cracking down hard on its leadership around the world, al-Qa'ida evolved into a far more geographically diffuse organization. It carried out or supported attacks in Indonesia, Morocco, Algeria, Tunisia, Turkey, Spain, Uzbekistan, Egypt, Saudi Arabia, Jordan, the United Kingdom, Yemen, and Pakistan. A succession of al-Qa'ida videotapes and audiotapes promised an unrelenting and costly continuation of *jihad* against the United States and its allies. Even with Bin Laden dead, it will be many years before Americans and Europeans might be able to emerge from the shadow of this threat (•**Figure 6.29**). Al-Qa'ida clearly will not hesitate to use the most devastating weapons, even against large numbers of civilians; this deadly goal is a tactical priority. There is also growing consensus that al-Qa'ida is no longer a solitary severe threat to U.S. interests; instead, the *jihadist* cause has spread to smaller groups and to "lone wolves" that are also keen to take on the West. There is little evidence that al-Qa'ida's message has any broad appeal, and it had virtually no role in the tumultuous Arab Spring (see page 196). But the danger remains.

Using GIS and a variety of other geographic tools, many geographers—along with U.S. intelligence agencies, of course—are at work interpreting the geographic dimensions of terrorism. One of the most critical questions for geographers is where future al-Qa'ida attacks are likely to take place. Rohan Gunaratna, probably the greatest authority on al-Qa'ida, estimates that al-Qa'ida maintains a reserve of at least 100 targets worldwide.[6] Where are most of these? Suleiman Abu Ghaith, one of bin Laden's top aides, offered a clue, along with a geographic answer to the question of what al-Qa'ida wants: "Let the United States know that with God's permission, the battle will continue to be waged on its territory until it leaves our lands."[7]

In 9/11 and other atrocities, a tiny minority of Muslims carried out terrorist actions that the vast majority of Muslims condemned. Islamic scholars and religious authorities point out that murder of civilians is prohibited in Islamic law and that the attacks have no legitimate religious grounds. Mainstream Islamist movements are not military or terrorist organizations and have distinguished themselves through public service to the needy.

The following section offers more insight into the peoples and nations of this vital region. It begins with a continuation of the geopolitical theme, examining one of the world's most problematic, persistent, and influential conflicts, the one between Israel and its Arab neighbors.

(a) **(b)**

• Figure 6.29 Bookends to Osama bin Laden's engagement with the United States. In the 9/11/2001 attack on the World Trade Center **(a)**, note United Flight 175 about to strike the South Tower. Belgian newspapers trumpet bin Laden's death on May 3, 2011 **(b)**.

6.6 Regional Issues and Landscapes

Israel and Palestine

The Arab-Israeli Conflict

The Arab-Israeli conflict persists as one of the world's most intractable disputes. It has not been resolved, in part because the central issues are closely tied to such life-giving resources as land and water, and to deeply held and unyielding religious beliefs. When you understand these issues, you should be proud and prepared. Among other things, you will have a strong foundation from which to follow the news about this conflict for years to come.

The Arab-Israeli conflict is above all a conflict over who owns the land—sometimes very small pieces of land—and is therefore of extreme interest in the study of geography. It is also a conflict that has repercussions far beyond the boundaries of the small countries and territories involved. As long as it simmers or boils, there are other countries and entities that will use the unresolved Arab-Israeli conflict to advance their interests at the expense of others; both al-Qa'ida and Iran, for example, derive much beneficial propaganda value from it. A United Nations–sponsored group called the Alliance of Civilizations argues that the Palestinian-Israeli conflict is *the largest force behind global tensions.* Let's look at this in different terms: when it is resolved, the world will be a much more safer place.

A geographic understanding of this conflict requires familiarity with the events leading up to the creation of the state of Israel and with the wars that have followed, particularly as they have rearranged the boundaries of nations and territories.

The modern state of Israel was carved from lands whose fate had been undetermined since the end of World War I. The Ottoman Empire, based in what is now Turkey, had ruled Palestine (roughly the area now made up of Israel and the Palestinian territories) and surrounding lands in the eastern Mediterranean since the sixteenth century. After the British and French defeated the Ottoman Turks in World War I and destroyed their empire, they divided the region between themselves (**•Figure 6.30**). The British received the "mandate" (authority to establish a government) for Palestine, Transjordan (modern Jordan), and Mesopotamia (Iraq), and the French received the mandate for Syria (now Syria and Lebanon).

During World War I, British administrators of Palestine had made conflicting promises to Jews and Arabs. They implied that they would create an independent Arab state in Palestine and yet at the same time vowed to promote Jewish immigration to

•Figure 6.30 The Middle East and North Africa in 1920. The victorious allies of World War I carved up the Middle East among themselves. Growing difficulties of administration would drive them from the region within a few decades.
Source: Based on John Haywood, ed., Atlas of World History (Barnes and Noble Books, 1997).

Palestine with an eye to the eventual establishment of a Jewish state there. The **Palestinians**—Arabs who historically formed the largest majority of the region's inhabitants—did not welcome the ensuing Jewish immigration and rioted against both the migrants and the British administration. Militant Jews attacked British interests in Palestine, hoping to precipitate a British withdrawal.

Placing themselves in a no-win position with these conflicting promises and under increasing pressure from both Jews and Arabs, in 1947 the British decided to withdraw from Palestine and leave the young United Nations with the task of determining the region's future. The United Nations responded in 1947 with the **two-state solution** to the problem of Palestine. It established an Arab state (which would have been called Palestine) and a Jewish state (Israel).

The two-state plan was deeply flawed. The states' territories were long, narrow, and fragmented, giving each side a sense of vulnerability and insecurity (Figure 6.29). When Israel declared itself into existence in May 1948, the armies of the neighboring Arab countries of Transjordan, Egypt, Iraq, Syria, and Lebanon mobilized. In what Israelis call "the War of Independence" and Palestinians call "the Catastrophe" (*al-Nakba*), the smaller but better-organized and more highly motivated Israeli army defeated the Arab armies, and Israel acquired what have come to be known as its **pre-1967 borders** (the 1949 Armistice Agreement borders shown in **•Figure 6.31**). The boundary separating Israel from the West Bank would later come to be known as the "Green Line" (seen in Figure 6.36).

Prior to and during the fighting of this **1948–1949 Arab-Israeli war**, approximately 800,000 Palestinian Arabs chose or were forced to flee from the new officially Jewish state of Israel to neighboring Arab countries (**•Figure 6.32**). The United Nations established refugee camps for these displaced peoples in Jordan, Egypt, Lebanon, and Syria. Little was done to resettle them in permanent homes, and both the local Arab governments and the refugees themselves continued to insist on the right of the refugees to return to Israel and the restoration of their property there. This Palestinian **right of return** is one of the central problems of the modern peace process.

Other countries assumed control of those parts of the proposed Palestine that Israel did not absorb. Egypt occupied the Gaza Strip, a piece of land on the Mediterranean shore adjacent to Egypt's Sinai Peninsula that was inhabited mostly by Palestinian Arabs. Transjordan, renaming itself simply Jordan, occupied the predominantly Arab hilly region of central Palestine on the west side of the Jordan River, known as the West Bank, Fig. 6.31
and the entire old city of Jerusalem, including the Western Wall and Temple Mount, Judaism's holiest sites. The Palestinian Arab state envisioned in the UN partition plan was thus stillborn.

The **Six-Day War** of 1967 fundamentally rearranged the region's political landscape in Israel's favor, setting the stage for subsequent struggles and the peace process. This conflict was precipitated in part when Egypt, by positioning arms at the Strait of Tiran chokepoint, closed the Gulf of Aqaba to Israeli shipping. Egypt's President Nasser and his Arab allies took sev- 177
eral other belligerent but nonviolent steps toward a war they

• **Figure 6.31** The 1947 UN partition plan for Palestine and Israel's original (pre-1967) borders. The 1948–1949 war, which began as soon as Britain withdrew from Palestine and Israel proclaimed its existence, aborted the UN plan and created a tense new political dynamic in the region.

Source: The Middle East and North Africa: A Political Geography by Drysdale & Blake (1985) Fig.9.7 The 1947 UN Partition Plan for Palestine and Israel's Original (Pre-1967) Borders p.280. By permission of Oxford University Press, Inc.

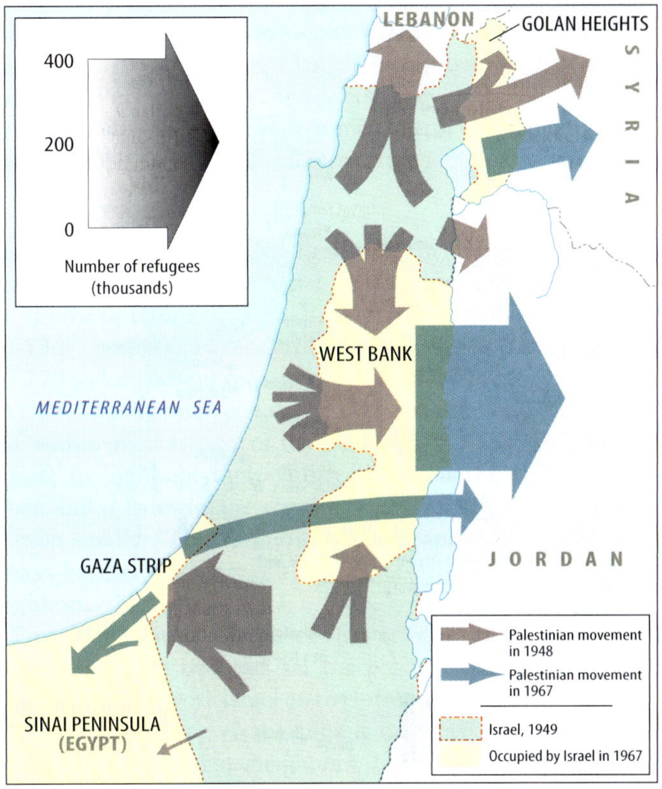

• **Figure 6.32** Palestinian refugee movements in 1948 and 1967. Many who fled in the first conflict relocated in the second.

Source: The Middle East and North Africa: A Political Geography, by A. Drysdale and G. Blake. 1985 Oxford University Press. Based on The Wandering Arabs, Geographical Magazine XLV (3) 1972 page 180.

were ill prepared to fight. Israel elected to make a preemptive strike on its Arab neighbors, virtually destroying the Egyptian and Syrian air forces on the ground. Israel gave Jordan's King Hussein an opportunity to stay out of the conflict. However, Jordan went to war and quickly lost the entire West Bank and the historic and sacred Old City of Jerusalem. The entire nation of Israel was transfixed by the news that Jewish soldiers were praying at the Western Wall. The Israeli army (Israeli Defense Forces, or IDF) also seized the Gaza Strip, Egypt's Sinai Peninsula (shutting down the Suez Canal), and the strategic Golan Heights section of Syria overlooking Israel's Galilee region. These three pieces of land would henceforth be known to the world as the **Occupied Territories**. Israel had tripled its territory in six days of fighting (•**Figure 6.34**).

The Palestinian refugee situation became more complex as a result of the war. Israel took over the areas where most of the camps were located. Many people in the camps again took flight, and Palestinians fleeing villages and towns in the newly Occupied Territories joined them as refugees (see Figure 6.32). Some later returned to their homes, but an estimated 116,000 either did not attempt to return or were denied permission by Israel authorities to do so.

The **1973 Arab-Israeli War** was a multifront Arab attempt to reverse the humiliating losses of 1967. On October 6, 1973, the Jewish holy day of Yom Kippur, Egypt and Syria launched a surprise attack on Israel, with the hope of rearranging the stalemated political map of the Middle East. Egypt's army initially showed surprising strength, penetrating deep into the Sinai Peninsula and overturning Israel's image of invulnerability. Israeli troops soon reversed the tide and surrounded Egypt's army, but in the ensuing disengagement talks Egypt won back the eastern side of the Suez Canal and by 1975 was able to reopen it to commercial traffic. In 1979, Israel agreed to return Sinai to Egypt under the U.S.-sponsored **Camp David Accords**, and Egypt recognized Israel's rights as a sovereign state. The Gaza Strip, West Bank, and Golan Heights continued to be held by Israel. Different Israeli administrations regarded them as lands that rightfully belong to Jews or as cards to be played in the regional game of peacemaking.

Arabs and Jews: The Demographic Dimension

In addition to issues of land, water, politics, and ideology, the Palestinian-Israeli conflict is about sheer numbers of people. Each side has wanted to maximize its numbers to the disadvantage of the other. To realize the Zionist dream of establishing a Jewish state, Jews began immigrating to Palestine around the start of the twentieth century. Jews made up 11 percent of

KHALED DESOUKI/AFP/Getty Images

• **Figure 6.37** Egyptian masses protesting against the government of Hosni Mubarak in Cairo's Tahrir Square, February 2011.

(•**Figure 6.37**). Egypt's despised secret police used deadly force even when there was no need. Do you remember the scenes of armed men mounted on camels and horses charging against the demonstrators in Tahrir Square (which became known as the "Battle of the Camels")? Rather then disperse, over a period of 18 days the protestors held their ground in Tahrir and Egypt's other public squares and outside government buildings. Without arms, the demonstrators took casualties of dead and wounded. In the end, their people power prevailed. The army, an institution that Egyptians long revered, took up their cause. Mubarak offered political concessions to Egyptians, but it was too late. As the army assumed power, he fled from Cairo, and was soon imprisoned and put on trial.

The Libyan Domino

A worse fate would befall Libya's dictator, Muammar Qaddafi. Libya was not like Egypt: with a population of just 6.4 million, Libya did not have an unmanageable youth bulge, nor was it propped up by American aid. With its considerable oil reserves, more than 3 percent of the world total, Libya should have been a prosperous country. For more than 40 years, it was instead the toy of Muammar Qaddafi, an eccentric man (what other world leader would call for Switzerland to be abolished?). Colonel Qaddafi ran the country as if it were his fiefdom, favoring and distributing money to those who were loyal to him. Tribal and ethnic politics were decisive. Qaddafi loyalists drew mainly from his own tribe on the central coast and from other tribes he favored in the center and west. King Idris I, whom Qaddafi overthrew in 1969, had favored tribes in the eastern region known as Cyrenaica, especially around the city of Benghazi. Those tribespeople never forgave Qaddafi for ousting their leader, and Qaddafi made sure they never earned any graces from him. Qaddafi also treated the ethnic Berber tribes of Libya's western mountains (whose livelihood but not ethnicity is
165 much like that of the Bedouin) as second-class citizens.

The people of Cyrenaica along with the Berbers of the west were inspired by the tide of the Arab Spring to rise up against Qaddafi, and a brief civil war emerged between them and Qaddafi loyalists. The opposition to Qaddafi was advantaged by the commitment of NATO air power, weapons, and intelligence, and it was only a matter of time until Qaddafi and his
power base would be eliminated. Qaddafi was captured and 89
killed execution-style in October 2011, eight months after the uprising began.

Syria's Minority Dynasty versus the People of Syria

As in Bahrain and Libya, Syria's leaders opted for violent repression against demonstrators inspired by the Arab Spring; at least 9,000 and probably many more died. The Syrian uprising began in January 2011 after another self-immolation.

In Syria, a minority Shiite group (making up 7 percent of population) called the **Alawites** and led by president Bashir al-Assad, son of the late president Hafez al-Assad, ruled the majority (74 percent) Sunni. Resentment of Alawite rule was nothing new in Syria. Four decades of the repressive al-Assad regime rankled the Sunni majority, the Christian minority, and others. In 1982, a brutal army assault against a Sunni Muslim Brotherhood uprising in the city of Hama resulted in tens of thousands of deaths. The 2011/12 revolt involved much broader participation in more cities (•**Figure 6.38**). Arab Spring demonstrations in most of the Arab countries were loudest in the capital cities, but Syria was an exception. The fighting came to Damascus rather late.

The challenge of taking down the Alawite regime in Syria was great. The Alawites controlled the key unit of the Syrian military, so it would be difficult to secure a broad military uprising against the regime. NATO and the United States would be far less inclined to become involved than they did in Libya; the military challenge would be much greater. Western countries stepped up their sanctions against Syria. The United Nations tried to pass resolutions against Syria, but Russia and China vetoed these. The Arab League moved to oust Syria from the organization and to impose its own sanctions. Turkey was more involved, providing shelter to forces of the Free Syrian Army, made up mainly of Sunni defectors

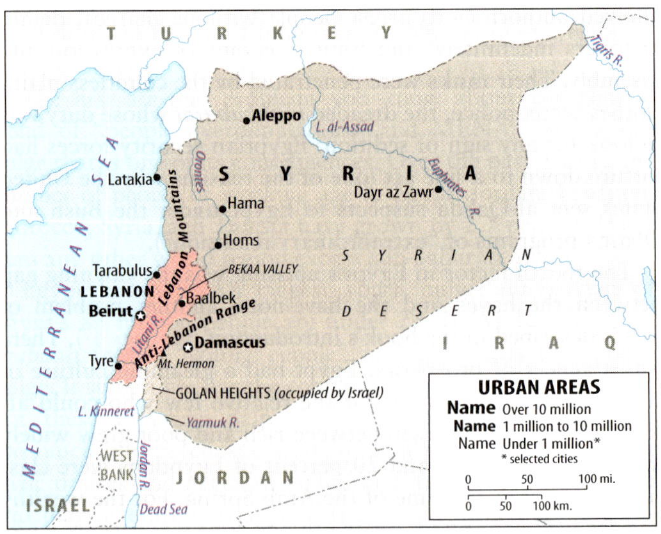

• **Figure 6.38** Principal features of Syria and Lebanon.

from the Syrian army. Israel did not become involved. Iran had much at stake: it would lose its only Arab ally should the Syrian regime fall.

Syria tried to keep foreign reporters out of the country, but a few sneaked in and were able to verify the authenticity of the video footage posted daily on YouTube. The savagery with which the Syrian army took on armed and unarmed civilians was extraordinary. Soldiers seemed to have no compassion even for children and women. It was unrestrained carnage, and there appeared to be no way to bring it to a stop.

Bahrain: A Pearl Is Crushed

In Bahrain, a small but oil-rich Gulf island shaikhdom (Bedouin monarchy) linked to eastern Saudi Arabia by a causeway, there is a Shiite majority of 70 percent that has long been ruled
174 by a Sunni monarchy (•**Figure 6.39**). Like Egyptians, Bahrain's Shiites were inspired by the Jasmine Revolution to challenge King Hamad bin Isa al-Khalifa.

Bahrain was the Arab Spring's best example of how repression can backfire by fueling resentment. The mainly Shiite protestors began by simply expressing wishes for democracy, public participation, and justice. They asked that King Khalifa take his place in a new constitutional monarchy. The monarch could have responded as a benevolent despot, yielding some ground to his subjects.

Instead, he ordered government forces to crush the rebellion with deadly force. As the casualties mounted, demonstrators' cries changed from reasonable petitions to "Death for Khalifa!" Worried that Bahrain's Shiites might overthrow the monarchy and put Bahrain into the embrace of his adversary Shiite Iran, the king called on Saudi Arabian forces to drive across
the causeway to Bahrain and helped put down the uprising. 203
Government forces destroyed the demonstrations' epicenter, Pearl Square, with the country's iconic pearl sculpture in the capital city of Manama, and turned it into a traffic intersection to ensure that crowds would not gather again. King Khalifa's opponents retreated, their hopes unrequited. Below we will consider Bahrain's future, especially in its geopolitical context.

Revolt in Yemen's Mountainous Redoubt

Although located on the Arabian Peninsula (**see Figure 6.39**), Yemen has only a small bounty of oil reserves. A stunningly beautiful country graced by mountains and deserts, and a distinctive urban architecture looking like gingerbread houses,

• **Figure 6.39** Principal features of the Arabian Peninsula.

Danita Delimont/Getty Images

• **Figure 6.40** The palace of the Imam in Wadi Dhahr, Yemen.

Yemen is cursed by poverty (•**Figure 6.40**). Loyalties are to clan and tribe, and Yemen has no natural sense of cohesion as a nation state. The country was previously divided into North and South Yemen. Al-Qa'ida has a strong foothold here; Yemen is in fact the ancestral homeland of Osama bin Laden's clan.

With the Arab Spring, a widespread revolt against the long-time ruler Ali Abdullah Saleh erupted. After surviving an assassination attempt, and just before resigning office, President Saleh offered concessions to Yemeni protestors. The country's established opposition political parties pleaded with the young people to stop their protests. Yemen's *shabaab* ignored that advice. They are a force to be reckoned with; the demographic profile of Yemen's *shabaab* is very large. With a population of 23.8 million, Yemen has one of the world's highest population growth rates, a staggering 3.1 percent (with a doubling time of 22 years). Poverty, unemployment, and repression of dissent made Yemen ripe for revolt. Its geopolitical context is also discussed here.

Hallmarks of the Revolution

In the Arab Spring's aftermath, Middle East observer Fareed Zakaria recommended a book on his Sunday morning "GPS" show on CNN. It was the United Nations' *Arab Development Report 2002*. Why recommend a decade-old book? To make the point that little had changed in those years. The chapter titles that so well summarized the region's conditions in 2002 did the same a decade later: "Development not engendered is endangered ... Bridled minds, shackled potential ... As development management of stumbles, economies falter ... The curse of poverty: denying choices and opportunities, degrading lives."[8]

The Arab Spring rumbled through Tunisia, Egypt, Libya, Bahrain, Syria, and Yemen—described in some detail here—and in Sudan, Algeria, Mauritania, Morocco, Jordan, the Palestinian Territories, Lebanon, and Oman. Many of Egypt's problems that we have considered—population growth, unemployment and underemployment, lack of political representation, and oppression by authorities—were shared by these countries. Here are some of the other notable characteristics of the uprising and its aftermath in the countries of the Arab Spring:

- The revolutions were much facilitated by Facebook, Twitter, 24/7 satellite television news (notably from Qatar's *al-Jazeera* network), and other social media. Governments tried to crack down, bringing down the Internet and suspending cell phone service, but to no avail. By the time the authorities woke up, detailed instructions on how, where, and when to gather and protest in public spaces had been widely posted and tweeted.
- Although new media made their debut as tools of revolution in the Arab Spring, the more traditional geography of revolt in public spaces played a prominent role. In the traditional Middle Eastern *medina*, with the exception of the mosque there was no large public space in which large numbers of people could gather. However, spacious traffic hubs that doubled as public squares were established in the post-colonial modern Middle Eastern cities. In the Arab Spring, Cairo's Tahrir Square, Tripoli's Green Square in Libya (renamed Martyrs' Square in recognition of the lives lost there in protest), and Bahrain's Pearl Square, served as central places where social media called upon demonstrators to gather. Many of the key events leading to success (and in Bahrain's case, the failure) of the revolutions took place in these locations.
- Women had an unprecedented strong role in the Arab Spring. In the conservative Muslim societies of the Middle East and North Africa, there is strict segregation of men and women (in some countries more so than in others) in public space. But in Tahrir Square and elsewhere, that convention was thrown aside as women and men stood shoulder to shoulder in protest. In a second round of revolt against Egyptian rule—this one against the Egyptian army late in 2011—the army struck back against women protestors with particular savagery.
- The traditional Islamic classification of sacred time played an important role in the Arab Spring. As often in Muslim history, Friday noon prayers and sermons incorporated politics with prayer. Prayer time gave way to protest time, often on such a huge scale that the authorities dreaded Fridays. This traditional day of prayer was often tweeted, Facebooked, and otherwise announced as the "Day of Rage" for which demonstrators should prepare themselves.
- Remarkably, religion and militant Islamism did not otherwise feature prominently in the Arab Spring. Mainstream and radical Islamists alike were conspicuous by their absence. In Egypt, the Muslim Brotherhood, a well-organized political group that could have promoted itself and its agenda, kept a low profile. Throughout the region,

Joe Hobbs

• **Figure 7.16** Spirit House in Laos.

has the following characteristic elements. It lacks a definite creed or theology. It is very absorptive, encompassing an unlimited pantheon of deities (all of which, most Hindus say, are simply different aspects of the one God, Brahman) and an infinite range of types of permissible worship (•**Figure 7.17**). Most Hindus recognize the social hierarchy of the caste system, deferring authority to the highest (Brahmin) caste. They practice rituals to honor one or another of the principal deities (Brahman the creator, Vishnu the preserver, and Shiva the destroyer) and thousands of their manifestations (*avatars*) and lesser deities, attending temples to worship them in the form of sanctified icons in which the gods' presence resides. They believe in reincarnation and the transmigration of souls and revere many living things (see A Closer Look).

Joe Hobbs

• **Figure 7.17** Hinduism has a complex and very colorful pantheon. This is Vishnu, preserver god, and his consort, Lakshmi, goddess of wealth and prosperity.

A closer LOOK

The Sacred Cow

Many attitudes, beliefs, and practices associated with cattle make up the world-famous but often poorly understood "sacred cow" concept of India. There are nearly 200 million cattle in India, representing about 15 percent of the world total and the largest concentration of domesticated animals anywhere on earth. India's dominant religion of Hinduism forbids the slaughter of cows but allows male cattle and both male and female water buffalo to be killed. Reverence for the cow is well founded. Indians favor cow's milk, ghee (clarified butter), and yogurt over dairy products from water buffalo. They value cows as producers of male offspring, which serve as India's principal draft animals. Both cows and bullocks provide dung, an almost universal fuel and fertilizer in rural India.

Reverence for the cow in particular and cattle in general pervades Hindu religion and mythology. The bull Nandi is associated with the Hindu god Shiva, the cow is linked to the goddess Lakshmi, and Krishna (an incarnation of the god Vishnu) was a cowherd. People allow cattle to roam the streets of Indian cities, and simply accommodate them in vehicle and pedestrian traffic (•**Figure 7.J**). As a symbol of fertility, cows are associated with several deities. The mother of all cows, Surabhi, was one of the treasures churned from the cosmic ocean. Hindus honor cows at several special festivals. They use cow's milk in temple rituals. They believe that the "five products of the cow"—milk, curds, ghee, urine, and dung—have unique magical and medicinal properties, particularly when combined. In India, there are about 4,000 *gaushalas*, or "old folks' homes," for aged and infirm cattle.

Remarkably, however, the subcontinent countries of Pakistan, Bangladesh, and India are major exporters of leather and leather goods. In India, the leather industry is mainly in the hands of Muslims, who do not share Hindus' restrictions on killing or eating cows. Muslims are forbidden to eat pork; therefore, pigs, a major food resource in many developing countries, are of little importance here. They are eaten mainly by Christians, very low-caste Hindus (some of whom are pig breeders), and tribal peoples.

Joe Hobbs

• **Figure 7.J** People and cows share street space throughout India.

They are supposed to be tolerant of other religions and ideas. They participate in folk festivals to commemorate legendary heroes and gods. To earn religious merit and to struggle toward liberation from the bondage of repeated death and rebirth, they make pilgrimages to sacred mountains and rivers.

The Ganges is a particularly sacred river to Hindus, who believe it springs from the matted hair of the god Shiva, and who call it "Mother" and "Goddess." The pilgrimage to celebrate the festival of Ardh Kumbh Mela at Allahabad, at the confluence of the Ganges and Yamuna rivers, drew 75 million pilgrims in January 2007. It was the largest pilgrimage, and probably the largest gathering of any kind, in human history. The most enduring Hindu pilgrimage destination is the Ganges city of Varanasi (Benares) in the state of Uttar Pradesh. Many elderly people go to die in this city and be cremated where their ashes may be strewn in the holy waters. Indian authorities have passed laws, so far ineffective, to clean up the Ganges, polluted in part by incompletely cremated corpses; many of the faithful poor cannot afford to buy the fuel needed for thorough immolation. Scavenging water turtles released into the river at Varanasi help dispose of the cadavers.

Buddhism is based on the life and teachings of Siddhartha Gautama, known as the Buddha or the Enlightened One (•**Figure 7.18**). The Buddha was born a prince in about 563 B.C.E. in northern India.[6] Although presumably born a Hindu, he came to reject most of the major precepts of Hinduism, including caste restrictions. When he was about 29 years old, he renounced his earthly possessions and became an ascetic in search of peace and enlightenment. The Buddha eventually settled for a "middle path" between self-denial and indulgence, and he meditated through a series of higher states of consciousness until he attained enlightenment (*bodhi*). From that point, he traveled, preached, and organized his disciples into monastic communities. In his sermons, he described the "Four Noble Truths" revealed in his enlightenment: life is suffering; all suffering is caused by ignorance of the nature of reality; suffering can be ended by overcoming ignorance and attachment; and the path to the suppression of suffering is the Noble Eightfold Path, made up of right views, right intention, right speech, right action, right livelihood, right effort, right mindedness, and right contemplation.

An important Buddhist concept is the doctrine of **karma**, a person's acts and their consequences. A person's actions lead to rebirth, in which good deeds of the previous life are rewarded (for example, by which one could be reborn a human) and bad deeds punished (for example, by being reborn as a resident of hell). The goal of the practicing Buddhist is to attain **nirvana**, a transcendent state in which one is able to escape the cycle of birth and rebirth and all the suffering it brings.

Buddhism evolved into two separate branches: **Theravada** ("Way of the Elders") and **Mahayana** ("Great Vehicle"). (Mahayanists sometimes apply the term *Hinayana*, "Lesser Vehicle," in a derogatory fashion to the Theravada.) Theravada Buddhism is strongest in Sri Lanka, Myanmar, Laos, Cambodia, and Thailand. Theravada Buddhists claim to follow the true teachings and practices of the Buddha. Mahayana Buddhism originated in India and then diffused along the Silk Road to

Joe Hobbs

•**Figure 7.18** The Buddha has many manifestations. This temple of the Theravada faithful is in Chiang Mai, Thailand.

central Asia, Tibet, and China and eventually into Vietnam, Korea, Japan, and Taiwan. Mahayana Buddhists accept a wider variety of practices than those espoused by the Theravada, have a more mythological view of the Buddha, and are interested in broader philosophical issues.

Confucianism is not as much a belief system as it is a sociopolitical philosophy that serves on its own or is blended with other religions, particularly among ethnic Han Chinese. It is based on the writings of Confucius (Kung Fu-tzu, or "Master Kung," who lived from 551 to 479 B.C.E.), collected primarily in his *Analects*. Mencius (c. 371–288 B.C.E.) later became a major force in the widespread diffusion of this belief system throughout China. Following a period in which Communist authorities disavowed them, Confucian ideals are making a major comeback in China today.

The major tenets of Confucianism are embodied in a system of ethical precepts for the proper management of society, emphasizing honor of elders and other authorities, hierarchy, and education. Confucius never proclaimed his beliefs to constitute a religion; he was simply attempting to create a social contract between different classes central to Chinese government and society. He viewed his philosophy as secular, and its diffusion to Korea, Japan, and Vietnam has been part of the cultural baggage taken abroad by a steady stream of Chinese emigrants to those places.

Another important Chinese school of thought, second only to Confucianism and also widely blended with Buddhism among ethnic Han Chinese, is **Daoism** (sometimes spelled **Taoism**). Its philosophy comes from a body of work known as the *Daodejing (Tao-te Ching)*, or in English, *Classic of the Way and Its Power,* ascribed to Chinese Lao-tzu (Laozi), who lived in the sixth century B.C.E. Daoism encourages the individual to reject Confucian-style social conformity and seek to conform only to the underlying pattern of the universe, the "Way" (Dao, or Tao). To follow that way, the individual should "do nothing," meaning nothing unnatural or artificial. One can become rid of all doctrines and knowledge to achieve unity with the Dao and thereby gain a mystical power. With that power, the individual can transcend everything ordinary, even life and death. Daoists hold the simple earthly life of the farmer in high esteem.

There are about 25 million Christians in India, most of whom live in the south of the peninsula; Buddhists, Jains, Parsis, and members of a variety of tribal religions make up the numerous smaller remaining religious minorities. Most of the estimated 7 million Buddhists are recent converts from among India's lowest castes. Numbering perhaps 75,000 and concentrated mainly in Mumbai, the famously entrepreneurial **Parsis** have attained wealth and economic power far out of proportion to their number. Their religion is the pre-Islamic Persian faith of **Zoroastrianism**, a monotheistic faith known mainly for its reverence of fire. Zoroastrians follow the teachings of the prophet Zarathustra, who was born around 600 B.C.E. in Persia (modern Iran). He preached that people should have good thoughts, speak good words, do good deeds, and believe in a single god, Ahura Mazda ("Wise Lord"). Persia's Zarathustras, as the Zoroastrians are also known, fled Persia in two waves, first during the conquest by Alexander the Great around 300 B.C.E. and then after Islam swept into Persia in the seventh century. They settled in China, Russia, and Europe, where they became assimilated with other cultures. In India, they maintained a separate existence until quite recently; now about one in three takes a spouse of another religion and abandons Zoroastrianism. Many scholars consider theirs an endangered religion, with only 124,000 to 190,000 followers worldwide.

India's 6 million **Jains** also have influence beyond what their numbers suggest, as they control a significant share of India's business. They live mainly in the western state of Gujarat. **Jainism**, founded on the teachings of Vardhamana Mahavira (a contemporary of the Buddha, c. 540–468 B.C.E.), is renowned for its respect for geographic features and animal life. Jains believe that souls are in people, plants, animals, and nonliving natural entities such as rocks and rivers. Jainism has taken the principle of nonviolence (*ahimsa,* also present in Hinduism) to mean they should not even harm microbes. Therefore, Jain worshipers often wear masks to prevent inhalation of microscopic organisms, and Jains cannot be farmers because they would have to destroy plant life and living organisms in the soil.

Effects of European Colonization

As in much of the developing world, Western colonialism reshaped many traditional geographic patterns in South and East Asia (•Figure 7.19). By the end of the fifteenth century, Portugal and Spain had begun to extend economic and political control over some islands and mainland coastal areas of South and Southeast Asia. In the eighteenth and nineteenth centuries, the pace of colonization and economic control quickened, and large areas came under European domination. By the end of the nineteenth century, Great Britain was supreme in India, Burma (modern

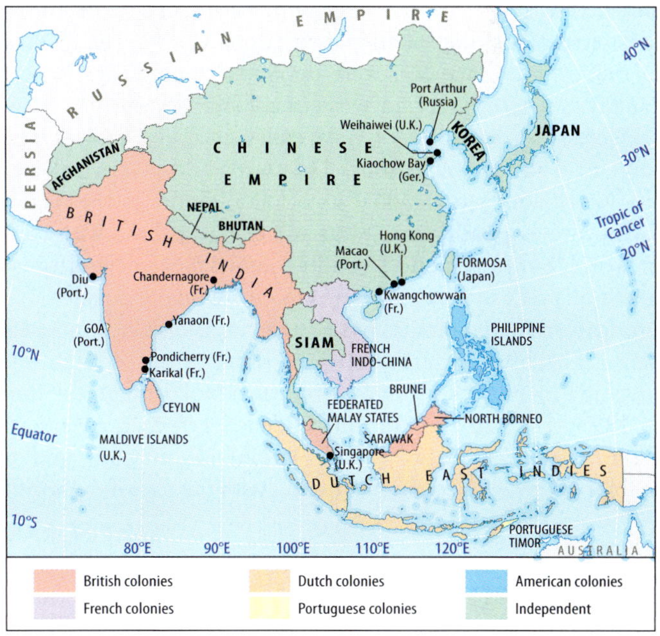

• **Figure 7.19** Colonial realms and independent countries in South and East Asia at the beginning of the twentieth century.

Source: Based on 'The World in 1900,' in R. R. Palmer, ed., Atlas of World History (Chicago: Rand McNally, 1957), pp. 169–171.

Myanmar), Ceylon (today's Sri Lanka), and Malaya and northern Borneo (now parts of Malaysia); the Netherlands possessed most of the East Indies (Indonesia); France had acquired Indochina (the region of modern Vietnam, Laos, and Cambodia) and small holdings on the coasts of India; and Portugal had Goa and Diu in India, Macau in China, and Timor (today's Timor-Leste) in the East Indies. Although retaining a semblance of territorial integrity, China was forced to yield possession of strategic Hong Kong to Britain in the 1842 Treaty of Nanking and to grant special trading concessions and extraterritorial rights to various European nations and the United States through the latter half of the nineteenth century. In 1898, the Philippines, held by Spain since the 1500s, came under the control of the United States.

A few Asian countries escaped domination by the Western powers during the colonial age. Thailand (historically known as Siam) formed a buffer between British and French colonial spheres in peninsular Southeast Asia. Japan withdrew into almost complete seclusion in the mid-seventeenth century but emerged in the late nineteenth century as the first modern, industrialized Asian nation and soon acquired a colonial empire of its own. Korea also followed a policy of isolation from foreign influences until 1876, when Japan began to colonize it. It is noteworthy that Japan, South Korea, and Thailand would emerge as very or relatively prosperous countries, whereas most of the former colonies lagged in economic development. Dependency theorists point to these discrepancies to affirm
44 their argument that colonization hampered development.

Asia was an extraordinarily profitable region for the colonizers. Western nations extracted vast quantities of tropical agricultural commodities and in turn found large markets for their manufactured goods. Westerners also invested heavily in plantations, factories, mines, transportation, communication, and electric power facilities. Some of the region's most important cities, including Shanghai in China, Kolkata (Calcutta) in India, and Singapore, were developed mainly by Western capital as seaports serving Western colonial enterprises.

Western domination of Asia ended in the twentieth century for a variety of reasons. The two world wars weakened the West's ability to conduct its colonial affairs in the region; Japan rose to great power status and challenged the West early in World War II; and effective anti-colonial movements arose in nearly all areas subject to European control. In the decades after World War II, all colonial possessions in Asia gained independence. The last to revert were Hong Kong (returned by Britain to China in 1997) and Macau (returned by Portugal to China in 1999). Until this era of independence, however, twentieth-century Asia was marked by revolution, war, turmoil, and an inability to gain an economic footing with the rest of the world.

7.4 Economic Geography

With some exceptions, this is a dynamic region with some of the world's strongest and fastest-growing economies. Their economic performance is often described as miraculous, given the region's tumultuous history. At the end of World War II, South and East Asia were poorer than Africa and Latin America. "The depleted states of East Asia, including Korea, Taiwan and Singapore, with few sources of money and practically no existing industry, seemed especially hopeless," writes journalist Michael Schuman in his book *The Miracle*. "Yet it was in these countries, supposedly the bottom of the international economic barrel, that the Miracle was born."[7]

If not literally a product of divine intervention, what explains Asia's extraordinary economic success? Many of the answers—government policies that nurtured the growth of industries, and government investments in industrial growth rather than in public amenities and services, for example—are explained in the following pages, in studies of the individual countries, including Japan and South Korea. It is also useful to look at the level of the individual and family to appreciate Asia's economic geography and economic success. To generalize, Asians are savers (whereas Americans are spenders!). These individual traits arguably played a significant role in the recent "Great Recession:" Americans went deeply into debt to finance consumption, while Asians stepped up their savings and exported more goods to Western consumers.[8] Savings are also important because most of the Asian countries provide few "safety nets" such as retirement pensions and health care. And most important of all, savings help to take care of the family, which is virtually sacred in many Asian cultures. It is always impressive to see how young people working even for low wages manage to save a significant share of those earnings to send home as remittances to Mom and Dad and Grandma. Needless to say, one of the great hopes in the West is that Asians will cut back on savings, loosening purse strings especially to buy Western exports—thus pulling Western economies and (the West argues) the world economic system out of the abyss.

The world's second and third largest economies are in East and South Asia: China and Japan, respectively. The recent global economic downturn slowed them down somewhat, but the economies of China and India have still been surging, with both goods and services inundating the global economic system. Their clout, their accomplishments, and their ambitions are enormous. Some economists and equities marketers have taken to conflating the two countries and calling them "Chindia," a double-barreled engine of economic growth.

It must be noted, however, that this growth is occurring against the backdrop of extraordinary poverty, and many hundreds of millions of people remain desperately poor. Although the region as a whole has taken great strides at reducing poverty rates, the gap between rich and poor is growing in many countries. This is particularly true in India, where the overall economy has grown about 8 percent annually since 2000; yet in that same period, there has been almost no reduction in childhood malnutrition, which afflicts almost half of India's young. Three-quarters of India's people live on less than $2 per day, which is the World Bank's benchmark for poverty. About 70 percent of Indians live in rural areas, but agriculture produces only about 20 percent of India's growth domestic product; no wonder then that rural people are poor, as they are a majority who live off a small portion piece of the economic pie. Poverty

tends to be deepest in the rural areas, fueling the classic push
56 factor of rural-to-urban migration. China overtook Japan to become the world's second largest economy in 2010 (and is projected to be the largest by 2020), yet on a per capita basis the country remains poor, with over a third of the people living on less than $2 per day. "For the first time, you have this odd combination," an economist in Hong Kong wrote as China reached this milestone. "One of the world's largest economies is also one of the world's poorest economies."[9]

The milestone of China becoming the world's second largest economy is one of the many that is described as a "China moment." With its galloping growth, China reaches one of these milestones after another. The world awaits the China moment of the country becoming the world's largest economy. There was a China moment recently when the country consumed so much coal that for the first time it began to import coal. There was the recent China moment of China becoming the world's second largest oil importer, after the United States. And so on.

Despite being generally a region of less developed countries, South and East Asia are home to the strong, industrialized, export-oriented economies of South Korea, Taiwan, Hong Kong, and Singapore, which late in the twentieth century came to be known as the **Asian Tigers** (or "Little Dragons"). Some economists consider the rapidly industrializing Southeast Asian countries of Thailand, Malaysia, Indonesia, the Philippines, and Vietnam the **New Asian Tigers**, or **Tiger Cubs**, following the path of rapid economic growth blazed by their neighbors. Indonesia's economy has been doing so well that some economists would like to include it as a second "I" in the exclusive "BRIC" club, which would then be Brazil,
133 Russia, India, Indonesia, and China, or "BRIIC."

Japan was the first Asian country to develop modern cities and modern types of manufacturing on a large scale. It has been a major industrial power for more than a century. China and India are much better supplied with mineral resources than is Japan. With the world's largest populations, they also have cheaper labor. But Japan's labor force is, at least for the time being, generally more skilled, and it leads Asia in high **value-added manufacturing**—the process of refining and fabricating more valuable goods from raw or semi-processed materials. Now, however, there are signs that Japan's preeminence in innovation is waning as China's intellectual resources grow. The biggest economic story in East and South Asia is the rise of China. Some highlights of that story are told elsewhere in Section 7.6.

China's Surging Economy

The statistics portray China as a steamroller. Among the China moments, China became the world's largest exporter, overtaking Germany in 2010. China overtook the United States as the world's largest energy consumer in 2010. China overtook Japan to become the second largest economy that same year.

China's economic growth soared at an average annual rate of 10 percent between 1990 and 2011. China's exports (along with imports) increased about 1,000 percent during those years, growing an average of 50 percent a year. China's share of world exports of goods grew 8 percent between 1981 and 2010 (compared with a decline of 3.5 percent in the United States during the same period).

Joe Hobbs

• **Figure 7.20** Mechanized silk production in China. Note the white cocoons in the vats below the worker's hands; the silk is spun from these.

Just a few years ago, China was notable for making low-value-added products—cheap toys, for example—that the richer countries really had no interest in making. There were shirts and shoes too, most of them manufactured in more than 20,000 factories in the Special Economic Zone (SEZ) of Shen-
zhen, adjoining Hong Kong. Now China is making at least a 267
little bit of almost everything, becoming the "workshop of the world" (•**Figure 7.20**). The country is making especially rapid progress in manufactures and exports of information technology (IT) hardware. Already China is the world's third largest manufacturer of personal computers, sold under such brand names as Dell and IBM. The roots of China's boom in IT may be traced to the dot-com bubble that burst in the United States in 2000. Many of the talented Chinese who worked in California's Silicon Valley at that time went back to China,
taking their skills with them in a kind of "reverse brain drain." 56

Advances in China's global status and in government economic policy have fueled the boom. China joined the World Trade Organization in 2001, prompting a surge in foreign investment. China sweetens the deal for investors with incentives like subsidized loans, tax exemptions, and 50 percent discounts on land prices. The addition of Chinese brawn and brains creates a nearly perfect investment climate. Chinese labor is much less expensive than is typical of most more developed countries, and China has a rapidly growing pool of university-trained engineers and other innovators.

China's Economic Impact

China's economic boom is having enormous consequences for other countries, particularly its Asian neighbors. Japanese electronics companies are cutting costs by moving increasing amounts of production to China. Surging investment in China is linked to disinvestment elsewhere, especially in Southeast Asia. That region's Tigers and Tiger Cubs—notably Malaysia, Singapore, and Thailand—dominated the 1990s with manufactures and exports of medium- and high-tech industrial products. With its growing consumer appetite, China was buying such goods from these countries. But recently the manufacturing shifted to China from those countries. South Korea's Samsung made China the main base for production of its flat screens and computers, for example. Japan's Toshiba makes televisions only in China. Dell recently moved some of its computer-making facilities from Malaysia to China.

With China's increasing gravity, Southeast Asian economies may at least temporarily be relegated to supplying low-end products, especially food and raw materials, to China. In turn, they may end up buying cheap Chinese manufactured goods, harming their own fledgling industries. They may try to counter this trend by developing higher-value niche products to meet demands in China. Thailand, for example, wants to boost manufactures of health care products, while Singapore focuses on biomedical products and financial services. Poorer Indonesia will probably be able to count on helping satisfy China's growing appetite for natural gas to fuel its southern industrial cities. Meanwhile, the large economies of Japan, Taiwan, and South Korea have lost ground to China in the market of foreign investment, as investors have been lured away from those countries by China's lower labor costs and its potential 1.3 billion consumers.

Cheap labor is unlikely to make China a permanent magnet for foreign investment, however. As affluence rises there, so will wages, and China's less prosperous neighbors will again attract investment. These figures are telling: in 2011, the average factory worker in Vietnam made $136 per month, in Indonesia $129 per month, and in China $413 per month. China is no longer cheap. China's one-child policy also means it will have fewer industrial workers in the future. Mindful of this demographic problem, farsighted multinational companies are boosting manufacturing investment in India, where the pool of young workers is likely to remain large and growing for decades to come.

In the long-dominant United States, there are rising fears of China's economic ascendance. Many Americans feel their country is becoming subservient to China because Chinese investments are funding U.S. federal deficits and because the United States is losing ground to China's industrial might. China's rise has been steadily chipping away at U.S. economic dominance in the region. In one country after another (including the most critical, Japan and South Korea), China has eclipsed the United States as Asia's most essential trading partner. The U.S.-China relationship is growing in China's favor. Of China's exports, about 18 percent go to the United States (its largest single trading partner), but its imports from the United States lag far behind. This U.S. trade deficit with China threatens to become a sore point in relations between the countries. Textile and other industries in the United States are pressuring the U.S. government to pass punitive tariffs on Chinese imports, a move that could precipitate a costly trade war. Problems with tainted food and toy imports 420
have already led to bans of some Chinese products in the United States.

The United States is also furious at China's reluctance to stop the counterfeiting or "piracy" of American products, particularly computer software and DVDs of Hollywood films. China is the epicenter of prolific Asian manufacture and trade in pirated products. In some countries like Malaysia, authorities try to crack down on sales of such goods on street corners, but the vendors are usually prepared to disappear at a moment's notice with their goods in tow, only to reappear moments later when police leave. In other countries like Vietnam, no one needs to run: there are countless permanent shops selling the latest counterfeited music, movies, watches, handbags, and more (**•Figure 7.21**).

Another great concern in the United States is how many jobs and profits are lost to outsourcing, especially to India (**see Insights, next page**).

Joe Hobbs

• Figure 7.21 Counterfeited goods made in China and sold throughout Asia bear the brand names of U.S. and other manufacturers but are sold to happy consumers at a fraction of the cost of the genuine product. This $4 digital audio player claims to be an Apple iPod, but you can be sure it is not.

INSIGHTS | Outsourcing

Outsourcing, also known as **offshore outsourcing** or **offshoring**, is the flight of technology and other jobs from countries with high manufacturing and service costs to countries with low manufacturing and service costs. India is the world's largest recipient of these jobs, and the United States is the leading outsourcer. India's outsourcing industry was a natural development after 1990, when the country abandoned its Soviet-style, centrally planned economic model of self-sufficiency and opened its doors to world markets and economic liberalization.

In the United States, there is growing worry and even alarm over the trend of outsourcing jobs to India. Many of these jobs are in computer programming and a wide variety of technical support, from computer problem phone calls to X-ray diagnosis. Many are mundane, like bill processing and order taking. Increasingly, however, high-paying white-collar employment is being outsourced to India, in such fields as investment banking, aircraft engineering, and pharmaceuticals research. Top Western corporations like the United States' Cisco Systems, the world's leading maker of communications equipment, are beginning to relocate substantial numbers of their senior executives to India, where much of the firm's work is actually done. The top destination is the south-central city of Bengaluru (Bangalore), which has come to be known as "India's Silicon Valley."

India, which has earned a reputation of being the West's "back office," is an especially attractive venue for information technology outsourcing for several reasons. It has a large population of well-educated people, most of whom speak excellent English. Those who work the telephone call centers typically receive training in vernacular American English and strive to perfect their American accents while shedding their Indian and British pronunciations. The bottom line is another major advantage: Indian skilled labor is cheap. The average computer programmer in the United States earns $75,000 per year, while his or her counterpart in India earns $9,000 per year.

Finally, the Internet and superb telephone communications, ironically combined with India's physical distance from the United States, provide some remarkable opportunities. While people in the United States sleep, the day shift is on in India. An American doctor can X-ray a patient in the afternoon and by early the next morning have the analysis—performed overnight in India. On the downside, when an American consumer places a technical service call for a computer problem, he or she may be speaking with a very tired Indian employee working the graveyard shift. There are many accounts of fatigue and burnout among Indian call center employees.

What about the loss of American jobs due to outsourcing? There is an expression that one's job may be "Bangalored." Some projections are alarming. A former official of the U.S. Federal Reserve described outsourcing as a "Third Industrial Revolution," predicting it would take jobs from as many as 42 million American workers in the coming years.[10]
Outsourcing companies generally argue that 420
reducing the costs of their services reduces inflationary costs for their consumers and improves efficiency and productivity. This rationale is, of course, cold comfort to the American worker whose job has moved offshore. Another trend that Americans may fear is that with its well-educated work force (including many graduates of U.S. universities), India is moving further up the value chain. There is new investment in research and development so that the country's information technology workforce will be innovating, and not just producing goods that Western thinkers invented.

There are surprising new twists in the economy of outsourcing. One is "reverse outsourcing," in which Indian and other companies hire American workers because their qualifications cannot be matched in India and other countries. Secondly, the pride in "made in America" products has grown great enough, and the costs of intellectual capital, wages, raw materials, real estate, and energy in India and China have risen high enough, that many American firms have begun to "onshore" work they had previously offshored. This countertrend, known as **insourcing**, onshoring, inshoring, or homeshoring, is especially popular in the United States at a time when jobs are scarce.

The Green Revolution

Most Asian countries do not want to improve their economies only by enhancing their outputs of services and manufactured goods. They also want to boost agricultural self-sufficiency and crop exports. Revolutionary changes in crops are at the heart of efforts to reshape Asia's agricultural economies. One of the most significant agricultural innovations in Asia during the past century is the development of new seed types and innovations in planting, cultivating, harvesting, and marketing of crops, efforts known collectively as the **Green Revolution**. These developments are not exclusive to South and East Asia, but much of the growth and investment have taken place here.

Ever since 1962, when the International Rice Research Institute (IRRI) was founded in the Philippines, there have been efforts to use science to increase food yields (particularly of rice) to stave off hunger and generate export income. Notable success has been achieved in breeding the new high-yield varieties of seed stock, and there has been a large upsurge of production in certain areas where the new strains have been widely introduced. Many of the new varieties are
bioengineered or genetically modified (GM) crops. This means 86
that precise manipulation of the genetic components of a crop can not only produce a higher yield but also a crop more resistant to drought, flood, or pests and which generates a higher amount of a desired nutritional component such as vitamin A. China, second only to the United States in the global biotechnology industry, has produced genetically engineered rice, corn, cotton, tobacco, sweet peppers, petunias, poplar trees, and more.

Because the underlying premise of the Green Revolution is economic—more crops will be produced at lower cost for higher revenue—most countries in the region are racing to

increase their biotechnological output, especially in an effort to catch up with China and ensure they do not lose positions in the global marketplace. Malaysia hopes to build Biovalley, a biotechnology research park near Kuala Lumpur, and Indonesia is planning a similar Bioisland. One of Biovalley's goals is to create palm oil trees genetically modified to produce the raw
253 material for specialized plastics used in medical devices. South Korea, Japan, and India have large and well-funded biotechnology research programs. Japan has so far resisted production of GM food crops out of fears of possible health hazards, a concern that most Europeans and their governments share.

For Asian farmers to capitalize fully on the Green Revolution, they must overcome many obstacles. Success requires levels of capital that are often beyond the means of peasant farmers, landlords, and governments. Such expenditures are needed for water supply facilities (for example, the tubewells that now number hundreds of thousands in the Indo-Gangetic Plain of the northern Indian subcontinent and that are severely depleting the groundwater), chemical fertilizers, and chemicals to control weeds, pests, and diseases. And as agriculture becomes more mechanized, considerable increases in the costs of machinery, fertilizers, and fuel have to be borne by farmers who have, in many cases, had little experience with the cash economy. Governments must improve transportation so that the large quantities of fertilizer required by the new seed varieties can be delivered in a timely fashion. Not only must there be these associated infrastructural changes to support this "revolution," but also the crop calendar of the farmer becomes much less forgiving because many of the newest seed grains demand more precise water, fertilizer, and cultivation requirements than traditional grains. Grain storage facilities, now subject to plundering by rats, must be improved.

There are other problems associated with the Green Revolution. Economic dislocations result when rice-importing countries become more self-sufficient, causing hardships for rice exporters. Large infusions of the agricultural chemicals associated with high-yield varieties have negative repercussions on natural ecosystems. India's government subsidizes urea fertilizer, and farmers are naturally tempted to overuse this cheap product. Unfortunately, the overuse of urea degrades the soil, and fertilized soil demands much more water. There is also concern that development of a limited number of high-yield crop varieties grown in vast monocultures will dramatically reduce the genetic variability of crops, in essence interfering
29 with nature's ability to adjust to environmental changes.

7.5 Geopolitical Issues

In the Middle East and North Africa, and in Russia and the Near Abroad, principal geopolitical concerns focus on the production and distribution of fossil fuel energy resources. In South and East Asia, by contrast, some of the most serious geopolitical issues are prospects for what may be done with weapons created from a particular energy source: nuclear energy. There are also concerns about seafloor resources (see Geography of Energy, next page), Islamist terrorism, and the security of shipping lanes. Of these, seabed resources may prove to be the most critical. There are several discussions in this chapter of seemingly insignificant small islands that are extremely critical because of the fossil fuels and other resources that may lie in nearby waters. 237,

The earlier discussion of the mushrooming economic clout 273 of China and India raises a big picture geopolitical reality: Asia is emerging as a center of gravity that will seriously challenge the century-long primacy of the United States in world affairs. Japan long stood alone as a pillar of strength in Asia, but with China and even India surging ahead, regional might has been reshuffled. Eyes in the region and around the world will be focused especially on China, which some analysts fear may use its fast-increasing military power to enhance its fast-increasing economic clout.

Nationalism and Nuclear Weapons

As we will see in Section 7.6, after its independence from Britain in 1947, India emerged as an avowedly secular democratic state. But after the 1998 victory of the Hindu nationalist Bharatiya Janata Party (BJP), hopes for new privileges emerged among India's Hindu majority. Among BJP supporters, there was hope for a "Hindu bomb," a counterweight to a long-feared "Islamic bomb" in neighboring Pakistan.

On May 11, 1998, much to the surprise of U.S. and other Western intelligence agencies, India conducted three underground nuclear tests in the Thar Desert. With the blasts, India's government seemed to be staking India's claim as a great world power, exhibit its military muscle to Pakistan and China, and increasing political support for the BJP. Initial reactions among India's vast populace were highly favorable; 91 percent of residents polled within three days of the event supported the tests. Outside India, there was alarm. India had defied an informal worldwide moratorium on nuclear testing that went into effect in 1996, when 149 nations (not including India and Pakistan) signed the **Comprehensive Test Ban Treaty** (also known as the **Nuclear Nonproliferation Treaty**, or **NPT**), which prohibits all nuclear tests.

The world's eyes turned quickly to India's neighbor to the west. Governments pleaded with Pakistan to refrain from answering India with nuclear tests of its own, arguing that Pakistan would have a public relations triumph it if exercised restraint: it would appear to be a mature and responsible power, whereas India would seem to be a dangerous rogue state. But Pakistan's leaders felt obliged by their population's demands for a tit-for-tat response to India's blasts, and soon followed with six nuclear tests. India and Pakistan thus joined just five other nations—the United States, Russia, China, the United Kingdom, and France—in acknowledging that they possess nuclear weapons. Israel neither confirms nor denies that it has nuclear weapons, but is thought to have about 250.

There is disagreement about what this recent regional and global shift in the balance of power means. Some analysts fear an escalating nuclear arms race that, perhaps ignited by a border skirmish in Kashmir or by a terrorist attack like the

GEOGRAPHY OF ENERGY | The Spratly Islands, Geopolitical Hot Spot in the South China Sea

About 60 islands make up the Spratly Island chain, which lies in the South China Sea between Vietnam and the Philippines (for its location, see Figure 7.37). They are an idyllic tourist destination where divers can hire luxury boats to explore the coral reefs and palm-lined beaches of remote atolls. However, the islands are much more significant for their strategic location between the Pacific and Indian Oceans. During World War II, Japan used the islands as a base for attacking the Philippines and Southeast Asia. Still more significant, as much as $1 trillion in oil and gas may lie beneath the seabed around the Spratlys.

Not surprisingly, many nations covet control of the Spratlys. Six countries claim some or all of the islands, and international law is not equipped to sort out these conflicting claims (see the UN Convention on the Law of the Sea, page 272). China, Vietnam, and Taiwan claim sovereignty over all of them, while Malaysia, Brunei, and the Philippines claim some of them. During the Cold War, the competing nations felt it was too hazardous to push their claims on the islands. As the Cold War drew to a close, however, the situation became more volatile. All the contenders except Brunei placed soldiers, airstrips, and ships on the islands. In 1988, the Chinese navy invaded seven of the islands occupied by Vietnam, killing about 70 Vietnamese soldiers. In 1992, China again landed troops in the islands and began exploring for oil in a section of the seabed claimed by Vietnam. In 1995, China moved to expand its territorial designs on islands already claimed by the Philippines and has since built what it calls "shelters," which look like fortifications to the Filipinos. Late in the 1990s, Malaysia occupied two disputed Spratly Islands and began building on one of them. In 2004, Vietnam began renovations of an airport on one of the islands, saying it was necessary to boost tourism there. China condemned the construction as a violation of its territorial sovereignty, an accusation it repeated as Vietnam negotiated natural gas and oil development with Indian and other foreign companies in the Spratlys. The situation moved to the brink of all-out conflict in 2011, when Chinese ships sabotaged Vietnamese oil exploration vessels operating within what Vietnam claimed was its own economic zone in the South China Sea (which Vietnam calls the East Sea—the same name the Koreans call the Sea of Japan).

As China's need for energy grows, so does its assertiveness over the Spratlys, the nearby Paracel Islands, and the Senkaku Islands, among others (see • Figure 7.K).

"China has indisputable sovereignty over the South China Sea islands and adjacent waters," a Chinese Foreign Ministry spokesman asserted in 2011.[11] Such a bold claim makes smaller powers in the region fear that China could turn small islands into "permanent aircraft carriers" that China could use to dominate them militarily. China even used a tiny submarine to plant a Chinese flag on the bed of the South China Sea. Not surprisingly, the United States took a position in this geopolitical chess match: in 2010, Washington declared that the South China Sea was of strategic importance to the United States. The United 248
States is solidifying its position to offset China: it has an alliance with the Philippines, and has drawn closer militarily to Vietnam. Clearly, an incident in these remote islands could trigger a 276
much wider and more serious conflict in Asia.

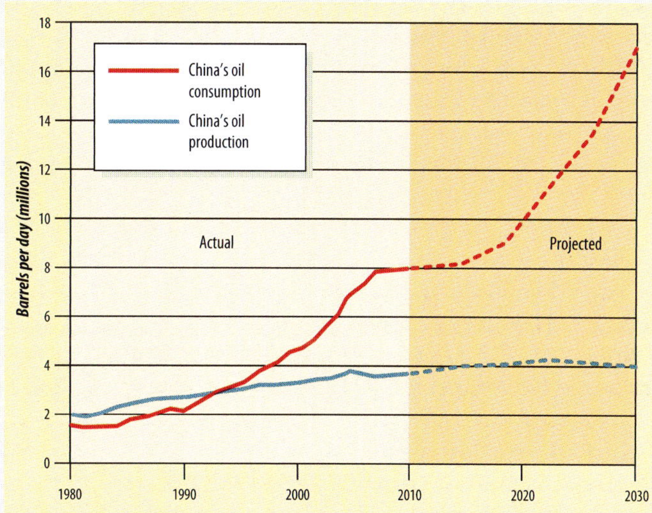

• Figure 7.K With China's oil appetite expected to grow enormously, every potential source of oil—like the Spratly Islands'—takes on unprecedented importance.

Source: Based on 'The World in 1900,' in R. R. Palmer, ed., Atlas of World History (Chicago: Rand McNally, 1957), pp. 169–171.

one in Mumbai in December 2008, could lead to the **mutually assured destruction (MAD)** of Pakistan and India. Others, particularly in South Asia, argue that the weapons represent the best deterrent against such a MAD conflict, as they did for decades between the countries of NATO and the Warsaw Pact. However, few people anywhere argue against the view that the nuclear rivalry between India and Pakistan has taken an enormous economic toll in two nations that need to wage war on poverty.

To the United States, both India and Pakistan are **pivotal countries**, defined by several influential historians as countries whose collapse would cause international migration, war, pollution, disease epidemics, or other international security problems.[12] The disposition of their nuclear arsenals is one of the main reasons for their being considered pivotal. If the government of Pakistan were to fall, Pakistan's nuclear weapons could come into the possession of a rogue element or a new government hostile to the West. One of those influential historians, Francis Fukuyama, declared that Pakistan is "the most dangerous place in the world."[13] India has much to fear from Pakistani scenarios. Either a right-wing or an Islamist government in Pakistan would be far more likely than the current regime to take India on in the disputed Kashmir region. Such a confrontation could set the stage for a nuclear exchange between Pakistan and India, not only decimating those countries but also sending economic shock waves around the world. India's recently surging economy has been a major force in India's recent diplomatic overtures to Pakistan; nothing discourages investment like war or the fear of it.

U.S.-Pakistan Relations since 9/11

Pakistan became even more pivotal with the events of 9/11. Up to that point, the United States had slapped tough economic sanctions against Pakistan (and less severe ones against India) for the nuclear weapons tests. The United States pursued a diplomatic courtship with India, both as a counterweight against China (India's longtime foe in the region) and as a means of expressing displeasure with Pakistan (China's ally) for helping Afghanistan's Taliban harbor Osama bin Laden. But the United States and Pakistan did an abrupt diplomatic about-face and embraced one another after 9/11. Pakistani President Pervez Musharraf instantly dropped support for the Taliban and quietly allowed the United States to use its territory to prepare for the assault on the Taliban and al-Qa'ida in Afghanistan. In return, the United States forgave much of Pakistan's debt to the United States and lifted its sanctions against Pakistan. To avoid isolating India, the United States also lifted its post-nuclear test sanctions against that country.

The United States has dramatically strengthened military ties with both countries, recognizing Pakistan in 2004 as a major non-NATO ally—one of just 14 countries to have that designation—and acknowledging India the same year as a "strategic partner." These designations opened the doors of both countries to major new shipments of American war material. But India would be favored: in 2007, Washington singled out India as its most critical strategic ally in the region by offering nuclear fuel and technology to New Delhi.

Under U.S. and allied attacks in Afghanistan after 9/11, Taliban and al-Qa'ida fighters retreated to and regrouped in western Pakistan, particularly in the semiautonomous federally administered tribal areas (FATA; see Figure 7.29). Here the populace is mainly Pashtun and is sympathetic to the causes of their Taliban ethnic kin and their al-Qa'ida spiritual kin (•**Figure 7.22**). This is an area where Pakistani government authority has long been kept at bay and where it was considered extremely difficult to insert Pakistani forces on a sustained basis. Bowing to American pressure, Pakistan's government periodically undertakes military operations in the FATA but considers a permanent or ongoing presence there as potentially detrimental to its survival.

The United States is deeply unpopular in Pakistan, so the sustained insertion of U.S. troops in Pakistan (in search of al-Qa'ida, for example) would be very destabilizing, perhaps even result in rebellion or coup against Pakistan's government. This could lead to the rise of an anti-American administration in Pakistan and raise questions about the disposition of Pakistan's nuclear weapons. To avoid that scenario, the U.S.

Joe Hobbs

• **Figure 7.22** The Pashtun of western Pakistan and Afghanistan are a tight-knit ethnic group with a long history of defiance against foreign rule. Most are strongly opposed to American interests in the region.

military has relied heavily on predator drones to stalk and kill al-Qa'ida and Taliban suspects in Pakistan.

Even those unmanned missions are politically risky for the United States, so imagine the risk involved in sending U.S. troops on the mission that killed Osama bin Laden, deep in Pakistani territory. The fact that bin Laden had lived not far from Pakistan's capital, and just around the corner from a Pakistani military academy, raised many questions on the quality and the loyalty of Pakistani intelligence forces. Many analysts and politicians in the United States asserted that elements within Pakistan's intelligence service, the ISI (Inter-Service Intelligence agency), had helped protect bin Laden. There are also suspicions that Pakistani intelligence and armed forces also continue to support the Taliban both in Afghanistan and Pakistan. The American military and intelligence communities often perceive Afghanistan and Pakistan as a single "theater" of operations, and had a neologism for it until the Pakistanis objected: *AfPak*.

What Does North Korea Want?

There are also major concerns about nuclear weapons in Northeast Asia. Ever since suffering Hiroshima and Nagasaki, Japan has had an official policy never to develop or use atomic weapons. But Japan worries about the potential nuclear threat from three adversaries, all of which possess nuclear weapons: Russia, China, and North Korea. The Japanese feel that the West dismissed potential threats from Russia too readily when the Soviet Union dissolved. Japan still has territorial disputes with Russia, particularly involving the four Kuril Islands of Kunashiri, Etorofu, Shikotan, and Habomai (lying just off the

northeastern coast of Hokkaido; see Figure 7.56), which the government of Josef Stalin seized at the end of World War II. To show that Russia was intent on maintaining its control, Russian President Medvedev visited one of the islands in 2010 and talked
147 about raising living standards there. The Japanese were furious.

Japan and China also have a territorial dispute concerning the East China Sea, and if, as expected, major oil reserves are discovered there, relations between the two countries could deteriorate. China continued to test nuclear weapons through 1996, adding to tensions in Japan. Finally, Japan fears reunification in Korea, which, as a former Japanese colony from 1910 to 1945, has a historical dislike of Japan. There is speculation that Japan may do the unthinkable and the officially disavowed—develop nuclear weapons—to counter the perceived threat of North Korea's nuclear weapons program.

The world looks nervously at the troubled relations between North and South Korea and between North Korea and the West. A crisis flared in 1994 when North Korea refused to permit full inspection of its nuclear facilities by the International Atomic Energy Agency (IAEA). The country was suspected of separating plutonium that could be used in making nuclear bombs. Over time, the United States, South Korea, and Japan worked out an agreement in which North Korea would agree to freeze development of nuclear weapons in exchange for the others providing fuel oil and assistance in building nuclear power plants. Those nuclear reactors would be of the light water variety, much less likely than North Korea's plutonium-based reactors to be "dual use" for both military and civilian purposes. The Clinton administration trumpeted the agreement as a success, but it had several flaws. One was that North Korea did not have to dispose of its existing nuclear fuel; it simply had to stow it safely away—meaning that it could quickly be reactivated.

President George W. Bush in 2002 proclaimed that North Korea was one of three countries making up an "axis of evil," along with Iraq and Iran. Late that year, North Korea dropped a virtual bombshell on the United States by admitting—after being confronted with evidence collected by U.S. intelligence agencies—that it did indeed have an active nuclear weapons program (developed, as it turns out, with Pakistani technical assistance). This admission followed earlier strident denials of such a program by North Korean officials.

North Korea's admission could have brought an end to the economic assistance in the form of food and fuel that the United States, South Korea, and Japan had been providing for a decade to the beleaguered country. What, then, was North Korea hoping to accomplish by revealing its nuclear weapons program? There are several possibilities. The least likely is that North Korea was instigating a military confrontation with the United States and its allies South Korea and Japan. In any war scenario, those allies would obliterate North Korea. The United States is obliged to defend South Korea in the event of a war with the North and would effectively wield its huge military advantage (•Figure 7.23). However, that victory would come at an enormous price for South Korea and possibly Japan. North Korea has a standing army of nearly 1 million soldiers, an impressive military arsenal including short- and medium-range missiles, a stockpile of chemical and biological weapons that it might not be shy to use, and presumably at least a few nuclear weapons that it could deploy. The casualties in South Korea would be enormous, whether from a conventional or an unconventional assault from the North.

Joe Hobbs

• **Figure 7.23** A little humor eases the tension for U.S. troops serving along what may be the world's tensest border, the "no-man's-land" separating North and South Korea. Even a minor incident here could touch off a conflagration that might take a huge toll in human lives.

It is much more likely that by disclosing its nuclear weapons program, North Korea sought guarantees that it could avoid war and also gain even more assistance from the West. For years, North Korea had only to mention the prospect of reactivating its nuclear weapons program to get more concessions from the United States and its allies—especially food aid during a succession of droughts and floods in the 1990s. In effect, North Korea was extorting money from the United States and its allies, which have been happy to ante up as a means of containing the rogue nation.

A series of on-again, off-again negotiations known as the Six Party Talks (held between North and South Korea, the United States, China, Japan, and Russia) culminated in 2007 with an agreement: The United States would help unfreeze North Korean funds in a Macau bank, remove North Korea from its list of countries supporting terrorism, lift trade sanctions in place since the Korean War ended in the 1950s, and along with its negotiating partners restart the flow of fuel oil to North Korea; in return, North Korea would close its nu-
274 clear plants and allow them to be inspected by international monitors. In the deal, North Korea did not promise to dismantle its nuclear facilities at Yongbyon, but only "abandon" or "disable" them, which it did do by the end of that year. North Korea is not required to destroy any existing stocks of nuclear weapons, and much uncertainty remains. The United States apparently hopes that the involvement of China and the other countries will make North Korea less likely to restart its nuclear program this time. North Korea watchers did not know how the policies of the country's new leader, Kim Jong-Un, might differ from those of his deceased father Kim Jong-Il.

Islands, Sea Lanes, and Islamists

As will be seen in Section 7.6, Indonesia—another pivotal country from the U.S. perspective—is an ethnically complex nation whose integrity has at times been threatened from a variety of secessionist movements. These disturbances range all the way from Aceh (pronounced "a-chay") in the country's far west to Papua (formerly Irian Jaya) in the east (see Figure 7.37). A possible **domino effect** scenario for Indonesia concerns foreign powers: If one province falls or proclaims its independence, many others will, and the largest country in the region will break apart. The United States and other countries fear what would happen to vital international shipping lanes should several new and perhaps militant states emerge from Indonesia's fragmentation. Washington is concerned that many states, some of them anti-Western, could emerge in Indonesia's place and threaten American interests in the country—notably its oil, natural gas, and copper resources and its location astride vital shipping lanes.

There have been scores of pirate attacks on vessels plying the strategic Strait of Malacca (see Figure 7.37), a critical chokepoint through which a quarter of the world's trade passes, including two-thirds of the world's shipment of liquefied natural gas and half of all sea shipments of oil (it ranks second only to the Strait of Hormuz as an oil shipping lane). Security in 177
the Strait of Malacca is a source of great concern to Japan, which imports 80 percent of its oil on ships that pass through the strait. Aceh's vast natural gas deposits and Papua's copper and gold are also seen as critical in the global economy, so like Indonesia, many world powers are anxious to see that they stay in the "right" hands. The United States gives generous amounts of foreign aid to Indonesia, but some Indonesians resent the United States, perceiving it as anti-Muslim in the wake of 9/11.

U.S. intelligence agencies have kept a watchful eye on Southeast Asia, and Indonesia in particular, fearful that it might emerge as an Islamist terrorist hearth. The country with the largest number of Muslims, Indonesia is a secular rather than a religious state, and a moderate form of Islam prevails in politics and in everyday life. What concerns Western intelligence authorities and Indonesia's moderate leadership are the more militant Islamist organizations that gathered strength after 9/11. One of these is **Laskar Jihad**, created in 2000 to mobilize Indonesian Muslims to fight Christians in Indonesia's Sulawesi and Muluku islands and Americans throughout the country. Another is **Jemaah Islamiah (JI)**, led by Abu Bakar Bashir, whose nominal role was preacher in an Islamic school. Western authorities are much concerned about such religious schools, known as *pesantrens* in Indonesia. Like the *madrasas* of Pakistan and the Arab heartland of the Middle East, these are venues for both religious and secular learning, and some of the schools' curricula are imbued with virulent anti-Western content. Baasyir admitted to having tutored 13 of the men arrested in connection with a plot to blow up Western targets in Singapore.

This concludes an introduction that has set the stage for further exploration of land and life in the subregions and countries of South and East Asia. The journey continues in the region's far west.

7.6 Regional Issues and Landscapes

South Asia

South Asia has an enormous range of ethnic groups, social hierarchies, languages, and religions. This is the most culturally complex area of its size on earth. Its cultures have roots that are deep in antiquity. Its modern engagements with the world are very important in the realms of security, economics, and population, among others.

Afghanistan: Graveyard of Empires

Like Sudan and Turkey, Afghanistan (•Figure 7.24) is a country that is transitional between regions; some geographers place it in Central Asia or the Middle East. It is a land of limited resources, poor internal transportation, and little foreign trade. Afghanistan is one of only five landlocked nations in South and East Asia. As we have seen, most landlocked countries are disadvantaged by their location and are poor. Afghanistan is 45

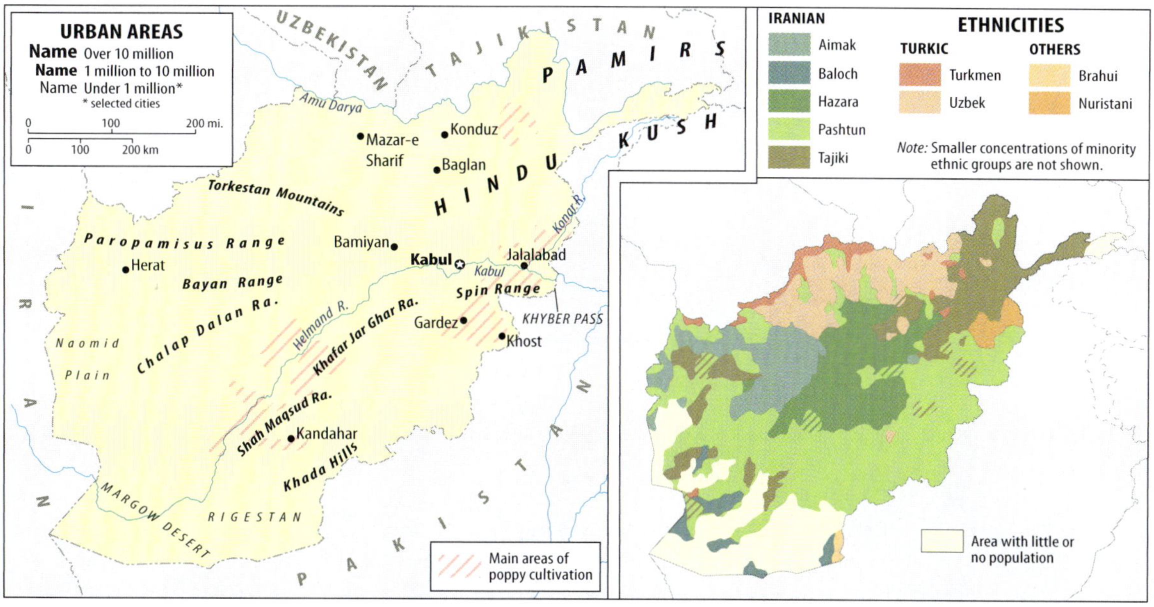

• **Figure 7.24** Principal features and ethnic groups of Afghanistan.

indeed one of the world's poorest countries and also ranks high as a failed state (seventh on that list).

Afghanistan is also a land of much conflict. It has repeatedly drawn the world's most powerful countries into conflict. Why? In large part because of its geographical situation: historically, it has occupied an important strategic location between India and the Middle East, and between India and Central Asia. Major caravan routes crossed Afghanistan, and a string of empire builders sought control of its passes. And although poor in natural resources, Afghanistan is neighbor to oil-rich Iran. This fact was critical to the Carter Doctrine discussed in Chap-
178 ter 6 and below.

High and rugged mountains dominate Afghanistan. Most of the country's 32 million people live in irrigated valleys around the fringes of the mountains. Northern Afghanistan borders three of the five central Asian countries, and millions of people on the Afghan side are related to peoples of those countries. The most populous area is the northeast, particularly the fertile valley of the Kabul River, where the capital and largest city, Kabul (population 2.3 million; elevation 6,200 ft [1,890 m]), is located. Most of the inhabitants of the southeast are Pashtuns (also known as Pushtuns or Pathans). Their language, Pashto, is
170 related to Persian. The Pashtuns are the largest and most influential of the numerous ethnic groups that make up the Afghan state; the next largest are the Tajik and Hazara, respectively.

The independent-minded tribal Pashtun people have always been loath to recognize the authority of central governments. In 1893, during the days of Britain's Indian Empire, British authorities and the Afghan King drew the "Durand Line" to separate their respective areas of influence. The line sliced right through traditional Pashtun lands, and Pashtun peoples have never accepted it. This border area saw warfare among Pashtun tribes, tribal raids on British-controlled areas, and British punitive expeditions against the tribes. In those days, the city of Peshawar—on the Indian (now Pakistani) side of the Khyber Pass into Afghanistan—became the hotbed for British and Russian agents playing what came to be known as the **Great Game** of vying (and spying) for strategic influence in Central Asia. The British were especially fearful that the Russians might use Afghanistan as a staging area to invade India, the "jewel in the crown" of Britain's colonies. To this day, Peshawar feels like a wild, lawless frontier city. It plays an important role in regional trade of weapons and drugs, especially heroin. For generations, the people of one of its outlying communities, Daraa, have specialized in replicating whatever foreign weapons a buyer desires, from old British muskets to modern anti-aircraft guns (•Figure 7.25).

Joe Hobbs

• **Figure 7.25** Pashtun *mujahadiin* in an Afghan refugee camp outside Peshawar, Pakistan. When I took this photo, these men were taking a break from fighting Soviet troops occupying Afghanistan. They and their sons could well have become Taliban fighting American troops in Afghanistan. Their bought their weapons in Daraa, the nearby factory and market for arms.

THOMAS J. ABERCROMBIE/National Geographic Stock

• Figure 7.26 The Bamiyan Valley and the Koh I Baba Range of north-central Afghanistan. This photo conveys some of the country's rugged beauty and pastoral character.

Afghanistan is overwhelmingly agricultural and pastoral. Its agriculture relies on traditional methods and produces low yields. The country is so mountainous and arid that only about 12 percent is cultivated (**•Figure 7.26**). The most successful crop has been the opium poppy, which routinely accounts for between one-third and two-thirds of the country's gross domestic product, and more than 90 percent of the world's opium (!). In the context of the global drugs trade, portions of Afghanistan, Pakistan, and Iran have come to be known as the **Golden**
257 **Crescent**. A succession of rebel armies and then Afghanistan's government under the Taliban used revenues from opium and heroin to obtain their arms. Most of the production takes place in the southern province of Helmand, where the Taliban have retrenched themselves most successfully. Principally to reduce funding to the Taliban, and also to crack down on illicit drugs, Britain and the United States have stepped up efforts to eradicate opium. A policy split has developed: the NATO mission in Afghanistan worries that in the absence of crop alternatives, elimination of the poppy crop could turn destitute farmers into vengeful fighters. One option rarely called for but perhaps most appropriate in Afghanistan's future would be to allow the country to produce opium legally for pharmaceutical needs, as a handful of countries, including Turkey and India, are allowed to do.

Through most of the twentieth century, this highland country was remote from the main currents of world affairs. After the Islamic revolution in Iran in 1979, however, Afghanistan's location next to that oil-rich country made it once again the
178 target of foreign interests. The Soviet military intervention of 1979 and the ensuing devastation catapulted the country into world prominence. The USSR's motives for its invasion of Afghanistan may have included a desire to prevent by force the spread of Iranian-style Islamic fundamentalism into Afghanistan (which is 80 percent Sunni and 19 percent Shiite) or into the central Asian Soviet republics deemed vital to the superpower's security. For its part, with the Carter Doctrine the United States warned the Soviet Union that it would not tolerate further Soviet expansionism.

The Soviet war in Afghanistan brought widespread killing and maiming of civilians, destruction of villages, burning of crops, killing of livestock, sabotage of irrigation systems, and sowing of land mines over vast areas. Soviet ground and air forces caused several million Afghan refugees to flee into neighboring Pakistan and Iran. Arms from various foreign sources filtered into the hands of the *mujahidiin*, the anti-Soviet rebel bands that kept up resistance in the face of heavy odds. The United States was one of the powers supporting the rebels and in that sense waged a proxy war against the Soviet Union in Afghanistan. In its waning years, the USSR recognized that it could not win its "Vietnam War," and its troops withdrew
from Afghanistan by 1989. It is estimated that of the 15.5 mil- 258
lion people who lived in Afghanistan when it was invaded in 1979, at least 1 million died, 2 million were displaced from their homes to other places within the country, and 6 million fled as refugees into Pakistan and Iran. About two-thirds of these refugees have returned to Afghanistan.

Support that the United States and moderate Arab states such as Egypt and Saudi Arabia lent to the *mujahidiin* and sympathetic Arab fighters soon came back to haunt those countries in a phenomenon that came to be known as **blowback**.

Emboldened by their victory against the Soviets, the most militant Islamists among the fighters turned their attention to the United States and its Middle Eastern allies. Chief among them was Osama bin Laden, who developed "the base" (*al-Qa'ida* in Arabic) of thousands of Afghans, Arabs, and other anti-Soviet war veterans he could call on to wage a wider *jihad*. Bin Laden's organization trained an estimated 10,000 fighters in al-Qa'ida camps in Afghanistan and from this unlikely, remote setting devised spectacular acts of terrorism: an assassination attempt on Egyptian President Hosni Mubarak, the bombing of U.S. military barracks in Saudi Arabia, the bombing of U.S. embassies in Kenya and Tanzania, the attack on the *U.S.S.*
184 *Cole*, and the 9/11 attacks.

The *mujahidiin* succeeded in overthrowing the Communist government of Afghanistan in 1992, but after that, rival factions among the formerly united rebels engaged in civil warfare. During this period, most other countries utterly neglected Afghanistan, inadvertently promoting the growth of militant movements (which tend to thrive in remote, underdeveloped, and ungoverned regions). By 1996, one of the rebel factions, the **Taliban** (backed by Saudi Arabia and Pakistan), gained control of most of the country, including the capital of Kabul. Proclaiming itself the sole legitimate government of Afghanistan, the Taliban imposed a strict code of Islamic law in the regions under its control and gained international notoriety for its austere administration. The Taliban removed almost all women from the country's workforce, forbade public education of girls, and outlawed "un-Islamic" practices such as dancing, flying kites, watching television, keeping birds, and trimming beards.

The Taliban continued to make advances against its opponents inside Afghanistan (particularly the **Northern Alliance**, whose leader, the Ahmed Shah Massoud—to his people a hero known as "the Lion of Panjshir"—was assassinated just days before and in apparent preparation for the 9/11 attacks) and, by 2001, controlled 95 percent of the country's territory. The neighboring central Asian countries, Russia, and even Iran grew increasingly fearful of the spread of the Taliban's extreme interpretation of Islam into their nations. Russia ended up supporting some of the rebels it had fought because they were now fighting the Taliban. Russia's once unlikely ally in supporting those rebels was the United States, which successfully lobbied the United Nations to levy economic sanctions against Afghanistan in 1999. The United States hoped that economic losses would pressure the Taliban into turning over bin Laden for prosecution, and it also announced a $5 million reward for information that would lead to bin Laden's capture or death.

The bin Laden bounty rose into the tens of millions of dollars after September 11, 2001. Named as the mastermind of the attacks against New York and Washington, bin Laden became the "world's most wanted" as U.S. President George W. Bush evoked a Wild West vow to have him captured "dead or alive." In crafting its pronounced war against terrorism (which it promised to carry anywhere in the world necessary), the U.S. administration identified the Taliban as al-Qa'ida's mentor and
184 targeted both organizations for elimination.

Within a month of the attacks on the United States, American warplanes and special forces struck in **Operation Enduring Freedom** against Taliban and al-Qa'ida facilities and personnel around Afghanistan. The British military assisted in the air campaign, and most of the ground fighting was carried on by the Northern Alliance, the Taliban's rival and nemesis. The United States successfully persuaded Pakistan to drop its support of the Taliban and join the effort, and much of the military campaign was based on U.S.-Pakistani cooperation. One after another of the Taliban's urban strongholds fell, including Kabul and the Taliban spiritual capital, Kandahar. Withering assaults from U.S. warplanes crushed the tunnels and other hideouts used by al-Qa'ida guerrillas, and many al-Qa'ida prisoners were taken to the U.S. naval base in Guantánamo Bay, Cuba. The elusive bin Laden, however, along with the Tal- 375
iban leader Mullah Omar, slipped away, presumably seeking safe harbor in Pakistan.

The reconstruction of Afghanistan began almost as soon as the Taliban were driven from power. Billions of dollars in aid went into the building of highways and other infrastructure, hospitals, and schools. The numbers of Afghans enrolled in primary schools rose from 25 percent in 1999 to over 90 percent, and girls and women are again being educated. Progress is slow because the country is so devastated, the security situation so tenuous, and the Afghan government so corrupt. Of all those billions of dollars in aid, some unknown amount has been siphoned off to line the pockets of those responsible for distributing it. U.S. Secretary of State Hillary Clinton said it was "heartbreaking" to see that billions of dollars of U.S. aid to the country had been wasted.

American forces and the Afghan government headed by Hamid Karzai effectively took control of only a small area around Kabul, and NATO troops were stationed in the troubled south of the country. Much of the country remains in the hands of warlords whose allegiance is up for grabs. Al-Qa'ida and Taliban fighters operate in the rugged countryside, sometimes carrying fighting into Kabul and other cities.

The mission of this war—the longest in U.S. history—changed substantially over the years. Initially, its purpose was clear: rid Afghanistan of al-Qa'ida and its Taliban supporters. With al-Qa'ida effectively displaced into Pakistan and elsewhere, the focus fell on the Taliban. American and NATO ground troops fought seemingly endless numbers of Taliban fighters, who fought with classic insurgent hit, run, and hide tactics. Western forces became engaged in a conflict that paralleled the long war of Soviets against the *mujahidiin* in Afghanistan. Although the Taliban and other Afghans, including civilians, bore heavy losses, the conflict also cost more than 1,800 American lives and over $400 billion (through 2011). Nearing the end of his first term in office, U.S. President Barack Obama steadily drew down the number of American forces in Afghanistan and announced that in 2014 the mission there would change from one of combat to one of support. U.S. military analysts came to the conclusion that the fight against the Taliban was unwinnable. For the losses of life and retreats suffered in earlier conflicts by the British and Russians

A closer LOOK

American Strategic Interests in Your Education

Arguably one of the most important answers to the question *Why has it been difficult for the United States to be involved in Afghanistan?* is something that cannot be seen in an atlas: the fact that the United States has been short on knowledge about the cultures, tribal allegiances, and other social complexities of Afghanistan, and in people fluent in some of the peoples' languages. The U.S. government is playing catch-up, in part by funding undergraduate and graduate-level college students to study the languages and become immersed for long periods in countries deemed critical to national security. See for example the Boren Scholarships, which "provide American undergraduate students with the resources and encouragement they need to acquire skills and experiences in areas of the world critical to the future security of our nation" (http://www.borenawards.org/). Boren Scholarships include language study of four of the five languages of Afghanistan depicted in Figure 7.10: Pashtu, Persian (Dari), Uzbek, and Turkmen. Scholarships like this can open many doors for you. They are also available for graduate study.

In the educational context, national security goes far beyond concerns with military thinking. Consider this definition of national security on the Boren Scholarship program overview: "It draws on a broad definition of national security, recognizing that the scope of national security has expanded to include not only the traditional concerns of protecting and promoting American well-being, but also the challenges of global society, including: sustainable development, environmental degradation, global disease and hunger, population growth and migration, and economic competitiveness." These concerns are virtually a definition of the field of geography. In a sense, your study of geography in this book is contributing to national security.

in Afghanistan, the country had come to be known as the "Graveyard of Empires." Historians began to add America to the list of empires bloodied by Afghanistan.

We can now check our answers to the geographic question that opened the book: *Suggest two or three reasons why it has been difficult for the United States to be involved in Afghanistan.* Even if you knew very little or nothing about Afghanistan, you might have looked at it in this book or in an atlas and proposed that it is mountainous, making supply logistics difficult, and that it is landlocked, also complicating those logistics. Now that you have read this section, you can supply many more details about its geopolitical and other contexts (see A Closer Look, this page).

Faith, Sectarianism, and Strife

The Indian subcontinent is one of the world's culture hearths, rich in a historical legacy and modern diversity of ethnicities, social practices, and faiths. Ironically, this cultural wealth often threatens the well-being and security of the subcontinent's peoples.

There are major social and political divisions within the subcontinent that relate to religion. Religious differences between the two largest religious groups—Hindus (whose religion is described on pages 228–230) and Muslims (whose religion is described in Chapter 6)—are often troublesome. Hinduism was the dominant religion at the time Islam made its appearance in this region. Muslims are significant minorities in the region's non-Muslim countries, including 13 percent of the population of India and 4 percent of Nepal. Muslims make up a seemingly small percentage of India's population, but they number over 160 million people, making India home to the world's third largest Muslim population, behind Indonesia and Pakistan. Muslims are majorities in Pakistan (97 percent) and Bangladesh (90 percent).

Seldom in history have two large groups with such differing beliefs lived in such close association with each other as Hindus and Muslims in the subcontinent. Islam holds to an uncompromising monotheism and prescribes uniformity in religious beliefs and practices. Hinduism is monotheistic for some believers but polytheistic for others, and asserts that a variety of religious observances is consistent with the differing natures and social roles of humans. The exuberant and loud celebrations of the Hindu faith are a striking contrast to the austere ceremonies of Islam. Islam has a mission to convert others to the true religion, whereas most Hindus regard proselytizing as essentially useless and wrong; one is born or reborn as a Hindu, so conversion is not an issue. Islam's concept of the essential equality of all believers is in total contrast to the inequalities of the caste system endorsed in Hinduism (see page 248). Islam's use of the cow for food is an anathema to Hinduism. And then there are religion-based politics, which often turn differences into discord and conflict.

During the colonial era, the colonizing power frequently favored one ethnic or socioeconomic group over another, sowing the seeds for eventual discord. This was the case in 339
British India, where the formerly subordinate Hindus came to dominate the civil service and most businesses. Many Muslims feared the results of being incorporated into a state with a Hindu majority, and their demands for political separation led in 1947 to the creation of two independent countries: the secular but predominantly Hindu nation of India (•Figure 7.27) and the Muslim nation of Pakistan (Figure 7.29). The respective secular and religious identities of both countries remain critical issues today. Pakistan had two parts, West Pakistan and East Pakistan, separated by 1,000 miles (c. 1,600 km) of Indian territory. In 1971, an Indian-supported revolt in East Pakistan led to the birth there of the new independent country of Bangladesh.

Immediately preceding and following partition in 1947, violence broke out between Hindus and Muslims, and hundreds of thousands of lives were lost in wholesale massacres. More than 15 million people migrated between the two countries. Pakistan today has a large population known as **Mohajirs**, or "migrants," the Muslim immigrants and their descendants who poured into the two Pakistans from India when partition occurred. Particularly in Karachi and elsewhere in the southern province of Sind, violent conflict frequently occurs between rival factions among the Mohajirs and between the Mohajirs and other ethnic groups, including Pashtuns (Pathans) and

• **Figure 7.27** Principal features of India.

Biharis. Large-scale migrations of ethnic groups like these are part of humanity's troubled experience.

The partition of colonial India precipitated all kinds of problems for the new countries. Some of them are still prominent today. The largest problem is the disputed Kashmir region. Since 1947, India and Pakistan have been in conflict over the status of Kashmir (called Jammu and Kashmir in India), a disputed province straddling their northern border (•**Figure 7.28**). Before independence, Kashmir was a princely state administered by a Hindu maharaja. About three-fourths of Kashmir's estimated population of 15 million is Muslim, which is the basis of Pakistan's claim to the territory. But under the partition arrangements, the ruler of each princely state was to have the right to join either India or Pakistan, as he preferred. Kashmir's Hindu ruler chose India, which is the legal basis of India's claim.

© Cengage Learning 2013

• **Figure 7.31** Principal features of Sri Lanka.

especially among the minority Tamils. Between 1983 and 2009, more than 70,000 people were killed and many more made homeless as the Tamils of the country's north have fought for autonomy or independence from Sri Lanka's majority Sinhalese government. Tamils complained that the Sinhalese treated them as second-class citizens and deprived them of many basic rights. Tamil fighters called the **Tamil Tigers**, officially the Liberation Tigers of Tamil Eelam (LTTE), wanted to establish their own homeland, Tamil Eelam, along the east, north, and west of the country. The rebels succeeded in holding wide swaths of northern Sri Lanka (see Figure 7.31). Unrest in rebel-held areas made it particularly difficult to rebuild northern sections of the area hit hard by the December
254 2004 tsunami. Government troops gradually pushed the Tamil forces back, and declared victory over them in 2009 (•**Figure 7.32**).

Joe Hobbs

• **Figure 7.32** The war between Tamils and Sinhalese has ended in Sri Lanka, but Tamil protests against Sinhalese rule continue where they can be practiced. This is a demonstration in Sydney, Australia. The depiction if of the Tamil Tiger founder and leader Velupillai Prabhakaran, who was killed just weeks before this demonstration. The red on the map on the right depicts parts of Sri Lanka held by the Tigers at their maximum advance. The Tamil flag, not fully in view, depicts a tiger jumping through a halo and crossed bayonets.

Arguably, one of the victors in the Sri Lankan civil war was China, which sided with the government and financed the construction of a port in the country's south. Some analysts argue that China will use this port to project its power into the Indian Ocean, where much of the world's oil and other valuable cargo is in transit. They see Sri Lanka as one pearl in China's **"String of Pearls"** strategy to gain critical allies in a line across the region. The largest "pearls" in this part of the string are China's allies Pakistan and Myanmar. India, which wants to see itself projecting power across these sea-lanes, views this as encirclement by China. Keep in mind that China and India are enemies and have been ever since they fought a high-altitude border war in 1962 (which India lost).

Known to early Arab seafarers as Seredib, the "island of Serendipity," Sri Lanka is a breathtakingly beautiful country. The return of peace to Sri Lanka should reopen the doors to international tourism.

The Caste System

One ancient Hindu belief is that every individual is born into a particular **caste**, a social subgroup that determines the individual's rank and role in society. Castes form a hierarchy, with the Brahmin caste at the top (comprising about 5 percent of all Hindus) and three others (Kshatriya, Vaisya, and Sudra) below. Caste membership is inherited and cannot be changed. Particular castes are associated with certain religious obligations, and their members are expected to follow traditional caste occupations. Marriage outside the caste is generally forbidden, and meals are usually taken only with fellow caste members.

At the bottom of the social ladder are the **Dalits**, or *scheduled castes*, once known as **untouchables**, accounting for about 20 percent of all Hindus. They are not part of the caste system but are "outcastes," and according to Hindu belief, they are not reincarnated. The untouchables were so called because they traditionally performed the worst jobs, such as the handling of corpses and garbage; therefore, their touch would defile caste Hindus. I will never forget the horrified look on my taxi driver's face when I shook hands with a man whose job was to cremate bodies on the "ghats" of Mumbai.

This ancient, rigid social system is changing. India's constitution outlawed the caste system in 1950, but a single decree could not undo thousands of years of tradition. Brahmin privileges are now being increasingly challenged and restricted by more laws. Indian law explicitly forbids recognition of Dalits as a separate social group. Confronting the fact that upper castes account for less than one-fifth of India's population but command more than half of the best government jobs, the Indian government has instituted an affirmative action program to allocate more jobs to members of lower castes. Indian leaders of high caste have championed the Dalits' cause for both moral and practical reasons. Politicians use caste divisions to their advantage to bring out voters, promising if elected to act in the interests of their respective large social blocs. In 1997, a Dalit was elected president of India for the first time in the country's history. Since then, Dalits have won more regional and local offices.

The Dalits are voting in greater numbers than ever before, with the conviction they can use India's democratic system to overcome ancient discrimination against them. They are making progress. Disintegration of the caste system has been especially rapid in the cities, where people of different castes must mingle in factories, in public eating places, and on public transportation. In India's vast rural areas, the caste system is more entrenched. There is also a north–south difference: lower caste southern Indians began agitating against the caste system early in the twentieth century and kept up their struggle. In northern India, by contrast, there was less effort to tackle the caste system. Caste is so important in northern politics that it is said that voters don't cast their vote; they vote their caste. Remarkably, partly because of the loosening of caste restraints, economic growth is much stronger in southern than in northern India.

Keeping Malthus at Bay

Poverty and human health are serious problems in South Asia, and the prospect of continued high population growth raises the question of whether growth in food supplies can avert the proverbial Malthusian crisis (see Chapter 3).

India's population has surged since independence, from 352 million in 1947 to 1.2 billion in 2011. In recent decades, the overall trend has been toward lower population growth rates. Fifty years ago, the average Indian woman had six children; now she has three. That rate is still well above the natural replacement rate, however, that would keep the population steady. The population base is already so vast, and the base of its age structure diagram so wide (about 50 percent of the population is younger than 25), that even modest growth will
53 add huge numbers. India is predicted to overtake China as the world's most populous country by 2040. There is uncertainty about how high India's population will grow. Estimates range from 1.5 to 1.9 billion. Most Indian decision makers think that even the low end of that range is too high. Unlike China, India is a democracy and cannot simply declare how many children a couple can have. Indian states rather than the central government establish laws and policies, and there is a wide range of them in the realm of population. Some states offer financial incentives, such as a $100 bonus if couples delay their first child's birth.

For many people, the name *India* invokes an image of grinding poverty, and with good reason: it is estimated that fully 40 percent of the population is "abjectly poor," defined as living on less than $1.25 per day, and almost half of India's children are malnourished. But perhaps in defiance of Malthus, an increased population has so far succeeded in producing more economic resources overall. A few people in India are very wealthy, and there is an emerging middle class of as many as 300 million people—the world's largest middle class (•**Figure 7.33**). India's economy has grown impressively in recent years. An apparent problem is that the benefits of the growing economy are unevenly distributed. Most of the economic growth benefits those who are already better off, widening the gap between rich and poor.

The issue of population growth is also critical for Pakistan, which has never mounted an effective family planning program, mainly because of opposition from Muslim religious authorities, who regard birth control as an intervention against God's will

Joe Hobbs

• **Figure 7.33** This Indian family visiting the Red Fort in Agra is among the growing ranks of the subcontinent's middle class. Note the Taj Mahal gleaming in the distance.

(Pakistan's birth rate in 2011 was 25 per 1,000, compared with 21 per 1,000 in India). Muslim Bangladesh, along with Nepal and Bhutan, also have high birth rates, but they are much lower in Bangladesh than they once were. In the 1970s, the average woman had seven children, whereas now she has two, and contraception use nationwide has increased from 4 percent then to more than 50 percent now. South Asia's success story in bringing population growth under control is Sri Lanka, with a birth rate of just 17 per 1,000 and an annual population growth of 1.3 percent.

Agricultural output in South Asia has increased since independence. Most notably, despite its huge and growing population, India had achieved self-sufficiency in grain production, with a fourfold increase in output since 1950. Almost half a million "ration shops" sell subsidized food staples to the country's poorest people. The successes of agriculture in the subcontinent have been due mainly to the increased use of artificial fertilizers, the introduction of new high-yield varieties of wheat and rice associated with the Green Revolution, more labor provided by the growing rural population, the spread of education, the development of government extension institutions to aid farmers, and better irrigation. "Mother Nature," however, has the last word. Sixty percent of India's farmland is rain-fed rather than irrigated, and when summer monsoons fail to deliver rainfall, as they did in 2009, the yields of valuable rice, sugarcane, and other crops plummet. This causes immediate difficulties for people (in the form of higher food prices) and the nation (in terms of trade and economic growth).

Many experts in agriculture and economics are calling for a "Second Green Revolution" in India to address agriculture in the context of climate change. The focus would be more nuanced than the Green Revolution of the 1960s. Now there is a need for nationwide cooperation to improve irrigation, capture rainwater and conserve groundwater while boosting crop yields. There is much at stake, not just food security but human well-being and equity. During droughts like the one in 2009, debt-ridden farmers resort to the dire options of selling their
wives or daughters (for $83 to $248) or committing suicide. 145

As the sale of wives and daughters suggests, one of the challenges confronting both rural and urban India is to improve the

status of women. Despite a 1961 ban on dowries (money and gifts given by a bride's parents to the groom), the practice is continuing. So is the rate of killing brides who do not provide enough dowry. The burden placed on the bride's family is one reason that many prospective parents choose to abort female fetuses, which
213 can be detected with ultrasound technology. India banned this use of ultrasound in 1996, but the practice continues. Female feticide—particularly prevalent among India's more educated and affluent, who can afford the ultrasound screening—and infanticide have taken tens of millions of young and unborn girls. In some of India's states, especially Haryana, Punjab, Delhi, and Gujarat, one result is a serious shortage of local brides; for every 1,000 men, there are 900 women. The gap is being filled by marriages
276 arranged with women from distant Indian states and Bangladesh.

Low-Lying Bangladesh and Maldives: Canaries in the Climate Change Coal Mine?

Formerly East Pakistan, Bangladesh is about the size of the state of Iowa or the nation of Greece—not an important fact until one considers that it is home to 150 million people (•**Figure 7.34**). With its limited resources, people-overpopulated Bangladesh is South Asia's third-poorest country on a per capita GNI PPP basis, after mountainous Bhutan and Nepal.

The Bangladesh portion of the Ganges-Brahmaputra Delta ranks, along with the Indonesian island of Java, as one of the two most crowded agricultural areas on earth. This area is subject to catastrophic flooding. Massive tropical cyclones (hurricanes) ravage Bangladesh regularly in September and October. Deforestation upstream on the steep slopes of the Himalayas has increased water runoff and sediment load in the Ganges, also contributing to Bangladesh's flood problems (the region's principal rivers may be seen in Figure 7.34).

Ironically, although there are concerns about climate change–related flooding in Bangladesh, there are even more worries about climate change–related water shortages and droughts. Climate change models point toward accelerating depletion of the

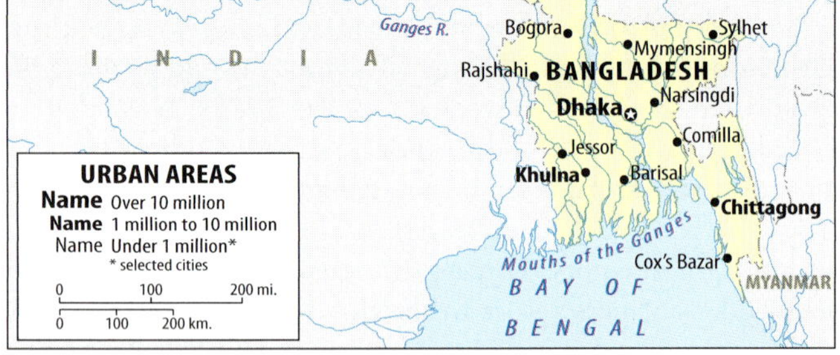

• **Figure 7.34** Principal features of Bangladesh, Nepal, and Bhutan.

• **Figure 7.35** Melting of glacial ice in the Himalaya and other great mountain ranges of South Asia threatens future water supplies for people downstream. This is Gokyo Glacier, close to Nepal's Mt. Everest.

so-called **"Water Towers of Asia"**—the glaciers and snow cover of the Himalaya, Karakoram, and other great mountain ranges that feed the region's 10 great rivers and scores of smaller ones. The perennial ice mass here is exceeded in volume only by the ice of the Arctic and Antarctic, earning the Tibetan Plateau and environs the name "The Third Pole." The region's roughly 18,000 glaciers, like most others in the world, are melting at a steady, relentless pace (•**Figure 7.35**). With temperatures on the Tibetan plateau rising much faster than the world average, a respected Chinese glaciologist predicts that two-thirds of them will be gone by 2050. 76, 358, 428
These glaciers supply water to about 40 percent of the world's population! The prospect of their melting away is so alarming that longstanding enemies India and China are putting away their differences and collaborating in scientific research on climate change.

Bangladesh and adjacent portions of India must also keep a wary eye on the potential for the sea level to rise if the world's temperatures increase, as predicted by many climate change models. The country's environment minister has forecast that 20 per- 33
cent of Bangladesh will be underwater by 2020 if nothing is done to control global warming. The Intergovernmental Panel on Climate Change recognizes this as one the world's most endangered areas, predicting a rise in sea level there of 23 inches (c. 58 cm) by 2100. On the Indian side of the delta, scientists are documenting the shrinkage of the river delta islands known as the Sundarbans. Renowned as habitat for the Bengal tiger, the Sundarbans are succumbing to a combination of sea level rise and deforestation that makes them very prone to flooding and erosion.

The roughly 1,100 islands that make up the Maldives (population 300,000; capital, Male, population 90,000), an independent country since 1965, appear to be the very essence of tropical paradise (•**Figure 7.36**). So prized are its palm-blessed and coral-fringed beaches that more than half a million tourists, mainly Europeans, visit the country each year. More than 60 percent of the country's foreign-currency earnings come from tourism, with fishing and clothing providing most of the rest. Some sources list the Maldives as South Asia's wealthiest country on a per capita basis. The greatest wealth is in the

Peter Essick/Getty Images

• **Figure 7.36** This aerial view of Male (pronounced *mah*-lay), the Maldives' capital, reveals how vulnerable the site is to storms and sea level rise. This is the world's most densely populated city.

tourist sector, where it is concentrated in so few hands that a potentially dangerous rich–poor divide has developed.

There may be other troubles in paradise. More than 80 percent of this country's very limited land area consists of limestone atolls less than 3 feet (90 cm) above the sea. If sea levels were to rise, as predicted by many common climate change scenarios, the entire country could rapidly be submerged. This prospect led to the Maldives' president's famous cry at a climate change conference: "We are an endangered nation!"

Southeast Asia

Deforestation of Southeast Asia

Southeast Asia today is composed politically of 11 countries: Myanmar (formerly Burma), Thailand, Laos, Cambodia, Vietnam, Malaysia, Singapore, Indonesia, Timor-Leste (formerly East Timor), Brunei, and the Philippines (**•Figures 7.37 and 7.38**). With the exception of Thailand (which was never colonized) and

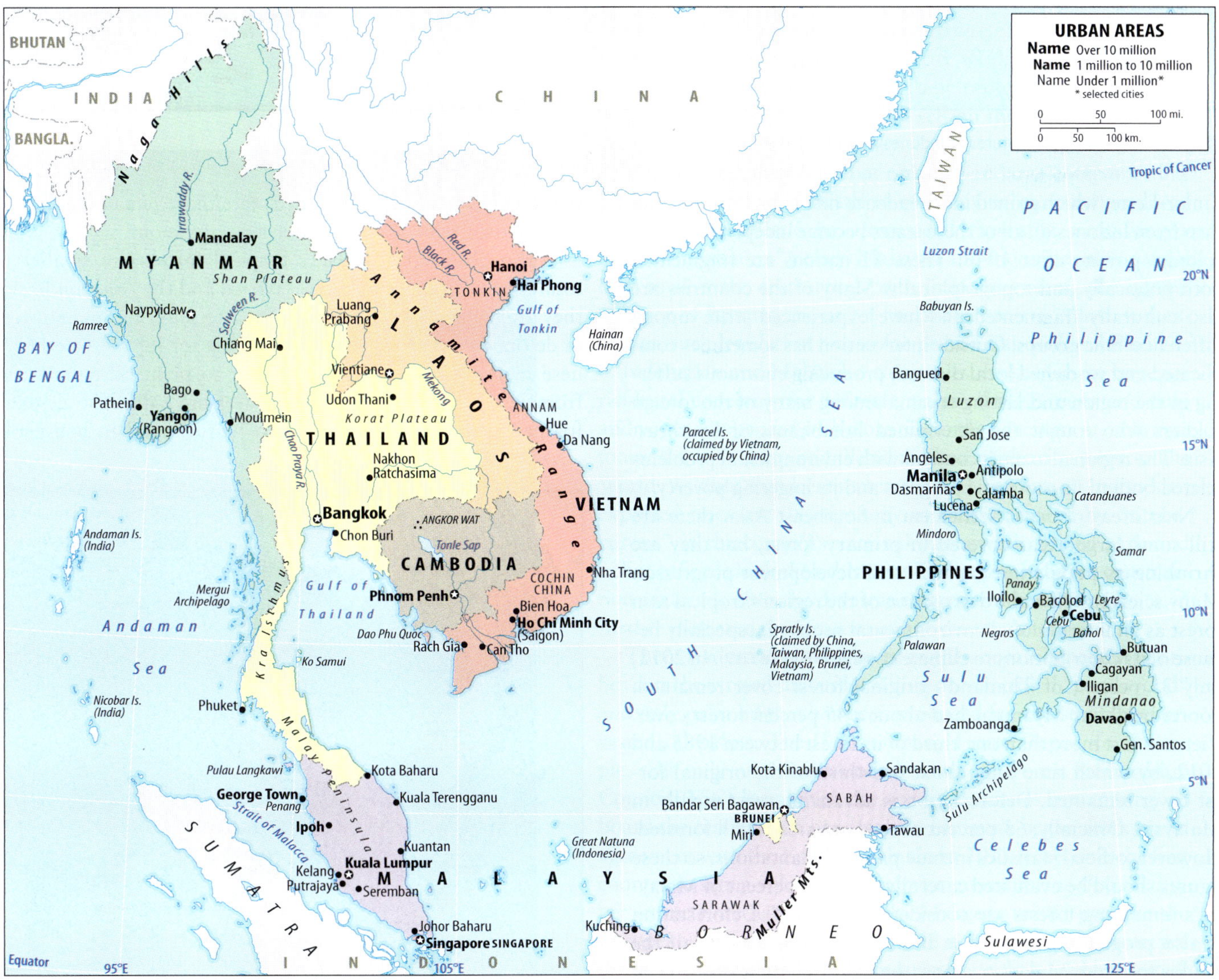

• **Figure 7.37** Principal features of Myanmar, Thailand, Cambodia, Laos, Vietnam, Malaysia, Brunei, Singapore, and the Philippines.

drought attributed to the warming of Pacific Ocean waters by the phenomenon known as El Niño turned the annual July–
358 October burning season into a conflagration of human origin. Deliberately set fires raced out of control over large areas of Sumatra and Borneo, resulting in a choking haze that shut down airports, closed schools, deterred tourists, and caused respiratory distress to millions of people in Indonesia, Malaysia, Singapore, the Philippines, Brunei, and Thailand. Less widespread but still severe smog crises struck Malaysia and Indonesia, particularly the island of Sumatra, in subsequent years, and Indonesian authorities have fought back by revoking the licenses of plantation companies.

Many of the plants and animals in these forests are **endemic**
344 **species** (found nowhere else on earth). Indonesia, which contains 10 percent of the world's tropical rain forests, is known, after Brazil, as the world's second most important **megadiversity country**, with about 11 percent of all the world's plant species, 12 percent of all mammal species, and 17 percent of
377 all bird species within its borders. The Southeast Asian region is also particularly significant in biogeographic terms because of the so-called **Wallace's Line**—named for its discoverer, the English naturalist Alfred Russel Wallace—that divides it. Nowhere else on earth is there such a striking local change in the composition of plant and animal species in such a small area on either side of this divide. East of the line (separating the Indonesian islands of Bali and Lombok, for example), marsupials are the predominant mammals, and placental mammals prevail west of the line. Similarly, bird populations are remarkably different on either side of the line.

The Great Tsunami of 2004

On December 26, 2004, the most cataclysmic natural event of modern times afflicted a vast portion of the earth. Beneath the Indian Ocean, just off the northwestern coast of the Indonesian island of Sumatra, a gargantuan "megathrust" earthquake measuring magnitude 9.3—with the force of more than 26 megatons of TNT, or 1500 times the power of the nuclear weapon that devastated Hiroshima—struck at 6:58 A.M. local time. In the Java (Sunda) Trench, as much as 750 miles (1,200 km) of fault line slipped 60 feet (over 18 m) along the subduction zone where the Indian Plate is sliding beneath the Burma Plate (part of the greater Eurasian Plate; see Figure 2.1).

As the seabed of the Burma Plate instantly rose several meters vertically above the Indian Plate, the massive displacement of seawater created a series of tsunamis (*tsunami* is Japanese for "harbor wave"; the phenomenon is sometimes popularly but incorrectly referred to as a "tidal wave") that pulsed across the Andaman Sea and the Indian Ocean at up to 500 miles (800 km) per hour. On the open ocean, such waves are barely perceptible, but they break on coastlines with the force of giant storm surges. The tsunami roared ashore, at heights up to 45 feet (14 m), in 14 countries: Indonesia, Malaysia, Thailand, Myanmar, Bangladesh, India, Sri Lanka, the Maldives, Seychelles, Somalia, Kenya, Tanzania, Yemen, and South Africa (•Figure 7.41).

These were the deadliest tsunamis in recorded history. The total number of dead exceeded 200,000. By far the greatest number of deaths (over 130,000) was in Indonesia, where Aceh's principal city of Banda Aceh was obliterated (•Figure 7.42). Sri Lanka lost 35,000; India, 12,000; and Thailand, over 5,000. Almost all the deaths occurred within a mile of the shoreline and were caused by blunt force from debris and by drowning. An estimated one-third of the fatalities were children. There were reports of children running out to investigate fish stranded by retreating seas—a characteristic precursor to a tsunami—only to be struck seconds later by the wall of water.

As many as 2 million people were made homeless by the disaster. An unprecedented international relief effort, designed to feed and house these refugees and to contain the spread of epidemic diseases, began within hours. In many cases, it was clear that recovery efforts were profoundly successful, sometimes in unexpected ways. In India's Tamil Nadu State, for

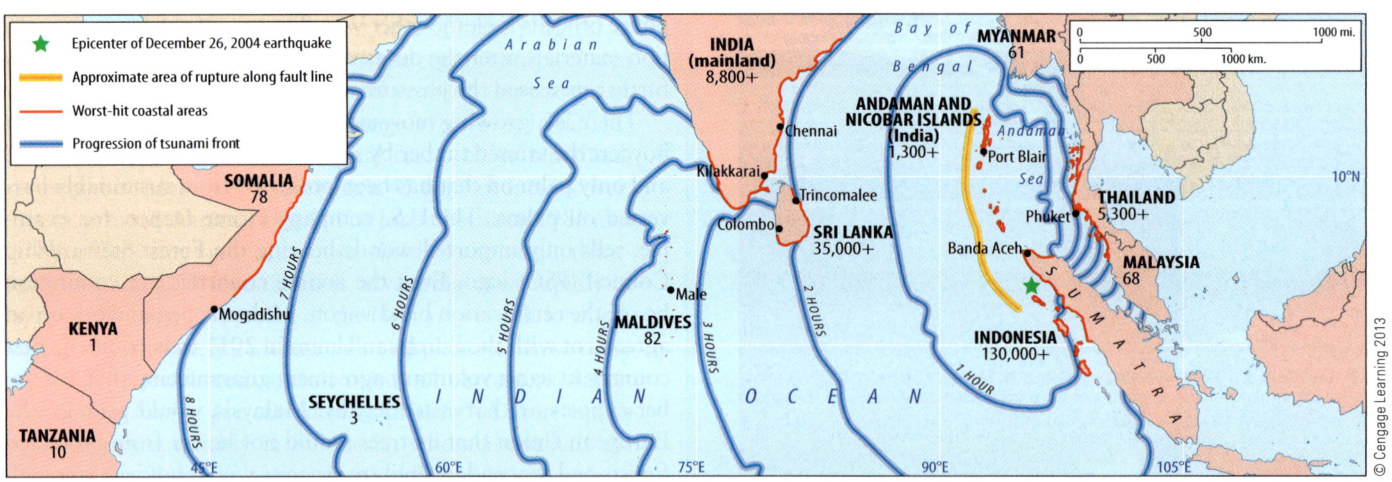

• Figure 7.41 The epicenter of the 2004 earthquake, which registered magnitude 9.3, was located off the coast of Sumatra, Indonesia. It created a tsunami of tremendous force and scope.

• **Figure 7.42** Banda Aceh before and after the tsunami.

example, there were new roads, housing and utilities, and jobs training. New homes were titled in women's names (because women were deemed more trustworthy than men), giving women an unprecedented rise in status as concerns respect. In the tsunami's wake, the social fabric was transformed.

Tsunami experts concur that countless thousands of lives would have been saved had the affected Indian Ocean region been equipped with the same kind of tsunami early warning system now in place in the Pacific Ocean.[16] That investment had not been made because historically about 90 percent of the world's tsunamis have occurred in the Pacific and because many of the nations afflicted lacked the financial resources for the wave sensors and other infrastructure required by the system. Installing an early warning system in the Indian Ocean basin became an urgent priority, and a system was up and running by 2006. Had this system been operational in 2004, Indonesia would have had just a few minutes' warning, but Thailand would have had two hours, and Somalia, seven hours to prepare for the disaster.

Japan's tragic tsunami in 2011 is discussed on page 16. Its mechanism was similar: an upthrust earthquake along a section of underwater plates. Its magnitude was slightly less (9.0 versus 9.3), and the loss of life in wealthy, tsunami-prepared Japan far less (15,000 versus 200,000). It was such a massive tsunami, however, that Japan's engineering works were simply overwhelmed.

Misrule in Myanmar

Myanmar (population 54 million, Figure 7.37) is centered in the basin of the Irrawaddy River and includes surrounding uplands and mountains.

Britain conquered what was then called Burma in three wars between 1824 and 1885. It became an independent republic outside the British Commonwealth in 1948. Civil war has been almost constant since independence in a country where ethnic minorities make up about 40 percent of the population. Communism, targeted persecution, and ethnic separatism have motivated rebellions by ethnic Karens and Shans, a number of less numerous hill peoples, and Muslims of the Rohingya ethnic group who fear the establishment of a pure Buddhist state. Beginning around 1999, however, Myanmar's ruling military government reached cease-fire agreements with most of these ethnic groups (except the Karens) and gave these groups what it called **contingent sovereignty**, offering them more civil rights and economic opportunities. The government also encouraged profits made by some of these groups in the drug business to be "invested" in national development.

The economy has been badly damaged by fighting and for years was mismanaged under a form of rigid state control promoted as the **"Burmese Way to Socialism."** Myanmar became the poorest non-Communist country in Southeast Asia, a sad contrast with its reputation in early independence years for having the best health care, highest literacy rate, and most efficient civil service in Southeast Asia. It used to be the world's largest rice exporter; it now exports less than one percent of the world total. Myanmar's government spends comparatively less on health care for its citizens than all but three countries in the world.

In the 1990s, Myanmar began to abandon socialism in favor of a free-market economy and enjoyed considerable economic growth, with China its main trading partner and military ally. Most of the fruits of this economic growth, however, lined the pockets of the ruling military. The **State Law and Order Restoration Council (SLORC)**, the military government that seized power in Burma in 1988 (and changed the country's name to Myanmar), has yielded little to popular pressure for democratization. Myanmar remains one of the world's most repressive places to live. Access to the Internet was prohibited until 1999 and is still strictly regulated. Foreign journalists are banned, and citizens may not allow foreigners into their homes. It is illegal to gather outside in groups of more than five. Giant green billboards across the country read: "Crush all internal and external destructive elements as the common enemy."

In an apparent move to gain strategic depth from its own population and from any outside threats, the government in 2005 established a new capital at Naypyidaw, in the country's remote central region. The move did nothing to quell popular discontent with the regime. Widespread anti-government protests, with broad participation of Buddhist monks, broke out all over the country late in 2007. Authorities responded with deadly force and the insurgency faded—at least temporarily.

In the wake of this political turmoil, Myanmar's isolated leaders turned a natural disaster into a humanitarian catastrophe. On May 2, 2008, a category 4 strength typhoon (hurricane) named Nargis slammed ashore across the Irrawaddy Delta. In this most agriculturally productive and densely populated region of Myanmar, more than 135,000 people lost their lives, and an estimated two and a half million lost their homes and livelihoods. The country's junta largely banned foreign aid agencies from providing medical, food, and other relief supplies. Foreign journalists were also barred entry, so the real impacts of the tragedy may never be known.

The Lady of Burma

A much publicized contest, followed around the world, has been raging between the government and one of its citizens, the Nobel Peace Prize laureate Aung San Suu Kyi, who had been placed under house arrest after leading her antimilitary political party to an overwhelming victory in the 1990 elections (which the military government nullified). In 1995, the government released "The Lady," as Suu Kyi is popularly known. She then called for dialogue with the military junta and appealed for the parliament that was elected in 1990 to be convened. The government responded by again placing her alternatively under house arrest and in prison, triggering strong economic sanctions by the United States.

Aung San Suu Kyi has called for would-be tourists to avoid Myanmar, which has abundant sites of interest but few visitors (•**Figure 7.43**). She argues that tourism revenue will help strengthen the repressive government. She has also been successful in gaining support abroad for boycotts of U.S. and other Western companies doing business with Myanmar. American corporations including PepsiCo and ConocoPhillips have withdrawn from Myanmar, and the U.S. government has banned new investments. The United States regards the country's government as illegitimate and therefore rejects the change of the country's name to Myanmar.

Gail Ludwig

• **Figure 7.43** The huge golden Shwedagon Pagoda is the holiest Buddhist shrine in Yangon, Myanmar, attracting multitudes of the faithful and the few tourists visiting the country. It is said that the relics or bones of three earlier buddhas and the eight hairs of Gautama Buddha are located within this spectacular golden architectural masterpiece.

The U.S. government was displeased when the American firm Unocal built a pipeline from the offshore Yadana natural gas field to a port west of Bangkok, Thailand, where the gas is used to generate electricity. Human rights groups complained that slave labor was used to build the pipeline and that profits from gas sales help prop up Myanmar's repressive government. The pipeline has to be well defended in view of the government's many domestic opponents who might try to destroy it as a means of weakening the regime. Other pipelines will serve China's energy appetite. One will carry Myanmar's offshore natural gas across the country to southern China. A parallel pipeline will carry Middle Eastern oil offloaded on the coast of Burma and piped to southern China. This will increase China's energy security, in part by avoiding the congested chokepoint of the Strait of Malacca. India is vying with China for access to 240,
Myanmar's natural gas, copper, and other natural resources. 251

The United States has been reevaluating its policies toward Myanmar, especially for an important geographical reason: its geographic *situation* is extremely important, as the country sits right between the two great powers of China and India. By withdrawing from Myanmar and imposing sanctions against it, the United States has been missing out on important commercial and strategic engagements with the country, pushing Myanmar into the arms of its rival China and enemy North Korea, and with sanctions effectively punishing the people rather than the government.

The long isolation of Myanmar may be ending, perhaps because Nargis forced the country to open its doors a little bit to allow food aid. Many changes came in 2011, when the 10 countries of the regional organization known as the **Association of Southeast Asian Nations** (**ASEAN**) agreed to allow its member state Myanmar to chair the bloc in 2014. In 2011, Myanmar made changes to its economy, released hundreds of political prisoners, eased some restrictions on the Internet, media, and political parties, stopped a dam project that was backed by China, and released Aung San Suu Kyi from house arrest again. In 2012, the ruling party allowed Aung San Suu Kyi to run in national elections. She won a seat in Parliament. In the same year the U.S. and Myanmar restored diplomatic relations; they had broken in 1990.

Sex, Drugs, and Health in Southeast Asia

Southeast Asia has long been one of the world's main source areas for opium (sometimes eclipsing Afghanistan) and its highly addictive semisynthetic derivative, heroin. For decades, 242
global concern focused on combating the drug trade originating in this region and on dealing with the problem of heroin addiction in the consuming nations. More recently, owing to the emergence of HIV/AIDS, the Southeast Asian industries of prostitution and tourism have woven drugs into a complex problem of global concern.

The region's drug production is centered in the so-called **Golden Triangle**, a vernacular region comprised of the

Joe Hobbs

• **Figure 7.44** Raw opium oozes from slices in a poppy pod. Heroin, which is derived from opium, has a hearth in Southeast Asia's Golden Triangle region.

borderlands where Laos, Thailand, and Myanmar historically exercised little control over their territories (see Figure 7.37). As in Afghanistan, for a long time the absence of strong government presence together with ideal growing conditions facilitated ex-
242 plosive growth in drug production (•**Figure 7.44**). But as Thailand and Laos exercised increasing authority in the area, drug production plummeted. Today, Myanmar is the only one of the three countries with a strong opium poppy economy, but even there, production has declined. The main reason is that neighboring China, with an eye to safe, conventional trade with Myanmar, is pressuring authorities there to crack down near their common border. As poppy cultivation slumped in the Golden Triangle, it spiked in the Golden Crescent of Afghanistan. This phenomenon, in which cracking down on a problem in one place causes it to pop up in another, is called the balloon effect and applies to
374 terrorism as well as to drugs.

In Asia, the U.S. Drug Enforcement Administration and other international antidrug authorities focus their concerns on Myanmar, where refining of heroin from opium is done in remote laboratories. In northeastern Myanmar, armed ethnic minority groups have produced both heroin and a potent methamphetamine known as *ya baa*. They smuggle raw opium into both China and India, and amphetamines into Thailand, providing investment capital to contribute to Myanmar's "development." Laundered drug money has paid for the country's recent construction boom, and the government encourages such investment from the drug sector to make up for shortfalls in legitimate export revenues.

Not all of the associated problems are exported. A flood of cheap and potent heroin hit the market in Myanmar after the current regime came to power in 1988. Heroin addiction has soared ever since. Primary delivery is by needles shared in tea stalls. The number of users continues to grow; perhaps 1 percent of the adult population is addicted. Accompanying the heroin use is an AIDS epidemic, with about 1 percent of the adult population infected. The best available figures indicate that the HIV infection rate among heroin users in Myanmar is around 37 percent.

The epidemic that began among heroin users spread quickly to Myanmar's sex industry. Recent estimates put the HIV infection rate among prostitutes at 26 percent, three times the rate in neighboring Thailand. Many young men with AIDS are shunned by their families and go to monasteries to die. Tragically, their virus has spread among the populations of Buddhist monks (there are now 400,000 in Myanmar), who share razors to shave their heads. AIDS prevention is virtually unheard of. The government had outlawed condoms until 1993. They are now legal, and over 30 million have been distributed by the government since 2001, but this is still short of what is needed. Practically no anti-HIV drugs find their way to Myanmar.

Myanmar's prosperous, cosmopolitan neighbor Thailand also has the scourge, with a 1.3 percent HIV infection rate among adults in 2011. This is a substantial reduction from the peak infection rate the country recorded in 1994. Thailand's liberal social climate has given rise to the unique phenomenon of sex tourism, with package tours catering to an international clientele (•**Figure 7.45**). Thailand's overall annual income from prostitution is estimated to be up to $4 billion, with prostitutes sending large sums as remittances each year to relatives in the villages from which they migrated. However, the spread of HIV infection among intravenous drug users and among prostitutes in the red-light districts of Bangkok and the popular beach resorts of Pattaya and Phuket threatens both the country's public health and its lucrative tourist trade. The Thai government mounted an all-out war on the drug trade in 2002 that resulted in the deaths of more than 2,000 people, most of them suspected drug dealers.

There are striking differences in how this region's countries are dealing with HIV/AIDS. Thailand, regarded globally as

Taimay Jones

• **Figure 7.45** Bangkok's red-light district was a breeding ground for HIV/AIDS until the government and nongovernmental agencies aggressively promoted a safe-sex campaign.

a leader in efforts to combat the problem, has run an aggressive and increasingly successful anti-AIDS public awareness campaign, resulting in a dramatic decrease in the number of new HIV infections. Thailand distributes antiretroviral drugs to all AIDS patients at extremely low cost. Other Southeast Asian countries lag far behind. The government of the Philippines has been unsuccessful in pursuing an anti-AIDS campaign because of opposition from the country's powerful and anti-contraception Catholic clergy. Church officials have denounced the government's anti-AIDS program as "intrinsically evil" and have set boxes of condoms on fire at anti-government demonstrations. Similarly, in the conservative and largely Islamic nations of Indonesia and Malaysia, Muslim clerics denounce their governments' anti-AIDS campaigns as efforts to encourage promiscuity.

Vietnam Then and Now

The Vietnam War, which is such an important chapter in the American experience, had roots that preceded U.S. interests in the region. This conflict profoundly affected Southeast Asia and has a lingering legacy there today.

France conquered Indochina between 1858 and 1907. The French first extinguished the Vietnamese empire, which covered approximately the territory of today's Vietnam, and defeated the Chinese, whom the Vietnamese called on for help. Later, the French took Laos (in 1893) and western Cambodia (in 1907) from Thailand. France's administration of Indochina, centered in Hanoi and Saigon, left a big cultural imprint. Its major economic success lay in opening the lower Mekong River area to commercial rice production, converting both Vietnam and Cambodia into important exporters of rice.

Japanese forces overran French Indochina in 1941, initiating five decades of warfare in the area. During World War II, a Communist resistance movement led by a Vietnamese nationalist named Ho Chi Minh carried on guerrilla warfare against the Japanese. When French forces attempted to reoccupy the area after the end of war, these forces fought to expel them. In 1954, Ho's forces destroyed an entrenched French garrison at Dien Bien Phu, in the northwestern mountains. With this defeat, France withdrew from Indochina, ending a century of colonial occupation.

Four countries came into existence with France's departure (see Figure 7.37). Laos and Cambodia became independent non-Communist states. North Vietnam, where resistance to France had centered, emerged as an independent Communist state. South Vietnam gained independence as a non-Communist state that received many anti-Communist refugees from the north. Vietnamese Catholics were prominent among those who fled southward.

The partition of Vietnam into two countries was largely the work of U.S. power and diplomacy. South Vietnam was from the outset a client state of the United States, whereas North Vietnam became allied first with Communist China and then with the Soviet Union. Warfare in Vietnam gained momentum. Between 1954 and 1965, the **Viet Cong**, an insurrectionist Communist force supported by North Vietnam, achieved increasing successes in the south in its bid to reunify the country. Increasing intervention by the United States in South Vietnam escalated in 1965 and eventually involved the commitment of half a million U.S. military personnel. Why did the United States get involved on such a massive scale in a small country half a world away? Remember that those were Cold War days, and fighting communism was a foreign policy priority for the United States.

North Vietnam responded to the U.S. military escalation by fielding regular army forces against American and South Vietnamese troops. The United States never invaded North Vietnam, although American air strikes there were devastating, with more ordnance dropped on Vietnam than on all of Europe during all of World War II. American bombs, napalm (an incendiary explosive), and defoliants such as Agent Orange (dioxin) caused enormous damage to the natural and agricultural systems of Vietnam, destroying over 8,400 square miles (nearly 22,000 sq km) of forest and farmland in an effort described in military parlance as **"denying the countryside to the enemy."** Agent Orange also left a legacy of birth deformities among Vietnam's postwar generation.

The United States avoided invading North Vietnam partly because of the risks posed by Chinese and Soviet support of the North. In limiting the theater of ground warfare, however, the United States found itself unable to expel from the South both the Viet Cong and the North Vietnamese army. Those combined forces were determined, skillfully commanded, increasingly better equipped by its allies, accomplished in guerrilla tactics, and willing to bear heavy losses. More than 3 million Vietnamese soldiers and civilians died in the conflict, and about 58,000 U.S. soldiers and support staff perished. In 1973, almost all American forces were withdrawn from the costly war, which was extremely divisive at home as growing ranks of Americans questioned its rationale and protested its conduct. There are still 1,948 Americans listed as missing in action in Indochina.

North Vietnam completed its conquest of South Vietnam in 1975. Saigon was renamed Ho Chi Minh City (although most of its people still call it Saigon), after the country's leader (who died in 1969). Vietnam entered a period of relative internal peace, but there were ongoing conflicts, including Vietnam's 1979 conquest of Cambodia to oust the brutal Khmer Rouge regime, continued warfare against Cambodian and Laotian resistance groups, and a brief border war with China. Within its borders, the Communist government of the reunited Vietnam unleashed a repression against those who had assisted the Americans, including the rebellious hill tribes known collectively as Montagnards and members of the Cao Dai religion. This backlash caused a massive 228 outpouring of refugees, including the "boat people" who risked much to flee. Most of the ethnic Vietnamese living in the United States today are those refugees and their descendants.

Vietnam (population 88 million), with its capital at Hanoi (•Figure 7.46), has restored parts of its war-torn landscape through large-scale reforestation, agricultural reclamation, and nature conservation programs, aided by international organizations such as the World Wildlife Fund. Animal species new to science have been discovered in the rugged Annamite Cordillera (Truong Son Mountains). Such natural treasures are attracting international ecotourists who provide much-needed revenue, and tourism in general is booming.

GEOGRAPHY OF THE WORLD'S GREAT RIVERS | The Mekong

The Mekong River, known as "Water of Stone" in Tibet, "Great Water" in Cambodia, "Nine Dragons" in Vietnam, and "Mother of Waters" (Mae Nam Khong, contracted to Mekong) in Laos and Thailand, is Southeast Asia's great river (its course may be seen in Figure 7.37). From its source on the Tibetan Plateau in the Chinese province of Qinghai, the Mekong snakes along a 2,600-mile (4,200 km) path through China's Sichuan and Yunnan provinces, forms the boundary between Laos and its neighbors Myanmar and Thailand, winds across Cambodia, and fans out across Vietnam's densely populated Mekong Delta region, before empting into the South China Sea.

Unlike the Chao Praya in Thailand, the Red River in Vietnam, and the Irrawaddy in Myanmar, the Mekong is not a great magnet for cities and civilizations. The biggest city on its banks is Cambodia's Phnom Penh, with just over a million people. The main factor limiting more settlement along the Mekong has been the difficulty in navigating the river above Phnom Penh. Historically, very little trade has been carried along much of its length; the oceans and overland routes have provided superior opportunities.

The Mekong is not economically unimportant, however. It is one of the world's most prolific fisheries, yielding about 2 million tons of fish each year, or twice the annual North Sea catch. There are more species of fish in the river (about 1,200) than in any other rivers but the Amazon and the Congo. Unique creatures inhabit its waters, including about 70 Irrawaddy dolphins and the giant catfish that attain 10 feet (c. 3 m) in length and 660 pounds (300 kg) in weight. Overfishing and the construction of dams have all but eliminated this superlative catfish throughout the Mekong Basin.

A big push is being made to capitalize on the Mekong's potential to produce hydroelectric power, with plans to build about 100 dams on the river and its tributaries. The first dam on the river itself (the Man Wan in China) was completed in 1993. China then went to work on five more. Vietnam is building five dams on the Mekong tributary called the Dak Hoyt, which forms part of its border with Cambodia. Thailand has dammed its main tributaries to the Mekong. Laos plans to build the most dams of all—20—and thereby become mainland Southeast Asia's biggest supplier of electricity. Laotian officials boast that their country will become "the battery of Asia." With funding from the World Bank—despite the criticism this organization has received for supporting large dams—Laos built the giant Nam Theun 2 dam. The World Bank insists it went to extraordinary lengths to minimize environmental impacts from this project.

These dams will nevertheless have enormous impacts, some of them unforeseeable. The fish catch will definitely fall dramatically 181, 265
because the main reason for the Mekong's productive fishery is its regular flooding, which will be dramatically reduced. Fishermen are already complaining of smaller catches since dam building began. There are concerns that dams will ruin the ecosystem of Cambodia's Tonle Sap Lake, which is drained by a river that actually changes direction when downstream Mekong floodwaters inundate its basin. More than a million people make their living by fishing this lake. Inevitably, the downstream countries suffer many ill effects of dams. In this case, for example, Vietnam's Mekong Delta will be deprived of the fertile silt that helps make it an agricultural "breadbasket."

As always with shared river systems, there are serious questions about how the Mekong waters should be divided. Vietnam, Cambodia, 181, 247, 438
Laos, and Thailand have joined to form the **Mekong River Commission**, whose biggest responsibilities are to fix the minimum amount of water each country must discharge downstream on the Mekong and its tributaries, and to agree on rules to ensure water quality. Both quality and quantity of water are threatened by upstream Myanmar and China, which have refused to join the Mekong River Commission. The downstream countries complain that China does not seem to care about what happens to them as a result of China's dam-building and other uses of the Mekong.

Joe Hobbs

• **Figure 7.46** Typical street scene in the Old Quarter of Hanoi, Vietnam's capital. Urban traffic in Vietnam is a cacophony of motorbikes, bicycles, cars, and pedestrian traffic.

Since its embrace in 1986 of free-market reforms in the policy known as *doi moi*, Vietnam's economy has improved markedly. The poverty rate fell by half in the 1990s, and in recent years Vietnam has had Asia's second-fastest-growing economy, after China's. The Communist government halted collectivized farming in 1988 and turned control of land over to small farmers under 20-year lease agreements. The results have been impressive; Vietnam since 1988 has become a major exporter of rice, second in the world after Thailand. This is a vulnerable commodity, however, especially because production is concentrated along the Mekong River, which is prone to flooding in the May–October rainy season, and to the specter of rising sea levels due to climate change (see following discussions of the World's Great Rivers and GIS). Vietnam is also pinning hopes on exports of its proven reserves of 2.5 billion barrels of oil, especially to energy-hungry China.

Vietnam in 1995 restored full diplomatic relations with the United States. Recently, the United States and Vietnam have been boosting military ties, as they share a joint interest in containing China. U.S. businesses have been scrambling for consumers in this large, promising market of mainly young consumers (about 45% of the Vietnamese are under age 25). After the two countries signed a trade agreement in 2000, Vietnam began exporting shoes, finished clothing, and toys to the United States. Both its imports and its exports boomed when the country joined the World Trade Organization (WTO) in 2007. The dual personalities of North and South are being perpetuated in the new economic climate, with most investment and infrastructural improvement focused in the South. The North is a more austere economic landscape, and northerners dominate the civil service. Along with a growing gap between the rich and the poor, the differences between North and South could slow the country's overall development.

Visualizing Climate Change with GIS

In 2010, I was invited by the government of Vietnam to advise the country's decision-makers on how best to deal with the anticipated impacts of climate change. My focus had to be on adaptation rather than mitigation (see page 34); Vietnam produces only a tiny fraction of the world's greenhouse gases, but has one of the world's longest and most densely populated coastlines. It is commonly ranked as one of the top 10 countries most threatened by climate change, particularly in the forms of rising sea levels and stronger typhoons (hurricanes; •**Figure 7.47**). How could Vietnam adapt to these threats?

Joe Hobbs

• **Figure 7.47** In recent years, Vietnam has experienced an unprecedented number of "super-typhoons," equating to the category 4 and 5 hurricanes of the western hemisphere. I arrived in the coastal city of Hoi An hours before super-typhoon Xangsane slammed ashore on October 1, 2006, and took this photo 3 days later as the city's residents struggled to recover from extensive flooding and other damage. Many climate change scenarios anticipate more frequent and severe typhoons in the country's future. Just as in Louisiana and in Bangladesh, higher sea levels would magnify the impacts of these storms. Engineering adaptation to climate change in the Mekong Delta must therefore consider sea level rise plus storm surge.

The best way to approach this question is by the basic geographical question: *Where* will the impacts of climate change take place? Using GIS, Vietnam's Ministry of Natural Resources and Environment (MONRE) has developed maps showing exactly where rising sea levels will inundate Ho Chi Minh City and the Mekong Delta (Vietnam's most densely-populated region, and one of the world's "breadbaskets," particularly for rice production). MONRE's Director, Dr. Tran 259
Thuc, told me that IPCC forecasts for sea level rise were far too low, because they did not take into account the melting of the world's great ice sheets in Antarctica and Greenland. Dr. Tran therefore commissioned maps depicting the impacts of 65-, 75-, and 100-centimeter (26-, 30-, and 39-inch) sea level rise by 2100—and told me he "fully expects" Vietnam to experience at least the 100 centimeter (one meter) rise.

MONRE used a GIS method known as "statistical downscaling" to derive large-scale maps of the Mekong Delta and Ho Chi Minh City from small scale global sea level models produced by the IPCC. The maps showing the one meter rise are breathtaking (the Mekong Delta map is shown in •**Figure 7.48**). Almost 40 percent of the Mekong Delta, and 25 percent of Ho Chi Minh City, would be underwater. About nine million people—10 percent of Vietnam's population—would have to be relocated. These GIS visualizations are enormously useful, especially in helping Vietnamese leaders appeal for outside foreign aid to help build sea walls, dikes, and other infrastructure to adapt to climate change, and in illustrating the potential impact of climate change on global agriculture and urban infrastructure.

Much more sophisticated GIS work lies ahead. In order for sea walls to be built in just the right places and according to the correct specifications, very detailed large-scale modeling and mapping must be done. Vietnam can look to the Netherlands,

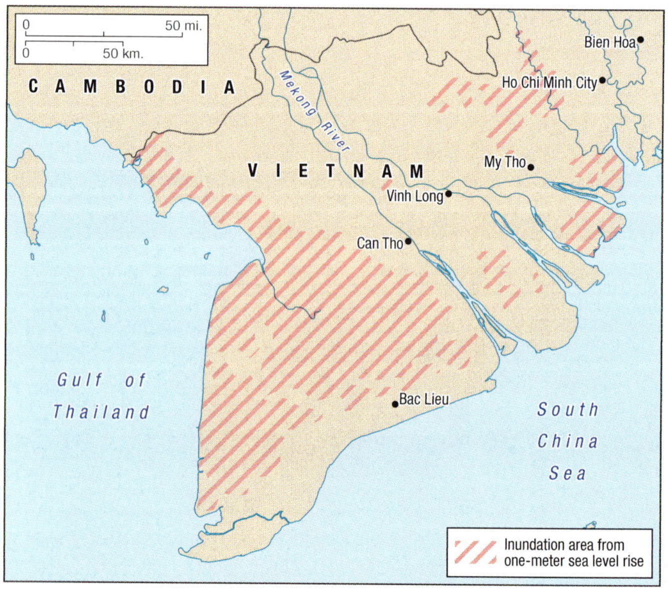

• **Figure 7.48** The impacts of a 1-meter (39-inch) sea level rise on the Mekong Delta, Vietnam. About 40 percent of this extremely productive farmland, one of the world's breadbaskets, would be inundated by the sea by the year 2100.

where Dutch geographers are using a GIS-based process of "spatial planning" to map out exactly where engineering works must be built to stave off the rising waters of the North Sea.

Indonesia: One Country, One People, One Language, 300 Ethnic Groups

The national credo of Indonesia (see map, Figure 7.38), seen on banners and placards throughout the country, is "One country. One people. One language." The government's constitution, in an effort to promote national unity, officially recognizes four faiths: Islam, Christianity, Hinduism, and Buddhism. But the presence of some 300 different ethnic groups, combined with religious frictions, physical fragmentation, and economic problems, has made it difficult to attain peace, order, and unity in Indonesia. There is an official national language (Malay, known locally as Bahasa Indonesian), but more than 200 languages and dialects are in use. The largest ethnic group is the Javanese, who make up 41 percent of Indonesia's population. Eighty-six percent of Indonesians are Muslim, 9 percent are Christian, and 2 percent are Hindu, living mainly in Bali.

Various groups in the outer islands have resented the dominance of the Javanese, who have traditionally asserted their power in colonialist fashion over the larger, more resource-wealthy outer islands. Such animosities have escalated at times to armed insurrections. The Suharto regime (1968–1998) countered these militant expressions with a state ideology called ***Pancasila***. Aimed mainly at suppressing militant Muslim aspirations, *Pancasila* was a pan-Indonesian nationalist ideology designed to neutralize all ethnic identities.

For a time, Indonesia's government vowed to break with the repressive ways of Suharto and appeased would-be separatists with promises of **liberal autonomy**, meaning that the provinces would have greater control over local administration and take more profits from the sales of local natural resources. With growing economic problems, pressure from nationalists, and accusations of corruption leveled against Suharto, however, the government shifted to using an iron fist against every Indonesian province that might aspire to follow the path the former enclave of East Timor took.

Now the independent country of Timor-Leste, East Timor was a Portuguese possession that Indonesia occupied after the collapse of the Portuguese colonial empire in the 1970s. The government carried on a long and bitter struggle against an independence movement led by East Timor's Catholics. In 1998, Indonesia and Portugal reached an autonomy agreement that would give the Timorese the right to local self-government.

Indonesia's incoming Wahid administration was bolder, however, and in 1999 allowed the people of East Timor to vote on whether they wanted outright independence. The result was overwhelmingly in favor of independence.

Since its birth, this tiny young country has had many woes: a breakdown of law and order, interethnic violence, and growing unemployment. The economic future of this poor nation could be brightened by resolution of an international legal wrangle over resources. Between Timor-Leste and Australia, in the portion of the Timor Sea known as the Timor Gap, lies a huge field of petroleum and natural gas. Because the distance between Timor-Leste and Australia is less than 400 miles (640 km)—making it impossible to apply the international standard of a 200-mile (320 km) offshore territorial limit—the two countries must come to an agreement on how to divide these undersea riches. Any deal is 272
likely to yield significant wealth to Timor-Leste.

From the West ... Of outstanding concern for Indonesia's unity is the province of Aceh (population 4.5 million), at the northernmost tip of Sumatra. In 1976, the inhabitants of Aceh, who are a predominantly Muslim people of Malayan ethnicity, began seeking independence from Indonesia. The central government had made them a promise of autonomy with Indonesia's 1949 independence from the Netherlands, but failed to keep it. The Acehnese insisted that the Jakarta government was too secular for their Muslim tastes and that too little of the revenue from sales of Aceh's abundant natural gas reserves remained in the province (which also has a wealth of oil, gold, rubber, and timber).

The secessionists' main voice was the **Free Aceh Movement** (known by the acronym **GAM**), which wanted to install a member of its own indigenous royal family as president of a free Aceh. Seeing Aceh as vital to the nation's unity and economic viability and fearful that Aceh might inspire separatism elsewhere in the country, the government refused to discuss independence for the province. Violent clashes persisted, and the death toll grew—to 12,000 between 1976 and 2004. But then an unlikely arbiter appeared. The tsunami of December 2004 concentrated its unspeakable sorrow on Aceh (see Figure 7.42). 255
The international community stepped in with a massive relief effort, and Indonesian authorities listened with unprecedented interest to Acehnese concerns. GAM dropped its demands for independence, and Indonesia agreed to allow GAM to participate in local elections. Nominally secular Indonesia allowed Aceh to become the only one of its 33 provinces to adopt Islamic *sharia* law. Redefining itself as the Aceh Transitional Commit- 176
tee, GAM laid down its weapons, and Indonesia scaled back its military presence in Aceh (•Figure 7.49). The government gave

HOTLI SIMANJUNTAK/EPA/Newscom

• **Figure 7.49** Following a victory in 2009 local elections, Aceh Party members who are former rebels of the Free Aceh Movement (GAM) hold a press conference in Banda Aceh.

amnesty to the rebels and, critically, pledged that 70 percent of the revenues from Aceh's natural resources would return to the provincial government.

Through a program called Aceh Green, founded by a former rebel, Aceh may emerge as a shining example of sustainable development. In one of Aceh Green's activities, hundreds of former rebels, who know the countryside very well, are being trained by the nongovernmental organization (NGO) Flora and Fauna International as rangers to look out for wildlife poachers and illegal loggers. If the rangers are successful in protecting the Ulu Masen rain forest, Aceh could net $26
35 million in carbon credits from the REDD funding mechanism.

... to the East Progress in Aceh raised hopes for a similar future for Papua, at the extreme eastern end of the Indonesian archipelago. Known formerly as Irian Jaya and West Irian, Papua is home to 3 million people of 200 different tribes and speaking 100 different languages. Most are of Melanesian origin, and most are at least nominally Christian. Collectively known as **Papuans**, they have little in common with the Javanese Muslims who control them from 2,500 miles (4,000 km) away in Jakarta. Their homeland on the western half of the island of New Guinea is of great importance to the economies of Indonesia and the United States because it contains the world's largest copper and gold mines. The American company Freeport-McMoRan, which provides more taxes to Indonesia's treasury than any other foreign source, operates these mines. There are also large assets of oil, natural gas, and timber.

Indonesia has never recognized the province's 1961 unilateral declaration of independence. It is little wonder that Jakarta refuses to let its poorest but most resource-rich province go; the Dutch, coveting this wealth, originally refused to relinquish the area when it granted independence to Indonesia. Pressure from the United States led to a UN-sponsored plebiscite on independence in 1969. The vote in favor of the region's becoming part of Indonesia was widely regarded as rigged. Recently, Indonesia has promised some autonomy and a greater share of resource revenues to Papua, but Papuans complain that these promises are unfulfilled.

China

Han Colonization of China's "Wild West"

China's growth as a land empire has involved both subjugation of people who are not ethnic Han and the colonization
226 of those ethnic areas by ethnic Han. There are at least 56 non-Han ethnic groups in China, in total about 100 million people living mainly in the west and southwest of the country. As the Soviets did in the Soviet Union, China's Communist government granted at least token recognition of the distinctive identity and rights of five large minorities by creating **au-**
127 **tonomous regions**. These are Guangxi, bordering Vietnam; Nei Mongol (Inner Mongolia) and adjacent, small Ningxia in the north; Xizang (Tibet); and Xinjiang in the far west (see **•Figure 7.50**).

• Figure 7.50 Principal features of China and Mongolia. On the inset map, the line represents the division between Humid China, or China Proper, to the east and Arid China, or Frontier China, to the west. Humid China has the great majority of the Chinese population and economic activity, and Arid China is home to many minority peoples.

The Hui of Ningxia are Muslims of many different ethnicities who absorbed Chinese language and many Han customs. In contrast, the Uighurs of Xinjiang and the Tibetans have defiantly resisted becoming Chinese—and so Han colonization, or **Hanification**, focuses especially on their territories. The Chinese government aims either to assimilate these minority peoples into a broader, Han-based Chinese culture or to establish a strong enough Han demographic presence among them to dispel any hopes of autonomy or independence.

Chinese authorities have also initiated what they call the **Western Big Development Project** with the stated objective of improving locals' livelihoods enough to diminish their desire for ethnic and political separatism. Focused on the autonomous regions and four other provinces in which there are large minority populations, this project seeks to improve infrastructure like airports and roads and to boost output from factories and farms. The Uighurs, Tibetans, and other minorities generally fear that this "development" is just a cover for demographic, political, and economic domination by the Han.

Buddhism reached Tibet in the middle of the eighth century, and its arrival was followed by decades of conflict between Chinese and Indian Buddhists. The Chinese were defeated in this struggle and expelled from Tibet at the end of the eighth century. China sometimes reexerted loose control over Tibet, but Chinese authority vanished again with the overthrow of the Qing (Manchu) Dynasty in 1911. From 1912 until the Chinese Communists' violent conquest of 1951, Tibet existed as an independent state, with its capital and main religious center at Lhasa.

After the Chinese completed roads to Tibet in late 1954, an increase in restrictive measures by the Communist government contributed to the rise of Tibetan guerrilla warfare. This culminated in a large-scale Tibetan revolt in 1959 and the flight of the **Dalai Lama** (the spiritual and political leader of Lamaism, or Tibetan Buddhism; •Figure 7.51) and many other refugees to India. Subsequently, the Chinese drove most of the monks from their monasteries, expropriated the large monastic landholdings, implemented socialist programs, and prohibited organized religion in the rest of China. Large-scale immigration by ethnic Han picked up; about half of Lhasa's population is now Chinese.

HARISH TYAGI/EPA /Landov

• Figure 7.51 The Dalai Lama is the enduring symbol of the nonviolent Tibetan quest for autonomy from China.

Beijing insists it is helping Tibet with major development investment, including the $4 billion Qinghai-Tibet railway recently constructed between Beijing and Lhasa. But Tibetan resistance to Chinese rule continues. From his place of exile in Dharamsala, India, and on frequent international road trips, the Dalai Lama continues to appeal for Tibetan freedom. He asks China to grant what he calls **genuine autonomy** to Tibet, meaning that Tibetans would have authority over most matters except foreign affairs and defense, which the Chinese government would control. His peaceful cause has widespread support. But some Tibetans want stronger, even violent, action. China's government has so far refused to negotiate with the Dalai Lama. For many years, India had promoted the cause of Tibetan autonomy, but as its trade ties with China have grown more vital, its advocacy for Tibet has quieted.

China's government regards Xinjiang's Muslims, mainly from the 8-million-strong Uighur ethnic group, who speak a Turkic language and are kin to the Kazakhs to the west, as the most problematic minority groups. Many Uighurs would like to pull at least a part of Xinjiang out of China's orbit to create an independent Chinese Muslim state. Two short-lived independent Uighur republics, both known as the Eastern Turkestan Islamic Republic, existed in 1933 and again from 1944 to 1949, and some Uighur separatists are fighting for one again. Beyond execution of those separatists deemed to be terrorists, China's response has been to press ahead with development projects and Han immigration.

At the outset of the Communist takeover of China in 1949, the Han population in Xinjiang amounted to 6 percent of the province's population. Today, it accounts for over 40 percent, and some 250,000 more Han Chinese immigrate to the cities (especially Urumqi and Yining) annually, enticed in part by relatively high salaries and other incentives. Many of these Han migrants are in Xinjiang only a short time; they stay long enough to make some money and then return to eastern China.

Human rights monitors report systematic Han abuses of the Uighurs. Scores of mosques have been torn down; Uighur literature has been burned; Muslim clerics have been forced into "reeducation" training; Islamic schools have been closed; public prayer is now forbidden; head scarves have been banned; and prodemocracy demonstrations are brutally quashed. China's struggle to maintain a grip on its ethnically distinct
regions is reminiscent of Soviet efforts in non-Russian areas. 134

The Three Gorges Dam

The Chang Jiang ("Long River"), also known as the Yangtze River, has been central to China's identity and welfare for thousands of years. The river delivers precious water and fertile soils, creating environments that have been intensively settled and farmed, especially for rice and wheat. But this great river has also delivered tragedy in the form of devastating floods. Flooding in August 1998 affected 300 million people along the river and did an estimated $24 billion in damage, for example. Inevitably, questions arose about the advantages to be gained—especially in flood control, drought relief, and hydroelectricity production—if the river were to be dammed. The concept of a giant dam on the river dates to 1919, when it was proposed by Sun Yat-sen. His vision has been realized on a scale far grander than he could have imagined in the massive Three Gorges (Sanzxia) Dam, begun in 1994 and completed in 2009 (•**Figures 7.52 and 7.53**; see also location on Figure 7.50).

As with any large dam project, like the Aswan Dam seen earlier (page 181), the Three Gorges may best be evaluated by considering its pros and cons. On the plus side, this largest dam ever built—1.3 miles (2.1 km) wide and 610 feet (186 m) high—is creating a reservoir 385 miles (620 km) long that will dramatically reduce the threat of downstream flooding. Stored waters will also help alleviate drought. The Three Gorges Dam will also improve navigation and therefore enhance

• **Figure 7.52** Changed landscape of the Three Gorges Dam region; NASA visualization.

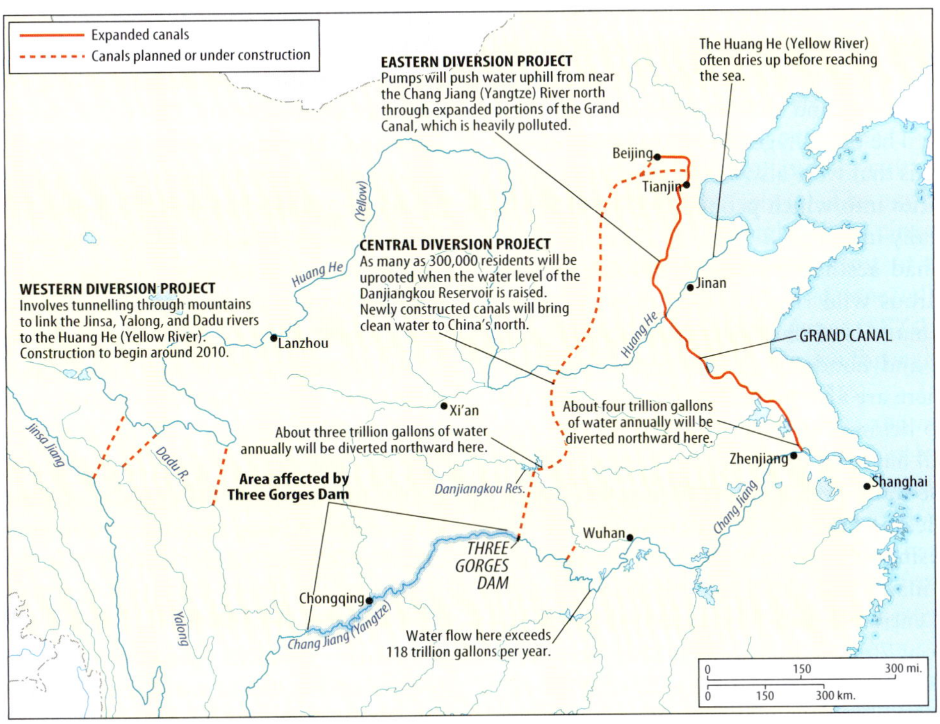

• **Figure 7.53** The Three Gorges Dam and the Chang Jiang Water Transfer Project.

Japan and the Koreas

Japan's Postwar and Its Costs

On the Way Up . . . Without colonies or empire, Japan (•Figure 7.56) became an economic superpower after World War II. The nation's explosive economic growth after its defeat was one of the most remarkable developments of the late twentieth century and, echoing the broader region's experience, is known widely as the Japanese "miracle." Japan watchers cite 232
different reasons for the country's economic success. Proponents of dependency theory argue that Japan, having never been colonized, escaped many of the debilitating relationships with Western powers that hampered many potentially wealthy countries. Some analysts believe that the country's postwar 44
economic miracle grew from an intense spirit of achievement and enterprise among the Japanese. Notably, many Japanese

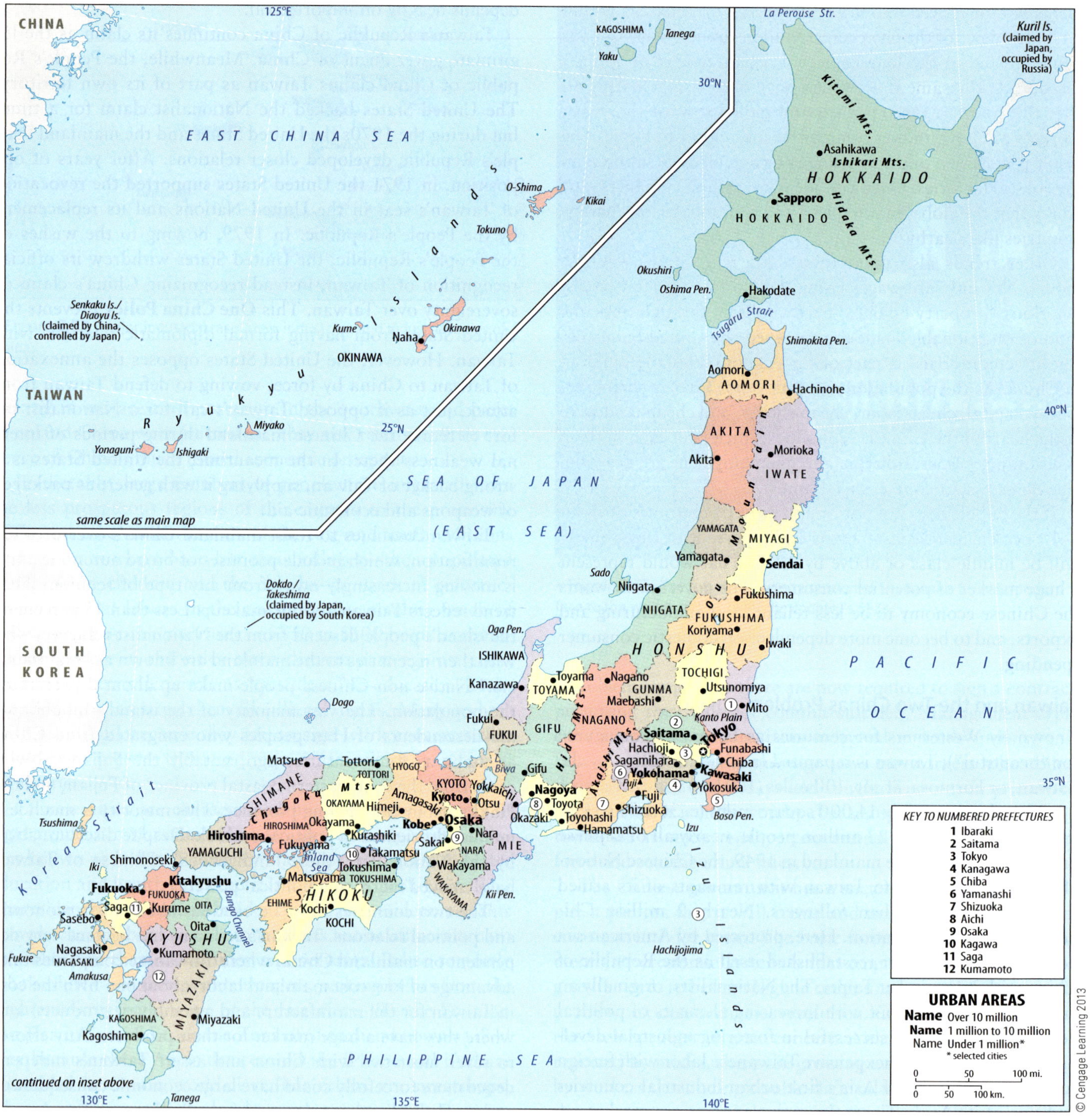

• Figure 7.56 Principal features of Japan.

attribute this industrious spirit to Japan's geography as a resource-poor island nation. To overcome the constraints nature has placed on them, the Japanese people feel they must work harder. Still others attribute Japan's postwar achievements to its association with the United States following World War II. Some of the postwar U.S.-imposed reforms worked well economically, and relations between the two countries since the war have also generally stimulated Japan's industrial productivity.

Economists also point to several unique features of Japanese management and employment to explain the country's meteoric postwar gains. One has been recruitment through an extremely challenging (some say brutal) educational system that emphasizes technical training. Although recent years have seen a move to a more lenient and more comprehensive education, a rigorous and stressful testing system still determines access to higher education. Japanese management strategies emphasize benevolence toward employees, encouragement of employee loyalty, and participation of workers in decision making. About 20 percent of Japanese workers enjoy guarantees of lifetime employment in their firms, and many large Japanese companies help provide housing and recreational facilities to their employees.

One essential factor in Japan's postwar economic growth was a high level of investment in new and efficient industrial plants. Investment capital was made available when government policies cut expenditures on amenities and services such as roads, antipollution measures, parks, housing, and even higher education. Japanese industrial cities grew explosively and became highly polluted areas of dense and inadequate housing, with few public amenities and snarled transportation. The money "saved" by not being spent on amenities and services was made available for investment in industrial growth, but the costs to Japanese society were high.

Some analysts cite elements of Japan's political culture to explain the country's economic successes. One political party, the **Liberal Democratic Party (LDP)**, has with few exceptions been elected to power since Japan regained its sovereignty in 1952. It is a conservative and strongly business-oriented and business-connected organization. The party has promoted Japanese exports with policies to keep the yen (the Japanese unit of currency)—and thus Japanese goods—inexpensive. The party has also cooperated closely with Japanese business in the development of new products and new industries. However, that very coziness between government and business—the "crony capitalism" seen in several other East and Southeast Asian countries—was partly to blame for the economic malaise that spread over Japan in the 1990s, and support for the LDP began to erode.

... and on the Way Down Despite all of these factors in Japan's economic favor, the Japanese miracle did not last. The peak of Japan's postwar success came in the 1980s. A powerful economic boom led to speculative rises in stock prices and land prices. At the end of the 1980s, the total value of all land in Japan was four times greater than that in the United States. The Tokyo Stock Exchange was the world's largest, based on the market value of Japanese shares. Then the **bubble economy** burst, and real estate and stock prices fell by more than 50 percent. Japan had near-zero economic growth through 70, most of the 1990s and into the early 2000s. Japanese industries 72, 88, whose growth had seemed unstoppable experienced an 421 increasing loss of market share to U.S. and European producers. Part of the reason the economy could not regain its footing was official Japanese commitment to prop up faltering industries and farms with huge subsidies. Because of the 420 disproportionate political influence of small towns and rural areas, the Japanese tried to modernize remote villages and islands with expensive and underutilized public works projects. Due to such projects, Japan is now the world's most indebted country. Overall, the system succeeds in subsidizing its past at the expense of its future.

Attitudes toward ethnic groups play a significant role in Japan's economic development, or more precisely the lack of development. Japan has one of the world's most homogeneous populations; it is 99.5 percent ethnic Japanese. This lack of racial and ethnic diversity has had mixed results for Japan. Many observers believe that it has helped the country achieve a sense of unity of purpose, allowing the Japanese to persist through periods of adversity, especially the postwar years of reconstruction. However, the Japanese have also earned a reputation for intolerance of ethnic minorities. This attitude may be costly for Japan's future economic development. One way to increase productivity in the face of Japan's declining population would be to import skilled workers from abroad, but most people in the country remain averse to immigration. 70, Inward-looking tendencies have also left Japan short of people 276, 393 capable of and interested in learning English, a language routinely sought by aspiring businesspeople in nearby South Korea and China.

There is definitely another upside to Japan's homogeneity. Capitalist Japan has had a remarkably egalitarian society, with about 80 percent of the population solidly middle-class. The richest third of the population has a total income just three times as great as that of the poorest third (compared with a fivefold differential in the United States). The Japanese have historically embraced the concept of *wa*, or harmony, based in part on the principle of economic equality. However, the recessions of the 1990s and early 2000s, followed by an economic rebound until the global financial crisis of 2008, resulted in growing joblessness and homelessness in Japan and a widening gap between haves and have-nots. The country's official unemployment rate stood at 4.7 percent in 2011, compared with 2.1 percent in 1990. There are growing ranks of "permanent temporary workers," mainly young people who live with their parents to keep their expenses low.

The government has balked at many suggestions for economic reform because of the threat they pose to a socially harmonious Japan. There have been loud international appeals for Japan to reform its economic system by allowing inefficient businesses to collapse (rather than prop them up with expensive subsidies) and to resist the temptation to satisfy marginal

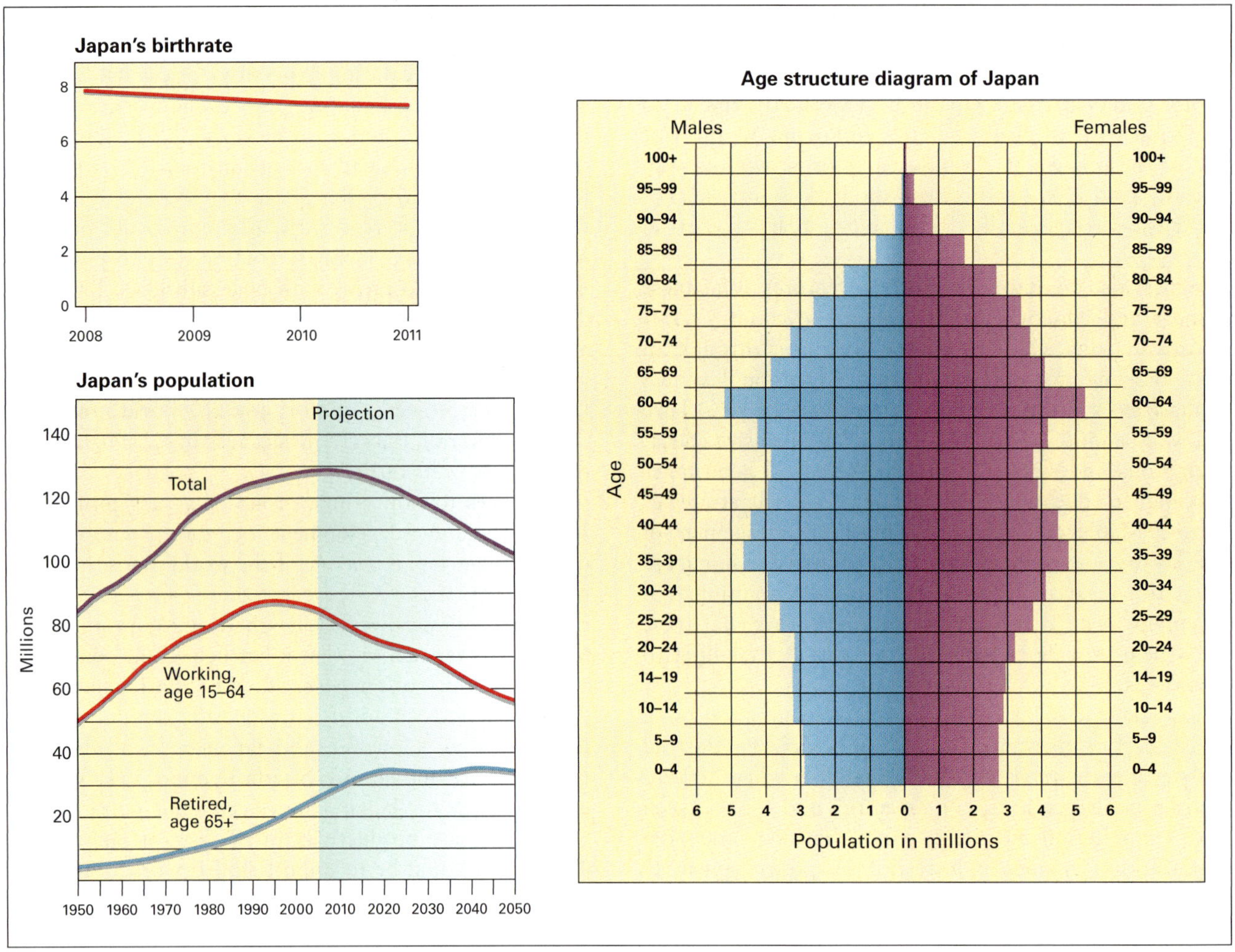

• **Figure 7.57** Many countries face problems of overpopulation, but Japan is concerned about not having enough people to keep its economic engine running in the future.

rural populations with costly education and public works projects.

The legendary Japanese work ethic has had its advantages and drawbacks. It has helped the Japanese create a prosperous country, but it has also created a nation of workaholics beset with the same problems—stress, suicide, depression, and alcoholism—experienced by workers in the world's other MDCs. There is a growing incidence of what Japanese call ***karoshi***, or death by overwork. The educational system encourages children to be highly successful but also to conform; critics say this stifles creativity and innovation. College graduates who are not hired during the annual recruiting season face difficult obstacles to entering the job market. And for those it affects, the practice of lifetime employment makes it difficult to change jobs. The country's welfare system, pensions, and social security are inadequate, compelling workers to work harder for savings. In their long hours at work, men are accustomed to spending little time with their spouses and children.

Women have not achieved parity with men in the workplace, and they complain increasingly of discrimination and sexism. In struggling to increase their footing in Japanese society, growing numbers of Japanese women are both working longer hours and marrying later, contributing to Japan's remarkably low birth rate (7 per 1,000 annually, one of the lowest in the world). This trend and the others make Japan's declining population stand out clearly in **•Figure 7.57**; note especially Japan's chimney-shaped age structure diagram, such a remarkable contrast with the diagrams of LDCs. Ironically, 53
Japan's falling birth rates will only increase the burdens of its workers. The population is both shrinking (it is expected to go from 128 million in 2011 to 100 million in 2050 and 67 million in 2100) and aging (already one in five people were over age 65 in 2011, and 1 million people a year join those ranks). The Japanese of working age will face an increasing load of taxes and family obligations to meet the needs of older citizens.

The Law of the Sea

Some 90 miles (145 km) between the shores of Japan and South Korea in the Sea of Japan (East Sea) lie 34 small, inhospitable islands known to outsiders as the Liancourt Rocks, as Takeshima to the Japanese, and as Dokdo (or Tokdo) to the Koreans (see Figure 7.56). South Korea has controlled the islands since 1956,

but with only a handful of Korean civilians and some Korean coast guard personnel living on the islands; Japan has periodically asserted its right to them. Since the 1990s, these remote rocks have been the focus of a dispute between Japan and Korea, not because of any riches they contain but because of a 1970s United Nations treaty known as the **Convention on the Law of the Sea**. This treaty would permit their sovereign power
237, to have greater access to surrounding marine resources.

427 The Law of the Sea began as an effort to apportion ocean resources as equitably as possible and to avoid precisely the kind of dispute that developed between Japan and South Korea. The treaty gives a coastal nation mineral rights to its own continental shelf, a territorial water limit of 12 miles (19.3 km) offshore, and the right to establish an **exclusive economic zone (EEZ)** of up to 200 miles (320 km) offshore (in which, for example, only fishing boats of that country may fish). The power that controls offshore islands such as Takeshima and Dokdo can extend its EEZ even further.

By early 1996, some 85 nations had ratified the treaty. South Korea ratified it late in 1995. Japan was preparing to ratify it in 1996, when news reached Tokyo that South Korea had plans to build a wharf on the islands. To avoid provoking Japan, North Korea, and China, South Korea avoided declaring an exclusive economic zone off its waters that would include the islands. Japan ratified the treaty. But fears that South Korea may build facilities and place more people on the islands as a step toward establishing an EEZ, and thereby excluding Japanese fishermen from the area, caused Japan to restate its claim to the islands in 1996. South Korean officials answered with military exercises near the islands, and South Korean civilians staged loud demonstrations outside Japan's embassy in Seoul. South Korea declared the islands a national monument in 2002. Long simmering, the issue boiled up again in 2006 when Japan announced it would send ships to survey the area and South Korea answered that it would form a naval blockade against the Japanese vessels. The problem may have to be resolved in the international legal arena.

Even more important than Japan's competition with South Korea over territorial waters and islands is its contest with China. Both lay claims to the islands in the East China Sea that the Japanese know as the Senkakus and the Chinese as the Diaoyu Islands (see the inset in Figure 7.56). Again, both are interested in staking out an EEZ in which oil reserves might be found. China does not accept the UN definition of the EEZ beginning from a country's coastline; it insists on a wider zone that starts at the continental shelf.

The Senkaku or Diaoyu Islands are uninhabited, but their control is championed by nationalists on both sides and is therefore a potentially explosive issue. In 2004, Chinese activists landed on one of the islands to stake China's claim there, but Japanese authorities promptly arrested them, and Japan took formal possessions of the islands the following year. Since then, there have been several encounters between Japanese and Chinese ships near the islands. The worst was the 2010 incident known as the "Senkaku Shock," which began when a Chinese fishing vessel rammed a Japanese patrol boat near the islands. Japan detained and refused to release the Chinese captain. China responded by halting exports of "rare earth" minerals needed in Japanese high-tech industrial products. By geological accident, and good fortune for China, a very large proportion of the world's supply of 17 rare earth elements are found exclusively or mainly in China. These are indispensable in cell phones, laptops, and many "green" technologies, like wind turbines and hybrid vehicles. They are also important in military technology, and the United States regards access to them as a matter of national security.

Japan's formal claim to the islands dates to 1895, when Japan declared that the islands had previously been *terra nullius*, or no one's property. But China insists that there is Chinese documentation about control of the islands dating to the Ming Dynasty in the fifteenth century. The islands are close to Taiwan, which also lays claim to the islands, so any move by mainland China to secure access to them could also precipitate conflict between the two Chinas.

North and South Korea: Night and Day

Just 110 miles (177 km) from westernmost Japan is the Korean Peninsula, roughly the size of Minnesota or Portugal (•**Figure 7.58**). The two Koreas have occupied an unfortunate

• **Figure 7.58** Principal features of the Korean Peninsula.

location in historic geopolitical terms. North Korea adjoins China along a frontier that follows the Yalu and Tumen Rivers; it faces Japan across the Korea Strait; and in the extreme northeast, it borders Russia for a short distance. These small countries are thus located near larger and more powerful neighbors—China, Russia, and Japan—that have frequently been at odds with one another and with the Koreans. China and Russia have traditionally feared Japanese rule in Korea because it might serve as a springboard for invasion of their countries. Meanwhile, to Japan, Korea has been seen as a Chinese or Russian dagger aiming at its heart. Japan therefore wanted to occupy or block Korea. For many centuries, the Korean Peninsula has served as a bridge between Japan and the Asian mainland. From an early time, both China and Japan have been interested in controlling this bridge, and Korea was often a subject or vassal state of one or the other.

Despite external ambitions of conquest and division, from the late seventh century to the mid-twentieth century, Korea was a unified state, sometimes invaded and forced to pay tribute but never destroyed as a political entity. The decline of Chinese power in the nineteenth century was accompanied by the rise of modern Japan, whose influence grew in Korea, and from 1905 until 1945, Korea was firmly under Japanese control. In 1910, it was formally annexed to the Japanese empire. A legacy of hostility and occupation continues to cast a shadow over relations between the Koreas and Japan, and disputes periodically emerge between them over issues such as control of islands and fishing rights (see page 272).

Japan lost Korea at the end of World War II, setting the stage for the division between North and South Korea. To understand Korea's split, it is important to consider the Korean War, sometimes in the United States called "America's Forgotten War." The conflict cost the lives of an estimated 2 million North Koreans, perhaps 300,000 South Koreans, and 33,629 Americans.

In the closing days of World War II, the Soviet Union entered the Pacific war as an ally of the United States against Japan. The two sides drew up plans to accept Japan's surrender on the Korean Peninsula, arbitrarily drawing a line at the 38th parallel. The Soviet Union would accept Japan's surrender north of that line and the United States south of that line. This was not meant to be a permanent boundary, but it became one. On either side of the line, the Soviet Union and the United States moved to set up governments that would be friendly to them. By 1948, the Soviet-style, Soviet-backed, Soviet-armed Democratic People's Republic of Korea was established in the north, while in the south, the Western-oriented Republic of South Korea was created under the auspices of the United Nations.

War came soon. On June 25, 1950, North Korean troops invaded the south and quickly took Seoul. In those Cold War days, the United States feared Soviet expansionism into and beyond South Korea. President Truman sent American forces in, and they soon fought under the United Nations flag with support from 16 other countries. Truman selected General Douglas MacArthur to serve as supreme commander of the effort. By September, MacArthur's forces had pushed the North Korean advance all the way back to the Yalu River on the Chinese border. The general proposed using nuclear weapons to press forward from there, but the president refused, and his protests over the issue cost MacArthur his job.

China, seeing enemy forces on its doorstep, reacted by sending huge numbers of forces across the border, driving the allies south across the 38th parallel and taking Seoul on January 4, 1951. Seoul was recaptured on March 15, and the battlefront stabilized generally along the 38th parallel. An armistice was signed at the border site of Panmunjom on July 27, 1953, by the Chinese, the North Koreans, and the United Nations command. It was only a cease-fire, and to this day, there has not been a full treaty or reconciliation between the Koreas.

The border between the two Koreas, often cited as the tensest boundary on earth, still follows this armistice line. The **demilitarized zone (DMZ)** between them, 150 miles (240 km) long and 2.5 miles (4 km) wide, is a virtual no-man's-land of mines, barbed wire, tank traps, and underground tunnels. The world's largest concentration of hostile troops faces off on either side of it. Remarkably, with the near absence of human activity, rare birds like the Manchurian crane and other endangered animal species have taken refuge in this narrow strip. The DMZ thus has the potential to become an international "peace park" like that taking shape along the old Iron Curtain
in Europe. 89

The Korean armistice line divides one people, united by their ethnicity and language, into two very different countries (•Figure 7.59; also see Table 7.1). South Korea is a republic that has fluctuated between attempts at democracy and a repressive military dictatorship. It has a capitalist economy heavily dependent on relationships with the United States and Japan. North Korea is a rigid and very tightly controlled Communist

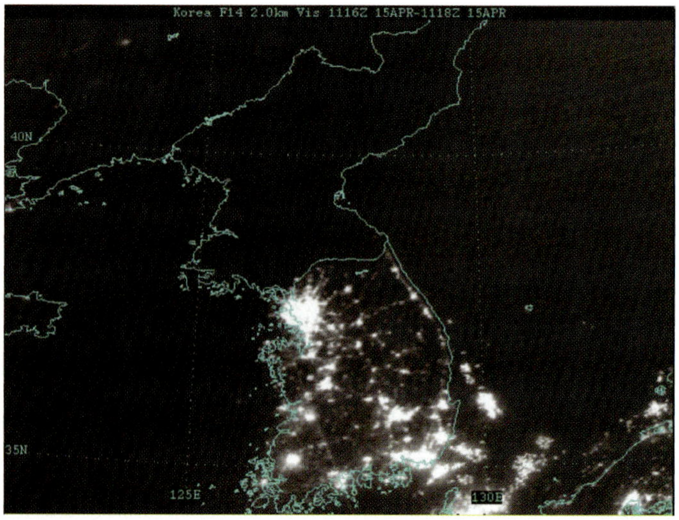

NASA

• Figure 7.59 The two Koreas are like night and day. A satellite image taken at night shows the profound economic distinctions, with prosperous South Korea awash with urban and industrial lights while North Korea is nearly dark.

state that promotes what it calls ***juche*** (pronounced "*djoo'*-cheh"), or self-reliance. North Korea's leadership has used this political philosophy's three principles of political independence, economic self-sustenance, and national self-defense to limit economic and other exchanges with the outside world. North Korea's principal Cold War ties shifted between China and the Soviet Union. With the collapse of the USSR, China became North Korea's main ally and benefactor, and the Russian Federation is emerging as one of North Korea's few trading partners.

From World War II until his death in 1994, the dictator Kim Il Sung, whom his citizens called "The Great Leader," governed North Korea. His son and successor, Kim Jong Il (dubbed "The Dear Leader"), perpetuated the country's militaristic character. The same is widely expected of Kim Jong-Un, "The Great Successor" and son of Kim Jong Il, who died in 2011. North Korea spends a staggering 23 percent of its gross domestic product on its military, compared with South Korea's expenditure of only 2.7 percent, China's 4.3 percent, and Japan's 0.8 percent. North Korea's lavish expenditures on defense have been one cause of its severe economic decline. North Korea's economy is only 3 percent the size of South Korea's. No data are available for North Korea, but South Korea's GNI PPP is $27,240, similar to that of Israel.

Beginning in 1995, successive waves of flood and drought brought famine to North Korea. The natural calamities only contributed to the already plummeting economic health of the country that came in the wake of the collapse of the Soviet Union, formerly North Korea's main benefactor and trading partner. North Korea's almost impenetrable veil of secrecy made it difficult to calculate the losses, but estimates of the number of people who died in the famine range from 900,000 to 2.4 million, or up to 10 percent of the country's prefamine population. As one crop after another failed, North Korea gradually and reluctantly sought food aid from abroad. It coaxed emergency supplies of wheat from the United States in part by threatening to withdraw from its 1994 agreement with the United States to halt its nuclear weapons program (see the discussion on pages 239–240). Although food aid from the United States, China, and South Korea helped alleviate the immediate crisis, the poor conditions of North Korea's health system, drinking water supplies, and electrical generation have perpetuated malnutrition and disease.

Another physical contrast between the two Koreas is that most of the nonagricultural natural resources are in North Korea. All of the peninsula's major resources—coal, iron ore, some less important metallic ores, hydropower potential, and forests—are more abundant in the North than in the South. North Korea was thus originally the more industrialized state, featuring a typical Communist emphasis on mining and heavy industry, together with hydroelectric production and timber products. Viewed strictly in terms of resource potential for development, North Korea should have become the more prosperous power. Its decision to close its doors to the outside world and insist on strict state control of industry and agriculture provides a fascinating opportunity to appreciate how different political systems produce different economic results.

While the North focused on defense and strict socialism, the South took over industrial leadership in the 1970s and 1980s by developing a dynamic and diversified capitalist industrial economy.[19] South Korea's economy first began to boom between 1961 and 1980 under a state capitalism approach rather than free-market and free-trade capitalism. This was a time during which the government favored and strongly funded a few small companies that it nurtured into giant conglomerates called ***chaebols***. Hyundai is a good example. Originally a small rice milk company, it diversified into trucks and buses. The government believed in its potential and supported Hyundai financially, supervising its growth into a world-class corporation renowned for the manufacture of automobiles and ships. Samsung, now one of the best electronics companies in the world, followed a similar path. Altogether about 15 of these gigantic, interlocking, family-controlled conglomerates came to dominate the economy. Large investments from Japan and the United States, as well as access to markets in those countries, aided in the explosive development of *chaebol* industries. Also important was the availability of inexpensive and increasingly skilled Korean labor. Finally, the factor South Koreans most frequently point to in their success has been an emphasis on education. They sell valued possessions if need be to afford their children's educations at home or abroad, and their culture prizes teachers and professors. South Korea has one of the world's highest numbers of Ph.D.s per capita.

With the world's 13th largest economy, South Korea is now among the top 10 countries worldwide in the production of automobiles, ships, steel, computers, and electronics. The development of its high-tech industries has concerned competitors in Japan and the United States; for example, Samsung has established an increasing market share in an industry dominated by Japanese firms. South Korea is the seventh largest trading partner of the United States, which has long subscribed to a "domino theory" that if South Korea's economy fell, so would Japan's and perhaps even its own. In 2007, the United States and South Korea signed a free-trade pact that will even more strongly enmesh their economies. The United States ratified this agreement in October 2011, and in November Korea did the same.

In sum, South Korea is one of Asia's success stories, an Asian Tiger that made an extraordinary postwar recovery, especially by using brainpower to overcome its resource limitations, and has enjoyed periods of rocketing economic growth. Although South Korea is an MDC by the strict technical GNI PPP definition, it is still something of an "in-between" country, not entirely an MDC or an LDC, not all urban but not all rural either, and both industrial and agricultural. The government still lavishes huge subsidies to protect its rice growers, for example, while also promoting the virtues of free trade in industry.

The transition from LDC to MDC has brought some growing pains, seen especially in the social changes of more women

kangaroos. These soft-hoofed marsupials are adapted to the Australian landscape and do not have such a damaging impact on the soil. Farmers and ranchers have traditionally eradicated about two million of them each year as pests. However, if more consumers at home and abroad could learn to appreciate kangaroo meat, there would be a strong incentive for ranchers to reduce their cattle and sheep herds and allow kangaroos to proliferate and be harvested. South Africa already imports about 20 tons of kangaroo meat yearly from Australia, and nearly 100 more tons go each year to countries including France, the Netherlands, and Germany. Russia was a big consumer until an *E. coli* outbreak led to a ban. Kangaroo exporters are now eyeing China, where there is a huge market for wild animal meats. Kangaroo skin is also used to make athletic shoes.

Australia is an unlikely exporter of another quadruped, the dromedary camel. Australia's nearly 1 million camels are feral descendants of those brought from South Asia and Iran early in the twentieth century to provide transport across the arid continent. Australian camels are exported even to Saudi Arabia, where their wild temperaments suit them to racing. Oddly, camels are part of Australia's "carbon farming initiative," which allows the country to cull (kill) the feral camels and claim carbon credits in the global carbon emissions trading scheme to mitigate global warming (Australia ratified the Kyoto Protocol). Every year a flatulent camel emits the equivalent of 99 pounds (45 kilograms) of methane, a greenhouse gas that is even more potent than carbon dioxide. The camel also eats about a ton of carbon dioxide–absorbing vegetation each year. On the carbon ledger, each camel's death earns an "emissions avoidance benefit" equal to about 15 tons of carbon dioxide.

From Australia, we move to the bottom of the world, to a continent twice its size, the white wilderness of Antarctica.

Antarctica: The White Continent

Antarctica is the world's fifth largest continent, with an area of 5.5 million square miles (14,245,000 sq km), lying south of the tip of South America and virtually filling the Antarctic Circle (•**Figure 8.24**). It is the setting of enormous human dramas in exploration, bravery, and foolhardiness as people have crossed the continent with dogsleds, on skis, on foot, and in airplanes and helicopters. Some, including all members of Britain's Scott Expedition of 1910–1912—which had hoped to beat the Norwegians to the South Pole but arrived five weeks too late—did not return alive.

Antarctica's distinctiveness comes in part from its winter of darkness, alternating with summer "whiteouts" caused by light refraction on snow and ice surface covering about 95 percent of the continent. Average temperatures during the summer months barely reach 0°F (−18°C). Winter averages are the coldest in the world, with winter mean temperatures averaging −70°F (−57°C). Winter's months of darkness of the polar night are an astronomer's dream, and many scientific teams, including one from China, have set up observatories here. Antarctica is also the world's windiest and driest continent and is very sunny during its half year of

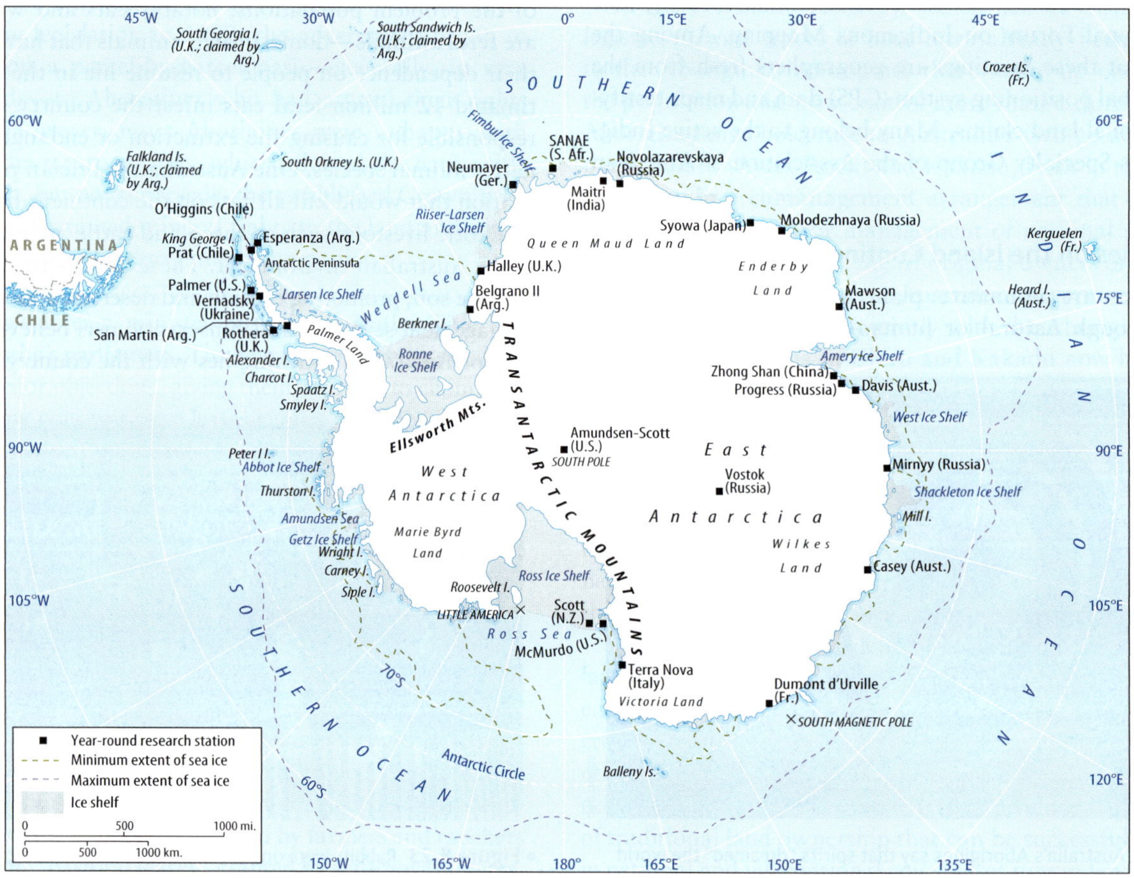

• **Figure 8.24** Principal features of Antarctica.

• Figure 8.25 Land and water on Antarctica's coast. The blue ice is part of a solitary iceberg that was "calved" from a nearby glacier. You are literally seeing the tip of the iceberg here. About 90 percent of the iceberg lies underwater, presenting a threat to navigation. The underwater portion of an iceberg fatally damaged the *Titanic*.

midnight sun; the South Pole on average receives more sunlight each year than the world's equatorial regions. Although the continent is capped with glacial ice up to 10,000 feet (c. 3,000 m) thick, only 2 to 10 inches (5 to 25 cm) of precipitation fall each year.

Antarctica has a prominent position in global concerns about climate change. The "ozone hole" in the earth's stratosphere is concentrated seasonally over Antarctica, and scientists keep constant watch as it dwindles, thanks to the Montreal Proto-
34 col treaty to phase out CFCs. Records show that the Antarctic Peninsula has warmed by 4.5°F (2.5°C) since 1944, causing both the peninsula's fringing ice sheet of sea ice and its inland glacial ice to begin to melt. Further melting of the land ice would lead inevitably to rising sea levels worldwide.

In 2011, the U.S. National Academy of Sciences published a report entitled *Future Science Opportunities in Antarctica and the Southern Ocean*. It identified key questions that should drive research in the region over the next two decades. Most of these questions focus on Antarctica's role in global climate change. Scientists see the southern polar region, like the northern polar region, as critical in understanding the global climate system and for creating models to predict future changes. Of particular concern is the Pine Island Ice Shelf, which is melting unusually rapidly. Scientists are using remote sensing to understand how relatively warm water is undercutting the ice and causing its melting. The fear is that accelerated melting will cause sea levels to rise more rapidly than anticipated.

Antarctica is a vast international scientific laboratory. There is virtually no human settlement beyond the research teams that have built facilities on small areas of exposed land that lie near the outer edges, on the island of Little America in the Ross Sea, and at the South Pole itself. About 1,000 researchers are on the continent in the winter and about 4,000 in the summer. The United States has blazed a 1,000-mile (1,600 km) "ice highway" from its McMurdo Research Station to the South Pole, angering Antarctic environmentalists.

Public, economic, and geopolitical interests in this distant world are increasing. Expensive voyages by sea bring about 40,000 visitors each year (up from 5,000 in 1990) to mingle with penguins and enjoy some of the wildest scenery on earth (**•Figure 8.25**). Seven countries have staked claims (some of which overlap) to portions of the continent as national territories: nearby Chile and Argentina, the Southern Hemisphere powers of Australia and New Zealand, the United Kingdom, France, and Norway. No other countries recognize these claims. Both the United States and Russia have avoided laying claim to parts of Antarctica but maintain that they have the right to do so. Such claims, which have been technically "frozen" since 1961, may be moot. The **Antarctic Treaty**, negotiated in 1959 and signed by 45 countries, including the seven laying claims, forbids any exploitation of the continent's natural resources until 2048. There has long been speculation about the potential for considerable resource wealth lying beneath the ice, but to date no mineral deposits of significant economic value have been found.

Antarctica is the world's largest nature preserve. But even at this end of the Earth there are pressures on resources. Some of Antarctica's waters are overfished. Japanese whaling vessels take about 1,000 whales from Antarctica waters
every year. Australia has taken on Japan in the International 99
Court of Justice to put an end to Antarctic whaling. In the meantime, conservationists argue that many more marine reserves like one created under the Antarctic Treaty in 2009 are needed.

From Antarctica, we move north (the only way to go) to Africa, another continent that boasts much wilderness and that is the cradle of humankind.

summary

- Oceania encompasses Australia, New Zealand, and the islands of the mid-Pacific lying mostly between the tropics. Tropical rain forest climates and biomes are most common, but Australia and New Zealand have several temperate climate and biome types.
- The Pacific islands are commonly divided into three principal regions: Melanesia, Micronesia, and Polynesia.
- Oceania is ethnically complex, having been settled by peoples of various Asian origins. Polynesia was the last to be populated. Papua New Guinea is the world's most linguistically diverse country. Christianity is the majority faith in this region.
- Although countless islands are scattered across the Pacific Ocean, there are three generally recognized types: continental islands, high islands, and low islands. Continental islands are either continents themselves (such as Australia) or were connected to continents when sea levels were lower (such as New Guinea). Most high islands are volcanic. Low islands are typically made of coral, a material composed of the skeletons and living bodies of small marine organisms that inhabit tropical seas.
- The island ecosystems of the Pacific region are typically inhabited by endemic plant and animal species—species found nowhere else in the world. Island species are especially vulnerable to the activities of humans, such as habitat destruction, deliberate hunting, or the introduction of exotic plant and animal species.
- Europeans began to visit and colonize the Pacific islands early in the European Age of Exploration and brought mainly negative impacts to island societies. However, a steady process of decolonization has accompanied a recent surge of Western interest and investment in the region.
- There are ethnic conflicts, related mainly to maldistribution of income, between Malaitans and indigenous Guadalcanalans on the Melanesian island of Guadalcanal and between indigenous Fijians and Indo-Fijians in Fiji. Interest in securing more income from minerals has pitted the people of New Caledonia against the ruling power, France.
- Aside from a few notable exceptions, the poverty typical of less developed countries prevails throughout most of the Pacific region.
- In general, the Pacific islands' economic picture is one of nonindustrial economies. Typical economic activities include tourism, plantation agriculture, mining, and income derived from activities connected with the military needs of occupying powers. Several countries are profiting from offshore banking and telemarketing.
- During the 1940s and 1950s, the United States used the Bikini atoll in the Marshall Islands as one of its chief testing grounds for nuclear weapons. Strong negative reaction arose throughout the region in the 1990s as the French resumed underground testing of nuclear weapons on the Mururoa atoll in French Polynesia. That testing has since ceased. The United States relies on the region for testing of its Strategic Defense Initiative missile technology. Many inhabitants of French and American military zones are fearful of the economic repercussions of the withdrawal of foreign military presence.
- Some of the low-island countries are fearful that global warming might raise sea levels and inundate them, and Tuvalu considered legal recourse against the United States and Australia for failing to ratify the Kyoto Protocol.
- Australia and New Zealand are products of British colonization. The British heritage is strongly reflected in the ethnic composition and culture of the two nations' majority populations.
- Both Australia and New Zealand have minorities and indigenous inhabitants. Native Australians are known as Aborigines, and the dominant indigenous group in New Zealand is the Maori.
- In Australia, one of the largest issues of contention between the Aborigines and the white majority is land rights. In 1993, the Native Title Bill addressed the Aborigines' major concerns and made several land rights concessions.
- Exotic species and human pressures have had catastrophic effects on native species and natural environments in Australia and New Zealand.
- Australia and New Zealand have been reorienting their focus toward the Pacific Rim and away from the United Kingdom. An important expression of this change in focus is the ongoing debate about whether Australia should become a republic, thus ending more than 200 years of formal ties with Britain.
- Antarctica is the world's coldest, driest, and least populated continent. There is no permanent population, only scientific researchers. The region's melting sea and land ice may reflect global warming, and the "hole" in the earth's stratospheric ozone layer concentrates over Antarctica. Seven countries stake claims to the continent, but no other countries recognize these claims, and so far they have not had any economic significance.

Review Questions

1. What three principal subregions comprise Oceania? What are some of the countries and colonial possessions of each?
2. Describe the major differences between high and low islands. According to the geographer Tom McKnight, how did the physical characteristics of the islands relate to social and political developments? What are the typical demographic and economic qualities of each?
3. What is a geologic hot spot, and how does it relate to Pacific islands?
4. What unique ecological attributes do islands generally have? What are the major threats to them?
5. What is an atoll, and how does it form?
6. What are the major ethnic groups, languages, and religions of Oceania?

7. How was Nauru able for a time to avoid the poverty typical of most Pacific islands? How is Nauru's economy diversifying now that its principal asset is dwindling?
8. What are the major economic activities in the Pacific islands, excluding Australia and New Zealand? Why and where have military interests been prominent?
9. What explains the peculiar jog in the International Date Line?
10. What fears do some Pacific countries have about potential global warming?
11. What features do Australia and New Zealand have in common? How are their identities changing?
12. What is the significance of the Dreamtime to Australia's Aborigines? What issues of land ownership concern them?
13. What advantages might kangaroo husbandry have over that of conventional livestock?
14. What are the major physical characteristics of Antarctica? What is the continent's political status?
15. What are "cargo cults?" Are they still active in the Pacific region?

Key Terms + Concepts

Aboriginal languages (p. 291)
  non-Pama-Nyungan (p. 291)
  Pama-Nyungan (p. 291)
Aborigines (p. 290)
Alliance of Small Island States (p. 297)
Antarctic Treaty (p. 305)
Asia-Pacific Economic Cooperation (APEC) group (p. 301)
atoll (p. 288)
Australia New Zealand United States (ANZUS) security alliance (p. 297)
Austronesians (p. 291)
Austronesian language family (p. 291)
  Formosan language subfamily (p. 292)
  Malayo-Polynesian subfamily (p. 292)
  Proto-Austronesian (p. 291)
cargo cults (p. 293)
Closer Economic Relations Agreement (p. 301)
"coconut civilization" (p. 289)
Compact of Free Association (p. 298)
continental islands (p. 285)
comanagement (p. 302)
counter-mapping (p. 302)
deterritorialized states (p. 297)
Dreamtime (p. 301)
environmental determinism (p. 287)
exotic species (p. 289)
"extinction capital of the world" (p. 289)
feral animals (p. 303)
Footprints of the Ancestors (p. 301)
geologic hot spot (p. 287)
guano (p. 295)
high islands (p. 287)
Indo-Fijians (p. 295)
low islands (p. 287)
Melanesia (p. 283)
Micronesia (p. 283)
millenarian movements (p. 293)
money laundering (p. 295)
Native Title Bill (p. 302)
Near North (p. 301)
Oceania (p. 282)
Operation Crossroads (p. 298)
pidgin (p. 292)
Polynesia (p. 283)
power projection hub (p. 297)
seamount (p. 287)
shell banks (p. 295)
Songlines (p. 301)
terra nullius (p. 302)
Torres Strait Islanders (p. 291)
Trans-Pacific Partnership (p. 301)
Walkabout (p. 301)
Way of the Law (p. 301)
"white Australia" policy (p. 301)
Wik case (p. 302)

Notes

1. Tom L. McKnight, *Oceania: The Geography of Australia, New Zealand and the Pacific Islands* (Englewood Cliffs, N.J.: Prentice Hall, 1995), p. 181.
2. Ibid.
3. Jeff Siegel, "Tok Pisin," http://www.une.edu.au/langnet/definitions/tokpisin.html.
4. An excellent article on Tanna's John Frum Cargo Cult is Paul Raffaele's "In John They Trust," *Smithsonian,* February 2006.
5. David Jolly, "Priced Off the Menu? Palau's Sharks Are Worth $1.9 Million Dollars Each, a Study Says." *New York Times*, May 2, 2011, p. A12.
6. Quoted in David Pilling, "How America Should Adjust to the Pacific Century." *Financial Times*, November 17, 2011, p. 9.
7. Quoted in Geoffrey Blainey, *Triumph of the Nomads* (Melbourne: Sun Books, 1987), p. 181.
8. Norimitsu Onishi, "Rich in Land, Aborigines Split on How to Use It." *New York Times*, February 13, 2011, p. 5.
9. Ibid.

Online Resources

CourseMate: Make the most of your study time by accessing everything you need to succeed in one place. Read your textbook, take notes, review flashcards, watch videos, complete activities, take practice quizzes, and more—online with CourseMate. Log in at **www.cengagebrain.com.**

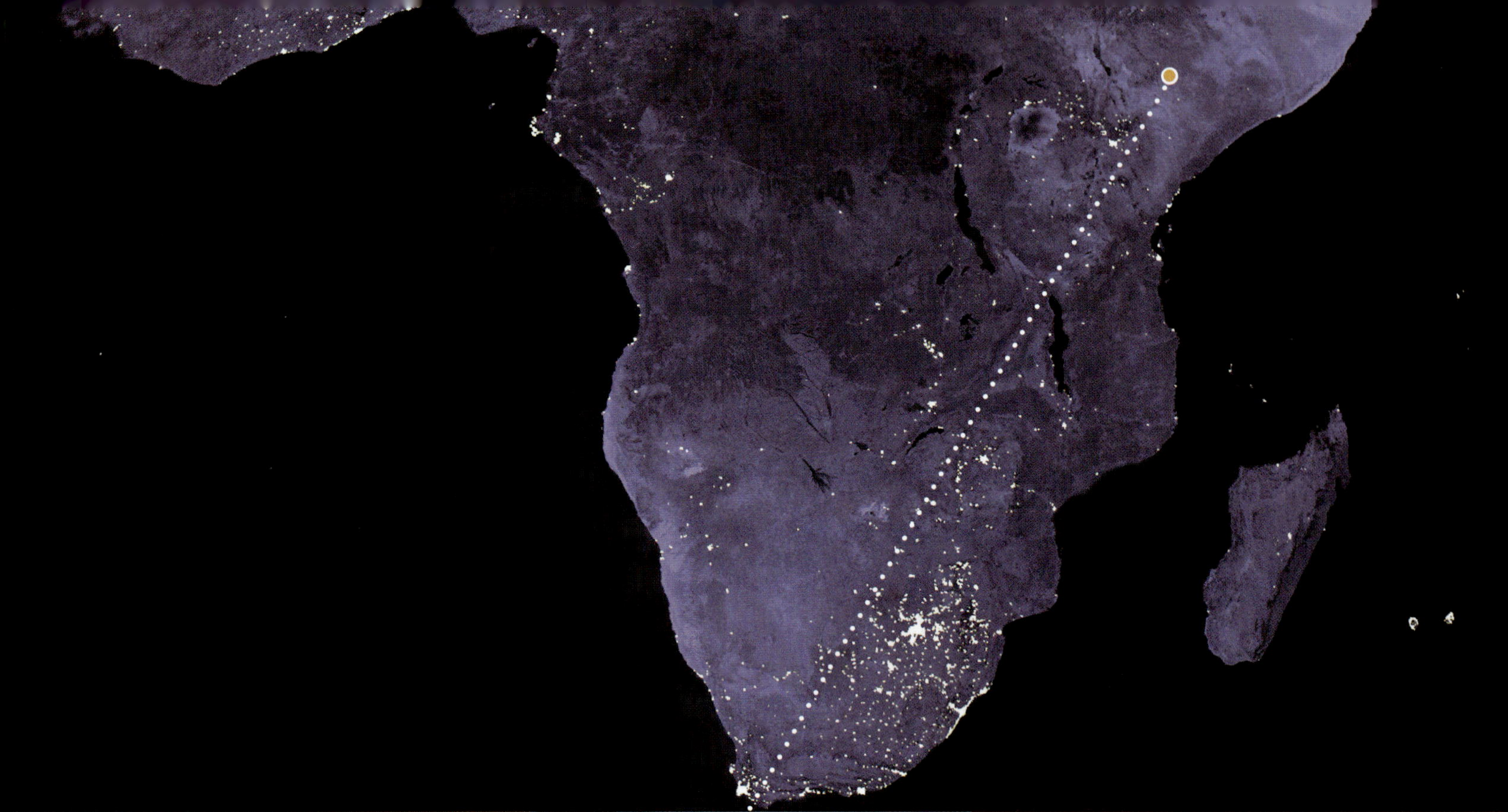

"I dream of the realization of the unity of Africa, whereby its leaders combine in their efforts to solve the problems of this continent. I dream of our vast deserts, of our forests, of all our great wildernesses."

—NELSON MANDELA

Africa at night. Note the lights of relatively prosperous South Africa on the world's poorest continent. Left: Doum palms rise from the banks of the Uaso Nyiro River in Kenya's Samburu National Park

C. Mayhew & R. Simmon (NASA/GSFC), NOAA/NGDC, DMSP Digital Archive

Joe Hobbs

9 Sub-Saharan Africa

Geographers typically recognize the continent of Africa as having two major divisions: North Africa (the predominantly Arab and Berber realm of the continent, described in Chapter 6), and ethnically diverse Sub-Saharan Africa (•**Figure 9.1**). This chapter sketches the broad outlines of the geography of Sub-Saharan Africa and examines issues in the major African subregions.

The region of Sub-Saharan Africa is culturally complex, physically beautiful, and problem ridden. This introduction to Africa surveys the region's diverse environments, peoples and ways of living, population distributions, European colonial legacies, and major current issues.

chapter outline

chapter objectives

This chapter should enable you to

- Understand what caused Sub-Saharan Africa to become and remain the world's poorest region.
- Know how the region came to have the highest HIV/AIDS infection rates in the world and how the epidemic could be reversed there.
- Appreciate the pressures on African wildlife and some unique approaches taken to protect the animals.
- Know what is uniquely African about African cultures.
- See how Africans have leap-frogged communications challenges with mobile phones, and use phones to overcome the "digital divide."
- Recognize why, after more than a decade at the sidelines, Sub-Saharan Africa is considered important again in geopolitical affairs.
- Recognize how European colonial favoritism of some ethnic groups over others sowed seeds of modern strife and warfare.
- Appreciate what corrupt leadership has done to impoverish people in countries with enormous oil and other natural resource wealth.
- Understand what causes the "resource curse."
- Understand how drought can trigger a process of environmental degradation.
- Consider the causes of the genocide and "ethnic cleansing" that marred Africa in the 1990s, particularly in East and East-Central Africa.

For many people, Sub-Saharan Africa is the most poorly understood or misunderstood of the world's regions. It seems to have so many countries, so much violence and disease, and so much poverty that it is simply too difficult to think about. Hollywood still perpetuates stereotyped images of an Africa rich in wildlife but with dangerous or backward peoples. This chapter will show that getting to know the real Africa is manageable, rewarding, and interesting. This region's diverse cultures and natural environments are sources of endless wonder. Conditions for many Africans are improving in a number of ways. There are many reasons for hope and much to celebrate in the study of this fascinating region. Just ask the people: a Gallup survey of 50,000 citizens around the world found that Africans are the most optimistic.[2]

9.1 Area and Population

Sub-Saharan Africa (including Madagascar and other nearby Indian Ocean islands, and the newest country of South Sudan) has the second largest land area of all the major world regions described in this book. It covers 8.7 million square miles (22.4 million sq km) and so is more than twice the size of the United States (•**Figure 9.2**).

Sub-Saharan Africa's population was 847 million in 2011. Even with the loss of population due to AIDS, the rate of natural population increase in Sub-Saharan Africa is 2.5 percent per year, or about five times that of the United States. The region's population is expected to double by 2050. As in most LDCs, African parents generally want large families for several reasons: to have extra hands to perform work; to be looked after when they are old or sick; and in the case of girls, to receive the "bride wealth" a groom pays in a marriage settlement. Large families also convey status. However, there are signs of significant change 50
in this pattern of population growth: birth rates have been dropping in every country in this region over the past two decades.

In spite of the large population, much of the region is sparsely populated. The region's average population density is 312
slightly more than that of the United States. Most of the people in this region live in a few small, densely populated areas (•**Figure 9.3**). The main areas are the coastal belt bordering the Gulf of Guinea in West Africa, from the southern part of Africa's most populous country, Nigeria, westward to southern Ghana; the savanna lands of northern Nigeria; the highlands of Ethiopia; the highland region surrounding Lake Victoria in Kenya, Tanzania, Uganda, Rwanda, and Burundi; and the eastern coast and parts of the high interior plateau of South Africa.

Sub-Saharan Africa is the world's most rural region (68%), with rural populations of most countries between 40 and 80 percent. The most rural are two East African nations with very fertile soils: Burundi (89% rural) and Malawi (86%). The most urbanized are Gabon (86%), where people are flocking to partake in an oil boom that is facilitating urbanization, and Djibouti (76%), where there is virtually no arable land and a small number of people are clustered in a port city. The region's major cities and large towns are magnets that attract many poor, rural people. The rate of urban growth is high; in 2011 there were twice as many cities with more than a million residents as there were in 1990 (•**Tables 9.1 and 9.2**). It is projected that 50 percent of Africans will live in cities in 2025. But for now, life in villages is the rule for the majority. A typical rural home is a small hut made of sticks and mud, with a dirt floor, thatched roof, and no electricity or plumbing (•**Figure 9.4**).

Africa has the world's youngest population, with 41 percent of the people under 15 years old (compared with about a third of the populations of Latin America and Asia). The median age is now 20 (compared with 30 in Asia and 40 in Europe). There are different ways of viewing this young, growing population. Some economists see the youthful median age as a "demographic dividend," as large numbers of workers will be able to contribute to the region's economic growth. Others fear that the Malthusian scenario will come about because of the gulf between population growth and agricultural output in these mainly agrarian societies. Since the 1960s, the population of Sub-Saharan Africa has grown at a rate of about 5 percent annually, whereas food production in the region has grown at only about 2 percent annually. This is the only world region where per capita food production declined in that period.

• **Figure 9.1** Principal features of Sub-Saharan Africa.

The wild card in Africa's population deck is the **human immunodeficiency virus (HIV)**. The disease it causes—**AIDS (acquired immunodeficiency syndrome)**—has inevitably been identified by some as the Malthusian "check" to the region's population growth (see Medical Geography, page 316).

9.2 Physical Geography and Human Adaptations

Sub-Saharan Africa is both rich in natural resources and beset with environmental challenges to economic development. It is home to some of the world's greatest concentrations of wildlife and some of the most degraded habitats.

The Landscapes of Africa

Most of Africa is a vast plateau, actually a series of plateaus, with a typical elevation of more than 1,000 feet (c. 300 m). Near the Great Rift Valley in the Horn of Africa and in southern and eastern Africa, the general elevation rises 2,000 to 3,000 feet (c. 600 to 900 m), with many areas at 5,000 feet (1,520 m) and higher (see Regional Perspective, page 317). The highest peaks and largest lakes of the continent are located in this belt. The loftiest summits lie within 250 miles (c. 400 km) of Lake Victoria. They include Kilimanjaro (19,340 ft/5,895 m) and Kirinyaga (Mount Kenya, 17,058 ft/5,200 m; **•Figure 9.5**), which are volcanic cones, and the Ruwenzori range (up to 16,763 ft/5,109 m), a nonvolcanic massif produced by faulting. Lake Victoria, the largest lake in Africa, is surpassed in area

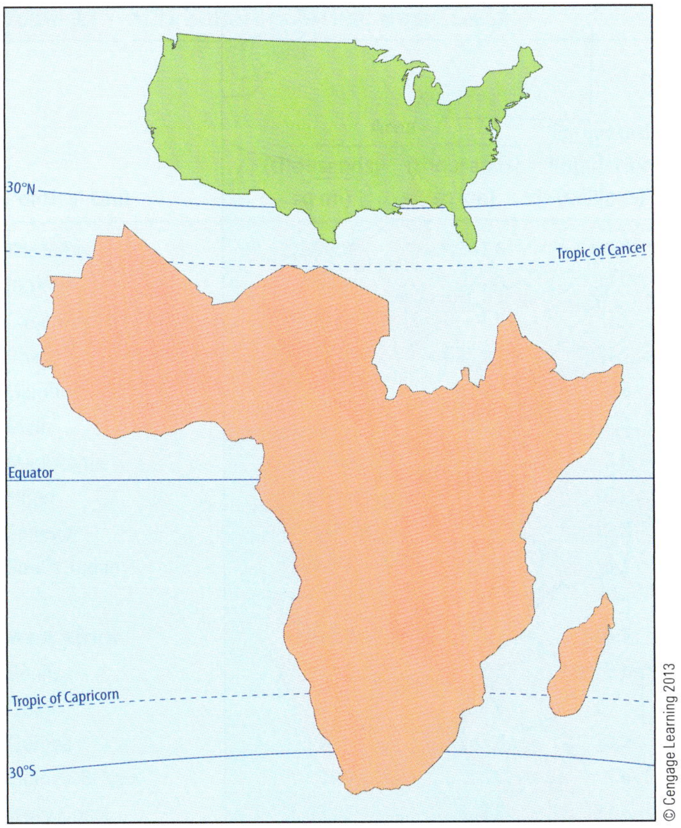

• **Figure 9.2** Sub-Saharan Africa compared in area and latitude with the conterminous United States.

among inland waters of the world only by the Caspian Sea and Lake Superior. It is relatively shallow, however, and the large numbers of people living on its shores are taxing its resources. Other very large lakes in East Africa include Lake Tanganyika and Lake Malawi.

The physical structure of Africa has influenced the character of African rivers. The main rivers, including the Nile, Niger, Congo, Zambezi, and Orange, rise in interior uplands and descend by stages to the sea. At some points, they descend abruptly, particularly at plateau escarpments, with rapids and waterfalls interrupting their courses. These often block navigation a short distance inland. The many waterfalls and rapids do have a positive side: they represent a great potential source of hydroelectric energy. There are major power stations on the Zambezi, Congo, Niger, Volta, and Blue Nile Rivers. But only about 5 percent of Africa's hydropower potential has been realized (compared to about 60% in North America). Many of the best sites are remote from large markets for power. In some cases, geopolitical considerations pose obstacles to dam construction. Downstream Egypt, for example, has expressed concern and even hostile rhetoric about dams and water diversions of the Nile and its tributaries by upstream Sudan, Ethiopia, Uganda, Kenya, and Tanzania (see page 180 in Chapter 6).

Africa's Biomes and Climates

The equator bisects Africa, so about two-thirds of the region lies in the low latitudes, having tropical climates and biomes; Africa is the most tropical of the world's continents (•**Figure 9.6**). Areas of tropical rain forest climate center on the great rain forest of the Congo Basin in central and western Africa. The forest merges gradually into a tropical savanna climate on the north, south, and east. This is the climatic and biotic zone supporting most of the magnificent large mammals of Africa. The savanna areas in turn trend into steppe and desert on the north and southwest. A broad belt of drought-prone tropical steppe and savanna bordering the Sahara on the south is known as the Sahel. There is desert on the coasts of Eritrea, Djibouti, and Somalia in the Horn of Africa. In South Africa and Namibia, a coastal desert, the Namib, borders the Atlantic. The Kalahari Desert, which lies inland from the Namib, is better described as steppe or semidesert than as true desert. Along the northwestern and southwestern fringes of the continent are small but productive areas of Mediterranean climate, while eastern coastal sections and adjoining interior areas of South Africa have a humid subtropical climate. Bordering the subtropical climate region is an area of marine west coast climate.

Total precipitation in the region is high but unevenly distributed; some areas are typically saturated, whereas others are perennially bone dry. Even in many of the rainier parts of the continent, there is a long dry season, and wide fluctuations occur from year to year in the total amount of precipitation. One of the major needs in Africa is better control over water. In the typical village household, women carry water from a stream or lake or a shallow (and often polluted) well. Use of more small dams would help provide water storage throughout the year.

Drought is a persistent problem in most of the countries. Although all droughts create problems, some last for years, with devastating effects in this heavily agricultural region. Droughts have been particularly severe in recent decades in the Sahel and in the Horn of Africa. The number of food emergencies, most of them tied to drought, has tripled since the mid-1980s in Sub-Saharan Africa.

The consensus among scientists is that this region will pay the highest price for global climate change, even though it contributes the fewest greenhouse gases to the atmosphere. United Nations studies predict even more severe droughts and crop losses, along with the destruction of coastal infrastructure that will accompany rising sea levels. Such dangerous environmental changes may be contributing to conflict in the region as well. Pastoralists and farmers already compete for scarce land and water resources in many places—this was one of the problems in Sudan's Darfur conflict, for example—and droughts intensify those struggles.

The patterns of Africa's land use (•**Figure 9.7**) reveal that the most productive lands are on river plains, in volcanic regions (especially the East African and Ethiopian highlands), and in some grassland areas of tropical steppes (notably the High

MEDICAL GEOGRAPHY

HIV and AIDS in Africa

AIDS is a global problem. The earth has not seen a **pandemic** (very widespread epidemic) like this since the bubonic plague devastated fourteenth-century Europe and smallpox struck the Aztecs of sixteenth-century Mexico. AIDS had killed over 25 million people worldwide by 2012, and there are almost 3 million new infections around the world each year. Authorities say that worldwide 7,400 people are infected every day. Sub-Saharan Africa is where the virus originated (apparently among chimpanzees; the first human case was reported in the Belgian Congo—today's Democratic Republic of Congo—in 1959) and is today the epicenter of this health crisis. The statistics are startling. In 2012, two-thirds of the world's estimated 33 million people infected with HIV lived in Sub-Saharan Africa.

The epidemic is most severe in southern Africa. In some countries, including Botswana, Lesotho, and Swaziland, as much as one-quarter of the adult population is HIV positive (•**Figure 9.A**). South Africa has more HIV cases (about 5.7 million) than any other country in the world. This scourge has been causing sharp reductions in life expectancy in Africa and, unless contained, will dramatically alter the region's population and development trends.

The good news is that HIV/AIDS is starting to be contained. The rate of new infections is down about 25 percent in the worst affected countries, mainly in southern Africa. Death rates, which were rising precipitously, are falling. This is brightening the prospects for development in this least developed of the world's regions. The incidence of HIV infection is particularly high among the region's most educated, skilled, and ambitious young urban professionals, including teachers, white-collar workers, and government employees. These are the people, in addition to truckers and merchants, who travel the most, have higher incomes, and are therefore the most likely to have many sexual liaisons.

What can be done to fight the AIDS epidemic? Two countries that have had remarkable success in stemming the AIDS tide provide some answers. Uganda and Senegal have used relatively inexpensive public health tools, especially education, in the fight against the disease. To reduce the huge stigma associated with AIDS and to encourage counseling, testing, and condom use, the countries' political and religious leaders have been outspoken about AIDS. The results have been impressive, with infection rates dropping in Uganda (from 18% to 6%) and staying low in Senegal.

In addition, wealthier countries outside Africa could invest more in combating and preventing the disease. The United Nations in 2001 established the Global Fund to Fight AIDS, Malaria, and Tuberculosis. Bono and other celebrity activists continue to pressure governments to pledge more monies to the Global Fund. But their efforts are flagging due to the global Great Recession and "donor fatigue."

The United States has its own effort to combat AIDS in 12 countries in Africa and 2 (Haiti and Guyana) in the Americas. In a program called **PEPFAR**, the President's Emergency Plan for AIDS Relief, the George W. Bush administration pledged over $45 billion to the effort. The United States gives more money to combat AIDS than all other donors combined. This is an enormous amount of money, and yet it is dwarfed by the scale of the problem: it is enough to treat only about one in four people infected with HIV. The Bush and Obama administrations spent most of those funds to put patients on life-lengthening **antiretroviral (ARV) drugs** and other drugs. About 5 million people worldwide were on these drugs in 2012, and another 10 million needed but could not get them.

The Obama Administration could not afford the escalating costs of treatment, and shifted the focus to prevention and mother/

CITES member states won the ivory ban over the strong objections of several southern African states, led by Zimbabwe, that were experiencing what they saw as an elephant overpopulation problem. At the 1989 CITES meeting, these countries appealed for an exemption from the ivory ban so that they could earn foreign export revenue from a sustainable yield of their elephant populations. The majority of CITES members rejected this appeal, arguing that any loophole in a complete ban would subject elephants everywhere to illegal poaching. Since 1989, CITES has periodically allowed Zimbabwe, Namibia, Botswana, and South Africa one-time sales of ivory under rigorous monitoring as a reward for their positive wildlife policies. All continue to cull (kill) their "excess" elephants, with the meat going to needy villagers and crocodile farms.

If elephant ivory is like gold, rhinoceros horns are like diamonds. Men in the Arabian Peninsula nation of Yemen prize daggers with rhino horn handles (•**Figure 9.10**). Although Western scientists deny the medicinal efficacy of powdered rhino horn, traditional medicine in East Asia makes wide use of it, including as an aphrodisiac and a cure for cancer. These demands, and the current black-market value of tens of thousands of dollars per horn ($45,000 per pound, more than gold, cocaine, or heroin), have led to a precipitous decline in population of black rhinoceroses in Africa. There were an estimated 65,000 black rhinos in Africa in 1982; in 2012, there were an estimated 3,600. Despite aggressive anti-poaching efforts, the Western Black Rhino became extinct in the wild in 2011, leaving three other subspecies with critically low numbers. Even rhinos mounted on display in European museums have had their horns poached!

9.3 Cultural and Historical Geographies

Many non-Africans are unaware of the achievements and contributions of the cultures of Sub-Saharan Africa. The African continent was the original home of humankind. Recent DNA studies suggest that the first modern people (*Homo sapiens*)

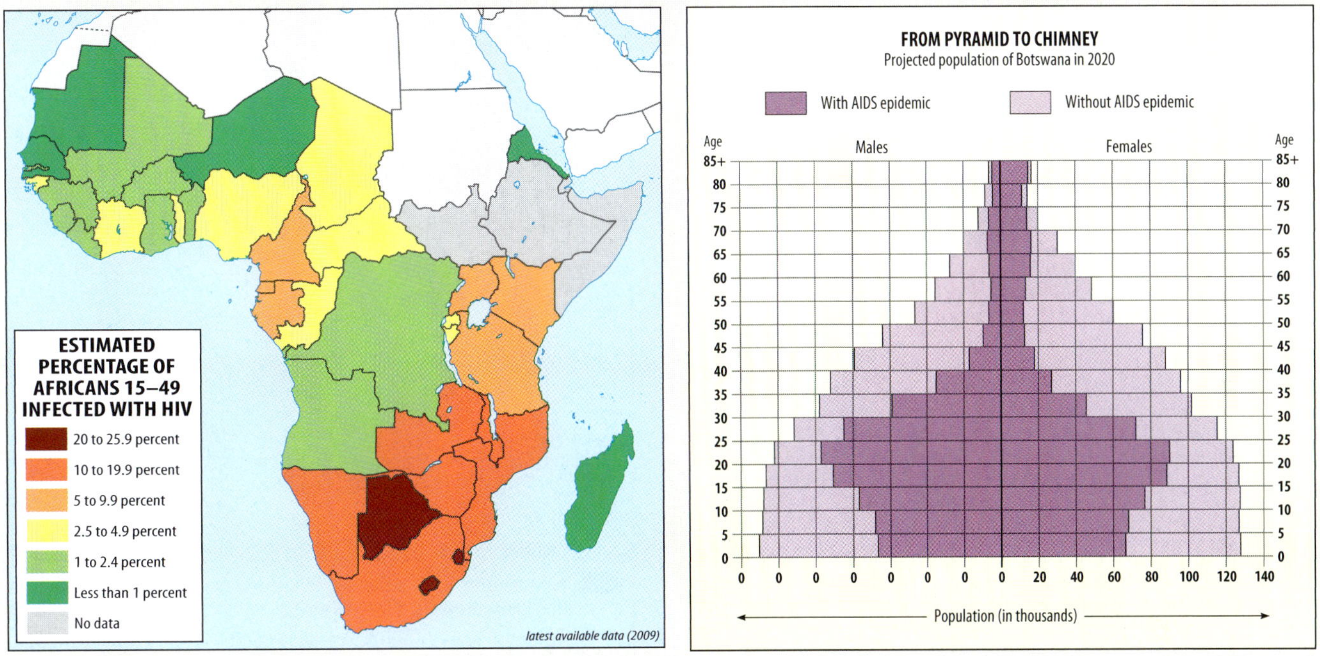

• **Figure 9.A** HIV/AIDS has had a devastating impact in Sub-Saharan Africa and threatens to redraw the demographic profiles of some countries in unprecedented ways.

child health. For sexual transmissions, the main weapons are male circumcision, a vaginal microbicidal gel, and of course, condoms. Needle-exchange programs are most important in preventing transmission among drug users. So far though, the math is heartbreaking. In the best-case scenario, the Global Fund and PEPFAR jointly will allow 10 million people to be put on treatment by 2014. An estimated thirty-six million will be infected by then. For every 100 people put on treatment, 250 more are newly infected. The overwhelming majority of the newly infected will die untreated. AIDS activists are angry about the spending focus of existing programs, saying that children will be saved now only to die later of AIDS.

to inhabit Asia, Europe, and the Americas were descendants of a small group that left Africa via the Isthmus of Suez about
49 100,000 years ago. After about 5000 B.C.E., indigenous people were responsible for agricultural innovations in four culture hearths: the Ethiopian Plateau, the West African savanna, the West African forest, and the forest-savanna boundary of West Central Africa. Africans in these areas domesticated important crops such as millet, sorghum, yams, cowpeas, okra, watermelons, coffee, and cotton. From Africa, these diffused to populations in other world regions. 82

REGIONAL PERSPECTIVE The Great Rift Valley

One of the most spectacular features of Africa's physical geography is the **Great Rift Valley**, a broad, steep-walled trough extending from the Zambezi Valley (on the border between Zimbabwe and Zambia) northward to the Red Sea and the valley of the Jordan River in southwestern Asia (see Figure 9.1 and also Figure 6.9, page 163 and Figure 2.1, page 22). Its relationship to the tectonic movement of crustal plates is still poorly understood. However, most earth scientists believe it marks the boundary of two crustal plates that are rifting, or tearing apart, causing a central block between two parallel fault lines to be displaced downward, creating a linear valley. This movement will eventually cut much of southern and eastern Africa away from the rest of the continent and allow seawater to fill the valley.

The Great Rift Valley has several branches. Lakes, rivers, seas, and gulfs already occupy much of it. It contains most of the larger lakes of Africa, although Lake Victoria, situated in a depression between two of its principal arms, is an exception. Some, like Lake Tanganyika—at 4,823 feet (1,470 m), the world's second-deepest lake, after Russia's Lake Baikal—are extremely deep. Most have no surface outlet. Volcanic activity associated with the Great Rift Valley has created Kilimanjaro, Kirinyaga, and some of the other great African peaks, along with lava flows, hot springs, and other thermal features. Faulting along the Great Rift Valley in Ethiopia, Kenya, and Tanzania has also exposed remains of the earliest known ancestors of *Homo sapiens*. 39

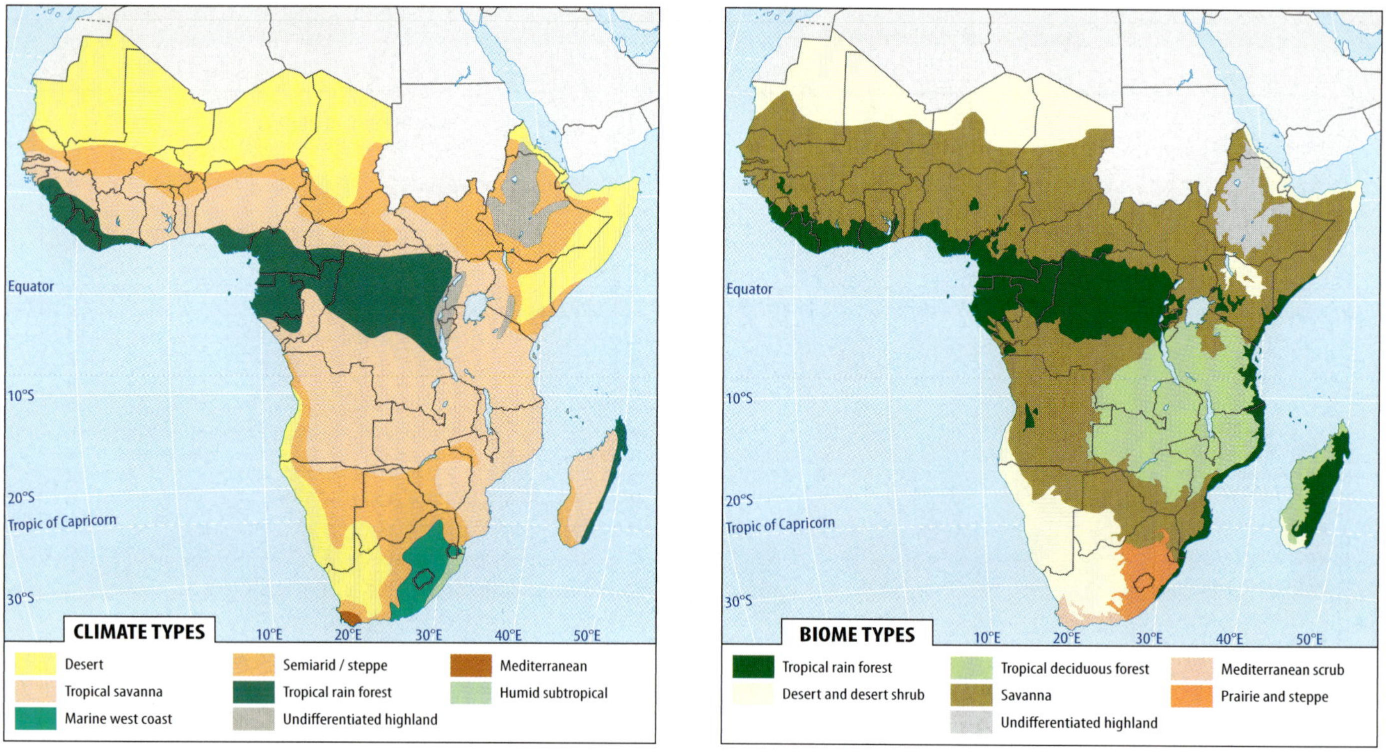

•Figure 9.6 Climates (left) and biomes (right) of Sub-Saharan Africa.
Sources: (left) Based on Rand McNally's Classroom Atlas, 2003. (right) Based on World Wildlife Fund ecoregions data, 1999.

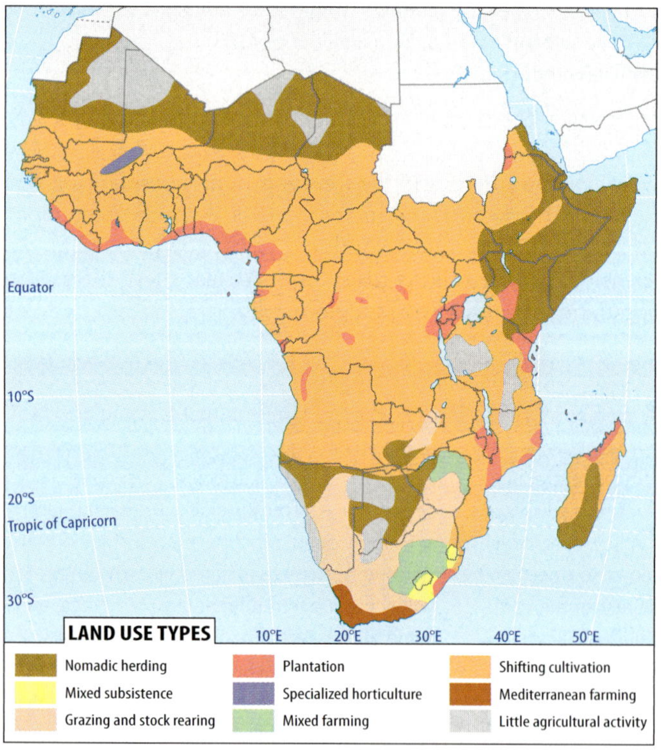

•Figure 9.7 Land use in Sub-Saharan Africa.
Source: The Oxford School Atlas edited by Patrick Wiegand (OUP, 1997), copyright © Oxford University Press. Used by permission of Oxford University Press.

Joe Hobbs

•Figure 9.8 Mother and child in Zimbabwe. Women plant and harvest most of Africa's food and care for most of its children.

Joe Hobbs

• **Figure 9.9** Zebu cattle are extremely important in the cultures and household economies of rural Madagascar. There are about 11 million cattle in this country of 21 million people.

Civilizations and empires emerged in Ethiopia, West Africa, West Central Africa, and South Africa. In the first century C.E., a Christian empire based in the Ethiopian city of Axum controlled the ivory trade from Africa to Arabia. Ethiopian tradition holds that a shrine in Axum still contains the biblical Ark of the Covenant and the tablets of the Ten Commandments, which
170 disappeared from the Temple in Jerusalem in 586 B.C.E. Several Islamic empires, including the Ghana, Mali, and Hausa states, emerged in West Africa between the ninth and nineteenth centuries. All of these agriculturally based civilizations controlled major trade routes across the Sahara. They profited from the exchange of slaves, gold, and ostrich feathers for weapons, coins, and cloth from North Africa. Three kingdoms arose between the fourteenth and eighteenth centuries in what are now the southern Democratic Republic of Congo and northern Angola. These included the Kongo kingdom, which had productive agriculture and was the hub of an interregional trade network for food, metals, and salt. In what is now Zimbabwe, the Karanga kingdom of the thirteenth to fifteenth centuries built its capital city at the site known as Great Zimbabwe. Its skilled metalworkers mined and crafted gold, copper, and iron, and merchants traded these metals with faraway India and China.

Joe Hobbs

• **Figure 9.10** Daggers are a nearly universal dress accessory for men in the Arabian Peninsula nation of Yemen. The most prized dagger handles are of rhino horn, a custom that has had a devastating impact on African rhinos thousands of miles away.

The Languages of Africa

Recent scientific studies have concluded that humans first began using language in southwestern Africa between 50 and 100 thousand years ago. Today, there is great linguistic diversity in Sub-Saharan Africa (•**Figure 9.11**). Africans speak about 2,000 of the world's 6,000 languages, belonging to four broad language groups.

The **Niger-Congo language family** (sometimes considered a subfamily of the Niger-Khordofanian family) is the largest. It includes the many West African languages and the roughly 400 Bantu subfamily languages that fall into seven branches: **Benue-Congo, Kwa, Atlantic, Mandé, Gur, Adamawan,** and **Khordofanian**. Most of these are spoken south of the equator. The *Bantu* language of the Benue-Congo branch is the most widespread.

The Afro-Asiatic language family includes Semitic languages (such as the *Amharic* language of Ethiopia and *Arabic*) and tongues of the **Cushitic** (*Oromo* and *Somali* of the Horn of Africa, for example) and **Chadic** (especially the *Hausa* of northern Nigeria) branches. People living in the area adjoining the Sahara, from West Africa to the Horn of Africa, speak these languages. Even some of the Niger-Congo languages originating south of the Sahara, such as the *Swahili (Kiswahili)* tongue spoken widely in East Africa, have borrowed much from Arabic and other languages with roots elsewhere. The prominence of Arabic words in Swahili reflects a long history of Arab seafaring along the Indian Ocean coast of Africa; in fact, Swahili means "coastal" in Arabic.

The **Nilo-Saharan language family** of the central Sahel region, the northern region of West Central Africa, and parts of East Africa includes about 100 languages in three branches: *Songhai* (a single language, spoken mainly in Mali) and the **Saharan** and **Chari-Nile** language groups.

The **Khoisan languages** of the **San** and related peoples in the western portion of southern Africa are nearly extinct. The

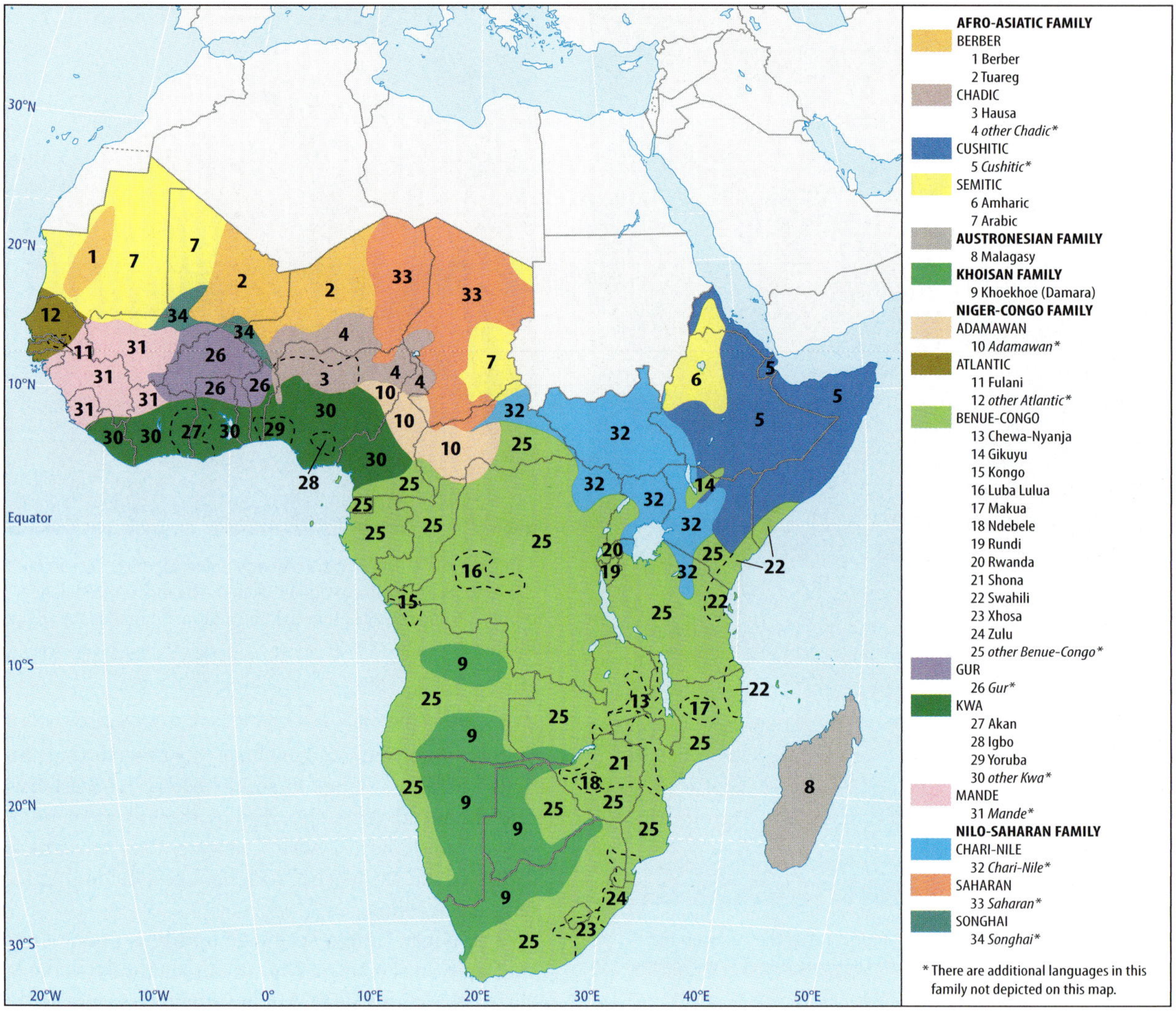

• **Figure 9.11** Languages of Sub-Saharan Africa.
Source: Adapted with permission from Bernard Comrie, et al., The Atlas of Languages, Revised Edition (New York: Facts On File, 2003). Reprinted with permission of the Publisher.

largest, *Khoekhoe (Damara)*, is spoken by fewer than 200,000 people in Namibia. Recognizable by their prominent "clicking" sounds, these may be humankind's oldest languages.

Quite distinct from those four groups, the people of Madagascar speak a language without African roots. This is *Malagasy*, an
291 Austronesian tongue that originated in Southeast Asia.

Africa's list of lingua francas—the languages most likely to be recognized on the continent and those that Africans can use to speak to the wider world—is provided by the six official languages of the African Union, the continent's chief supranational
85 organization. They are English, French, Portuguese, Spanish, Swahili, and Arabic. Use of English is growing most rapidly, even in the formerly French-dominated "Francophone" countries. Countries that formerly were British colonies have a distinct advantage in the world of outsourced services: South Africa, Kenya, and Ghana, for example, have already opened call centers that they hope will rival those of India. Some Francophone countries also have call centers catering to French industries and clients.

The dominance of foreign languages in computers and other digital technologies could endanger African languages, many of which are already in peril because they are spoken by so few people. As elsewhere, many languages in Africa are threatened by language death. Many linguists, however, see the computer 226
as a tool for *saving* the roughly 3,000 endangered languages of the world. They argue that the computer will make it much easier for people to learn and preserve their traditional tongues. Australian linguist James McElvenny uses software to revitalize vanishing languages (look up the Enduring Voices Project on the web). He gives young language learners portable electronic reference tools that give the definitions and sounds of words that are otherwise no longer spoken. Dr. Tucker Childs, a linguist working in the West African nation of Sierra Leone, is using a geographic information system (GIS) to plot locational variations of Kim and other endangered languages. With these findings, linguists can decide where to focus their efforts.

Africa's Belief Systems

The religious landscape of Africa is complex and fluid. Spiritualism is extremely strong, but spiritual affiliations and practices are more interwoven and flexible than in most other

INSIGHTS Africa's Greatest Conservationist

Not much larger than the common housefly, the tsetse fly of Sub-Saharan Africa packs a wallop. This insect carries two diseases, both known as **trypanosomiasis**, which are extremely debilitating to people and their domesticated animals. People contract **sleeping sickness** from the fly's bite, and cattle contract nagana. Since the 1950s, widespread efforts have been made to eradicate tsetse flies so that people can grow crops and herd animals in currently fly-infested wilderness areas (•**Figure 9.B**). Where the efforts have been successful, people have cleared, cultivated, and put livestock on the land. The results are mixed. Although people have been able to feed growing populations in the process of opening up these lands, they have also eliminated important wild resources and in many cases caused erosion, desertification, and salinization of the land. Where the tsetse fly has been eliminated, so has the wilderness. The diminutive tsetse fly thus may be characterized as a **keystone species**—one that affects many other organisms in an ecosystem. The loss of a keystone species—in this case, a fly that keeps out humans and cattle—can have a series of destructive impacts throughout the ecosystem. For its role in maintaining wilderness in Sub-Saharan Africa, some wildlife experts call the tsetse fly "Africa's greatest conservationist."

Joe Hobbs

• **Figure 9.B** Strenuous and successful efforts to eradicate tsetse flies have opened vast new areas of Africa to human use. This is a tsetse fly trap with two components: a liquid chemical called "simulated cow's breath" that attracts the fly and a pesticide-soaked tarp that kills it.

world regions. It is not uncommon for family members to follow different faiths or for an individual to change his or her religious beliefs and practices in the course of a lifetime.

Broadly, however, some dominant patterns of religious geography can be recognized (•**Figure 9.12**). Islam is the dominant religion in North Africa and the countries of the Sahel on the southern fringe of the Sahara. The **Ethiopian Orthodox Church**, closely related to the Coptic Christian faith of Egypt, makes Ethiopia an exception to the otherwise Islamic Horn of Africa region (•**Figure 9.13**). Islam is also the prevailing religion of the East African coast, where Arab traders introduced the faith. Muslims are a majority or strong minority in rural northern Nigeria and Tanzania, and there are minority Muslim populations in cities and towns across the continent. Christians are a majority in southern Nigeria, Uganda, Lesotho, and parts of South Africa. Both Christianity and Islam are strongest in the cities, whereas traditional religions prevail or overlap with these monotheistic faiths in rural areas.

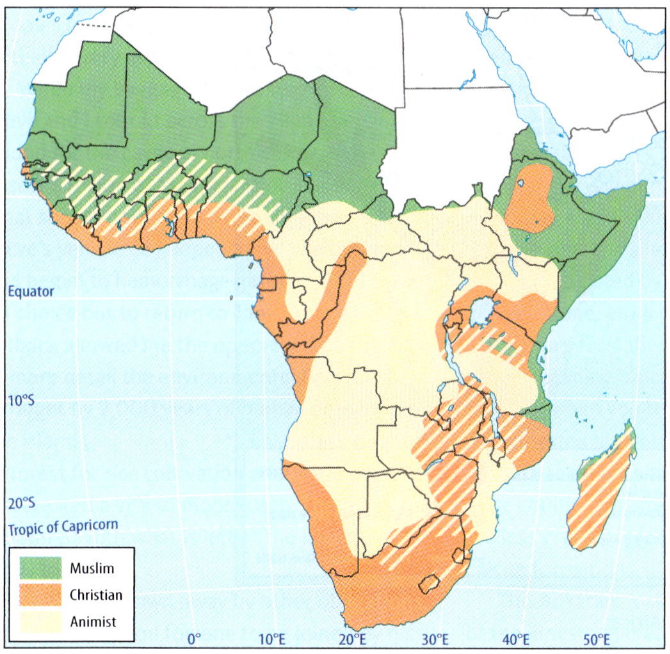

© Cengage Learning 2013

• **Figure 9.12** Religions of Sub-Saharan Africa.

Christians and Muslims have mounted efforts to dispose of "heretical" notions in African traditional belief systems, but these concepts are very resilient. Even where the large monotheistic faiths nominally prevail and their followers are devout, a strong substrate of indigenous beliefs has persisted and usually comfortably merged with the "official" faith—a syncretism also present in other cultures of the world. One example 228,
is the mingling of indigenous *vodun* ("voodoo") beliefs with 365
Christianity and Islam in West Africa.

Indigenous African religions have many animistic elements, emphasizing that spiritual forces are manifested everywhere in the environment. Many spirits are tied to particular places in the landscape. Such beliefs contrast with the introduced Christian and Muslim faiths, which tend to see

of those pledged by Africa's traditional financiers in the West. There is an important difference, however. Western institutions tend to have strings attached to loans and aid, insisting that African countries adopt democratic and other reforms to be eligible for the monies. China, in contrast, has a "no strings attached" policy. It is not insisting on political reforms, is not expressing concern about environmental or social responsibilities, and much to the chagrin of the West is even engaging with repressive governments that the West is trying to isolate. At the same time, Western governments and financial institutions
237, 335, 367, 373 are increasingly alarmed that China is making rapid inroads into African markets at the expense of Western economies and geopolitical interests.

Even Africans look with some trepidation of China's advance. Some complain about China's record on not hiring locals, instead bringing in Chinese nationals for virtually all labor and management positions. In other cases, Chinese firms have reportedly mistreated African workers and paid them what have been described as "slave wages." Some Chinese projects have been documented as having negative environmental impacts.

China is unabashed about its interests in Sub-Saharan Africa: it needs and wants the region's raw materials—especially oil, iron ore, copper, and cotton—to feed its insatiable industrial appetite. To many Africans, China is building the same kind of exploitive mercantile relationship that characterized the colonial West: it takes Africa's raw materials, transforms them into value-added products, and sells them mainly to its
367 own advantage. South Africa's president warned in 2007 that Africa risked becoming an economic colony of China. But desperate for assistance, even the wary are inclined to accept China's terms. The most typical arrangement is the "mines for infrastructure" deal, in which the African country grants mining concessions to the Chinese, and the Chinese build railways, roads, schools, and hospitals. There is especially great appreciation for the schools—there are simply not enough of them in many African countries.

China's model of state-led capitalism has proven attractive to many African governments. South African government officials go to Beijing to learn how to run state-owned companies more profitably. China is helping several African countries to build special economic zones, just as China did to spur its own
268 economic growth.

China and the West do not call all the shots. There are a growing number of African institutions designed to encourage economic growth and political stability. In 2002, the 53-member **African Union (AU)** was formed to replace the old and often inconsequential Organization of African Unity (OAU). One of the African Union's main articulated goals is to implement self-inspection or "peer review" to help promote democracy, good governance, human rights, gender equity, and development. It has formed an agency responsible for this monitoring, known as the **New Partnership for Africa's Development (NEPAD)**. NEPAD's premise is that improvements in such human affairs will improve the climate for international investment and assistance to Africa. The African Union has also created a 265-member Pan-African Parliament to make regional laws.

9.5 Geopolitical Issues

Sub-Saharan Africa has waxed and waned as a theater of geopolitical interest since the end of World War II. During much of the twentieth century, the great powers were interested in the region. To boost their competing aims during the Cold War years, the Soviet Union and the United States played African countries against one another, arming them with weapons with which to wage **proxy wars**. The superpowers also extended aid generously to many African nations, but not simply for altruistic reasons. 178, 242

Great power concerns about Africa changed with the end of the Cold War around 1990, and the nature of conflict in Africa changed. The great powers withdrew support, but their weapons remained to prolong smoldering conflicts. The United States helped build the arsenals of eight of the nine countries involved in the recent Democratic Republic of Congo conflict, for example. Cold War-era weapons like AK-47 assault rifles were 339
"dumped"—sold at low prices—in Africa, where they were not considered obsolete and where they fanned the flames of conflict. Increasingly, fighting began to spread across international borders. Previously, the United States and the Soviet Union had maintained a kind of security balance that kept warfare from becoming internationalized, but that restraint no longer exists.

The scenario of regional or even Africa-wide wars was realized with Africa's first "world war," centered on the Democratic Republic of Congo (see page 340). In the absence of regional or international powers to keep the peace, some countries (for example, Sierra Leone) simply imploded, fragmenting into fiefdoms run by factions that claimed to be revolutionaries but were essentially profiteers. Other national governments (as in Ethiopia) spent vast sums on expensive military aircraft while their people suffered from malnutrition.

The end of the Cold War constricted (and in the case of Russia, ruptured) aid pipelines. The United States cut its economic assistance to Sub-Saharan Africa by 30 percent between 1985 and 1992 and reduced it another 10 percent between 1992 and 2001. Why the waning concern? Former U.S. President George W. Bush was quoted as saying, "While Africa may be important, it doesn't fit into the national strategic interests, as far as I can see them."[5]

Then came 9/11. Al-Qa'ida's attack on the United States and the U.S. counterstrike against al-Qa'ida in Afghanistan raised fears that the Islamist organization might find new recruits and training grounds in Africa. Al-Qa'ida had already bombed U.S. embassies in Kenya and Tanzania in 1998 and in 2002 struck again in the Kenyan city of Mombasa. The United States maintains a focus on **terrorism hot spots** in the region, including Kenya, Somalia, Djibouti, Niger, Chad, and Mali, all of which are regarded as real or potential training grounds for al-Qa'ida-affiliated groups. As related on the next page, U.S. forces have made the tiny Horn of African country of Djibouti a major military center for potential action against nearby Somalia, Yemen, and other countries where Islamist militants are active. So far, U.S. interest in the hot spots has translated into military expenditures and partnerships with African governments, but not into wider development assistance.

In addition to the terrorism hot spots, and the Somali piracy that occasionally threatens commercial traffic in the region's vital sealanes, the United States has strong geopolitical interests in the region's oil-producing countries. These too became more critical after 9/11 and the U.S. war in Iraq, which heightened the sense that Middle Eastern oil supplies were
415 vulnerable. The United States is seeking more stable oil supplies and has pinned many hopes on Africa. Petroleum-rich Nigeria, Cameroon, Gabon, Congo, Angola, and Equatorial Guinea together in 2011 supplied 14 percent of the oil used by the United States, a figure projected by the U.S. Central Intelligence Agency to rise considerably in the coming years.

Increasingly, the United States will be competing with China for these dwindling finite fossil reserves, and given current trends, the geopolitical landscape of Sub-Saharan Africa will see Asia on the rise and American clout in decline. Countries that are not pleased with Washington's wishes or demands will find partners elsewhere, particularly in Asia. Some military analysts describe the emerging U.S.-China rivalry in Africa as a modern "Great Game", comparable to the British-Russian
241 contest over central Asia in the nineteenth century.

For many years following the deaths of U.S. service personnel in Somalia in 1993 (the "Blackhawk Down" incident), there was great reluctance to project U.S. military power in African trouble spots. Even Al-Qa'ida's simultaneous attacks on U.S. embassies in Kenya and Tanzania in 1998 did not change American military commitments in the region. The 9/11 attacks did, however. Since 2001, the United States has maintained a military base at Camp Lemonnier in Djibouti. Along with bases in Ethiopia and the Seychelles, it has been very useful in projecting U.S. power into Somalia and Yemen. Unmanned drones equipped with laser-guided weapons have been the weapons of choice against suspected al-Qa'ida (and in
183, 238 Somalia, al-Qa'ida's affiliate al-Shabab) targets.

The United States views South Africa as the region's economic powerhouse and as the key to regional stability. South Africa's military strength, which included nuclear weapons until they were decommissioned in 1990, contributes to its global significance. It is strategically important for its frontages on both the Atlantic and Indian Oceans, its possession of Africa's finest transport network (vitally important to many African countries that trade with and through South Africa), and its diversified mineral wealth.

Finally, HIV/AIDS in the region is also a geopolitical concern. Despite Africa's apparent remoteness from the United States, the two are linked by hundreds of air traffic routes traveled by thousands of people each day. Hundreds of HIV-positive Africans, perhaps most of them unaware of their virus, pass through U.S. airport gateways daily. Some African strains 316
(subtypes) of the virus are more easily transmitted by heterosexual contact than the strains prevalent in the United States, so the risk to the population at large is greater. There are other dimensions of the potential burden on the United States. As long as the epidemic rages in Africa, there may be large flights of "medical refugees" seeking asylum in the United States. There could also be AIDS-related political instability or civil wars that invite U.S. military intervention in the region. How the world should respond to poverty and recurrent crises is one of the key questions facing the huge, resource-rich, problem-ridden but promising region of Sub-Saharan Africa. The following section examines some of these challenges in greater detail.

9.6 Regional Issues and Landscapes

The Sahel

Drought and Desertification in the Sahel

The Sahel region extends eastward from the Cape Verde Islands to the Atlantic shore nations of Mauritania, Senegal, and the Gambia, and inland to Mali, Burkina Faso ("Land of the Upright Men," formerly Upper Volta), Niger, Chad, and the new nation of South Sudan (•Figure 9.20).

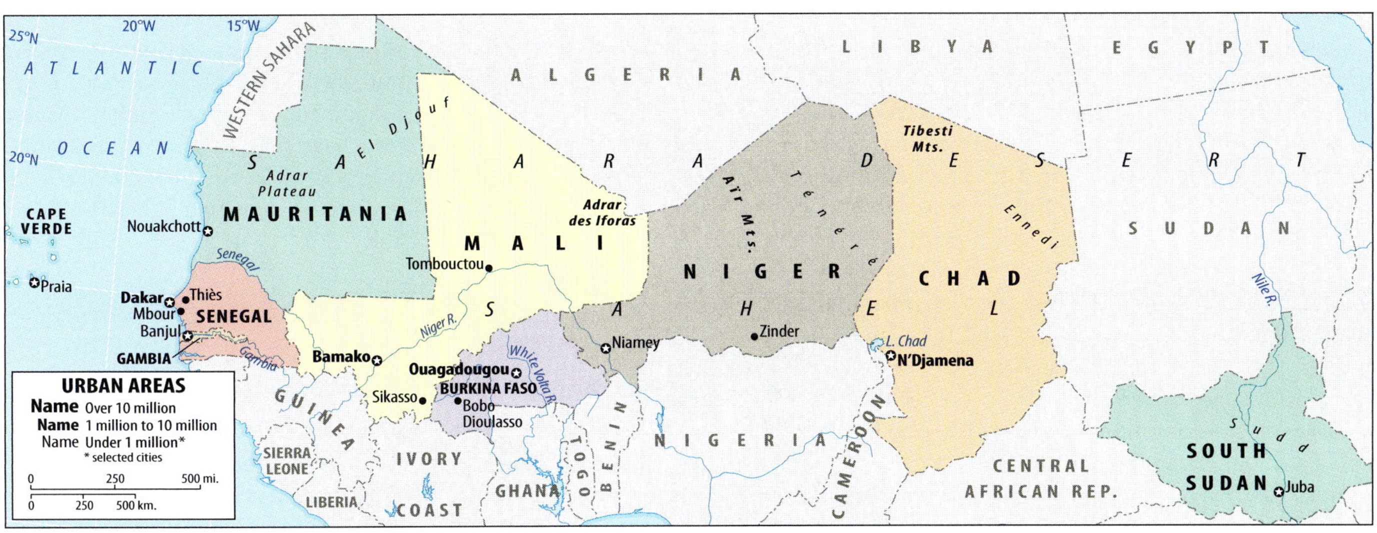

• Figure 9.20 Principal features of the Sahel.

The Sahel region, like much of Sub-Saharan Africa, experiences periodic drought. Drought is a naturally occurring climatic event in which rain fails to fall over an area for an extended period, often years. When rain does return, the arid and semiarid ecosystems of the Sahel come to life with a profusion of flowering plants, insects, and herbivores like gazelles, whose populations climb when foods are abundant.

These ecosystems have high **resilience**, meaning they are able to recover from the stress of drought and have mechanisms to cope with a natural cycle that includes periods of dryness and rain.

Desertification is the destruction of that resilience and the biological potential of arid and semiarid ecosystems. It is an unnatural, human-induced condition that has afflicted the Sahel periodically and severely since the late 1960s, and thus it is different from the natural phenomenon of drought. However, the recent history of the Sahel suggests that drought can be the catalyst that initiates the process of desertification. During decades of good rains prior to the late 1960s, the Fulani (Fulbe), Tuareg, and other Sahelian pastoralists allowed their herds of cattle and goats to grow more numerous. The animals represent wealth, and people naturally wanted their numbers to increase beyond the immediate subsistence needs
160 of their families. However, due to declining death rates, the numbers and sizes of families keeping livestock were also very large. Thus, the unprecedented numbers of livestock that built up during the rainy years made the Sahelian ecosystem vulnerable to the impact of drought on an unprecedented scale.

Drought struck the region in 1968, persisted through 1973, and returned over many years through 1985. With the drought, annual plants failed to grow, so the large herds of cattle and goats turned to acacia trees and other perennial sources of fodder. They ate all of the edible vegetation. This destruction was a critical problem because plants play an important role in maintaining soil integrity by helping intercept moisture and funnel it downward through the root system. Plant litter and decomposer organisms working around the plant contribute to soil fertility and help stabilize the soil. With hungry livestock in the region eating the plants, the landscape changed. No longer anchored and replenished, good soils eroded. "Junk plants" like Sodom apple (*Calotropis*) replaced palatable plants, and an almost impermeable surface formed on the land. This degraded ecosystem lost its resilience so that even when rains returned, vegetation did not recover promptly as it usually does.

Meteorologists studying the Sahel drought discovered another important link between precipitation and the removal of vegetation. They noticed that when people and livestock reduced the vegetative cover, they increased the **albedo**, the amount of the sun's energy reflected by the ground. Fewer plants meant more solar energy was deflected back into the atmosphere, lowering humidity and reducing local precipitation. This connection is known as the **Charney effect**, named for Jules Charney, the scientist who documented it in the early 1970s. It was a tragic sequence: People responding to drought actually perpetuated further drought conditions.

The 1968–1973 drought devastated the great herds of the Sahel. An estimated 3.5 million cattle died. Two million pastoral nomads lost at least 50 percent of their herds, and many lost as much as 90 percent. The Niger and Senegal rivers dried up completely. The cost in human lives was estimated at 100,000 to 250,000. Environmental refugees poured into towns unprepared to deal with such a large and sudden influx. Many pastoral nomads gave up herding and settled down to become wage earners and, where possible, farmers.

This crisis resulted in an international effort to combat desertification, using recommendations developed by the United Nations in 1977. The U.N. report, however, soon became a case study in how *not* to deal with an environmental crisis in the developing world. Most of the recommendations were universal, high-technology solutions that proved impossible to implement in the villages and degraded pastures of the Sahel. For example, many international agencies attempted to relieve the suffering of Sahelian pastoralists by using techniques such as digging deep wells to water livestock. But these agencies failed to anticipate that these wells would act like magnets for great numbers of people and animals; although there was plenty to drink, many animals starved to death when they decimated what vegetation remained around the new water supplies.

Since the disappointment of the 1977 U.N. plan, and with the lessons relief organizations learned as a result, there has been greater focus on sustainable development, with its local rather than universal solutions. In Niger, for example, hit by a 61
devastating drought as recently as 2005, farmers have ceased the traditional practice of clearing trees before planting crops and are now protecting trees and growing crops under and between them. They are also harvesting and selling many useful tree products. A key factor for the farmers' new concern for trees is that Niger's government released its long-held claim that all trees belonged to the state; now farmers recognize the trees as private property. New attitudes and practices like these, and a sustained period of good rains (which unfortunately have prompted periodic locust swarms), have combined to produce a remarkable regreening of much of the Sahel since 1985. Satellite imagery shows significant regrowth of vegetation in southern Mauritania, northern Burkina Faso, northwestern Niger, central Chad, and southern Sudan. Annual millet and sorghum harvests have risen in some cases as much as 70 percent above their pre-1985 levels. This change in the Sahel's fortunes is an important lesson that desertification is not inevitable or irreversible.

West Africa

The Poor, Oil-Rich Delta of Nigeria

West Africa extends from Guinea-Bissau eastward to Nigeria (•**Figure 9.21**). Its nine political units make up about 800,000 square miles (c. 2 million sq km), or nearly one-fourth of the area of the United States. These nine countries are Guinea-Bissau, Guinea, Sierra Leone, Liberia, Ivory Coast, Ghana,

• **Figure 9.21** Principal features of West Africa.

Togo, Benin, and Nigeria. The spatial, demographic, political, and economic giant of the region is Nigeria.

Oil is the economic lifeblood of populous Nigeria. Africa's largest oil producer, the country is well-endowed, ranking 10th in the world's proven reserves. Most of the production is concentrated in the Niger River Delta, home to about 12 million mostly Christian people (•**Figure 9.22**). They belong to several different ethnic groups who have one thing in common: they say they have derived few benefits and have suffered greatly from oil development in their homeland. They complain that for more than 30 years, thousands of oil spills have tainted their croplands and water, destroying their crops and fisheries, while the flaring off of natural gas has polluted their air and caused acid rain. They note that despite the enormous revenue generated from oil drilled on their land, little money has returned to the area; most goes to foreign oil companies like Shell and Chevron and to the Muslim-dominated government in the north. Meanwhile, most of them live in palm-roofed mud huts; half of the delta region lacks adequate roads, water supplies, and electricity. Schools have few books, and clinics have few medical supplies.

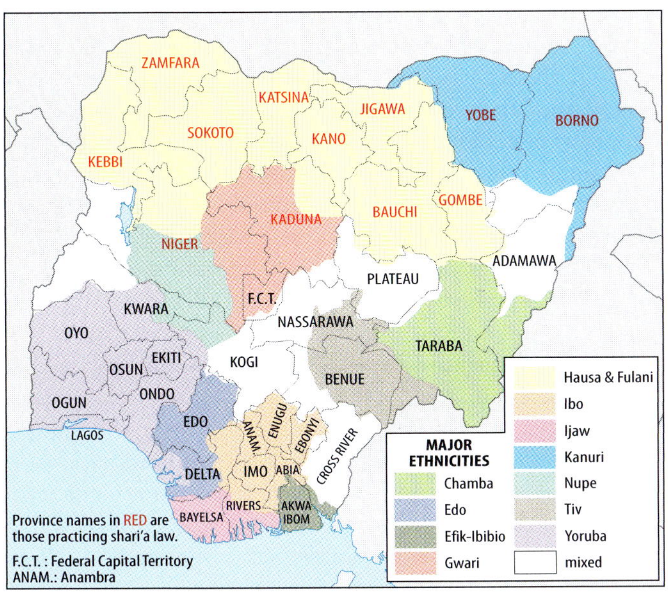

• **Figure 9.22** Nigeria's ethnic map.

Where does Nigeria's oil money go? There is not enough transparency to answer this question. Incredibly, until June of 2011, it was illegal in Nigeria to publish official government data and statistics, including government accounts. Recently, $22 billion from the government's Excess Crude Account simply "went missing." An estimated $400 billion has been pilfered from Nigeria's treasury since the country became independent in 1960.

Since 1990, several ethnically based organizations have been agitating for change. Ethnic **Ogoni** activists founded the **Movement for the Survival of the Ogoni People (MOSOP)**, issuing a bill of rights in which they declared the right to a safe environment and more federal support of their people. One of the movement's leaders, a popular author, playwright, and television producer named Ken Saro-Wiwa, also called for self-determination for the Ogoni. Despite international condemnation, Nigeria's government executed him after a questionable murder trial.

Despite the return to civilian rule in Nigeria that came after Saro-Wiwa's death, dissent in the delta continued to grow. Ogoni defiance and contempt of government and the oil companies spread to the region's other major ethnic groups: the **Ibo (Igbo), Yoruba**, and **Ijaw**. The Ijaw (who make up about 5% of Nigeria's population) have been especially strident in demanding that some of the oil money be used to fund roads, electricity, running water, and medical clinics. Ijaw activists have spoken of secession and self-determination if their demands are not met. They have periodically registered dissatisfaction by seizing onshore stations and offshore rigs belonging to foreign oil companies, temporarily disrupting Nigeria's oil exports on both occasions. In 2006, a mainly Ijaw group calling itself the **Movement for the Emancipation of the Niger Delta (MEND)** escalated these tactics into frequent kidnappings of foreign oil workers. Opportunistic profiteers without well-established political views also joined in the action, and kidnapping became big business. An oil worker is typically freed unharmed upon payment of a ransom of several hundred thousand dollars.

This militancy in the delta has sent shockwaves through the world economy. Nigeria had touted itself as the safe alternative to Middle Eastern oil but has had to cut production by as much as 25 percent when tensions have been particularly high. The resulting shortages on world markets helped send oil and gasoline prices to record levels.

Questions of equity in a potentially rich country thus pose a threat to national, regional, and international stability. The peoples of the eight oil-producing states in the delta have been promised 13 percent of all petroleum revenues (compared with 5% previously), and have been told that the region's foul gas flares will be shut off. But locals say the promised revenues have been embezzled, while the flaring off of gas continues.

One of Nigeria's ironies is that in this oil-rich country, there are continuous shortages of gasoline and electricity. Neglect, mismanagement, corruption, and theft are all responsible. One of the cruelest and most repetitive Nigerian news stories is the immolation of scores and even hundreds of poor delta villagers who, in an activity known as "scooping" or "bunkering," illegally puncture gasoline or oil pipelines in an effort to collect and sell the fuels, only to inadvertently ignite the fires that consume them.

• **Figure 9.23** Principal features of East Africa.

East Africa

No More Divisionism?

Five countries make up East Africa: Kenya and Tanzania, which front the Indian Ocean, and landlocked Uganda, Rwanda, and Burundi (•**Figure 9.23**). Their total area is 703,000 square miles (1.8 million sq km), roughly the size of the U.S. state of Texas, with a total population in 2011 of 143 million (nearly half the population of the United States).

Episodes of bloodshed have punctuated East Africa's history. This section and the next section on west-central Africa tell a complex and tragic tale of genocide. These details are related here because the potential for renewed conflict is high and because the world community vowed not to forget what happened here.

Rwanda and Burundi have had tragic disputes between their majority and minority populations. About 85 percent of the population in Burundi and 90 percent in Rwanda are composed of the **Hutu (Bahutu)**, and most of the remainder of the two populations is **Tutsi (Watusi)**. These were not originally separate tribal or ethnic groups; the peoples speak the same language, share a common culture, and often intermarry. The distinction between Hutu and Tutsi was instead based on socioeconomic classification. Historically, the Tutsi were a ruling class that dominated the Hutu majority. Tutsi power and influence were measured especially by the vast numbers of cattle they owned. Tutsi who lost cattle and became poor came to be identified as Hutu, and Hutu who acquired cattle and wealth often became Tutsi.

Colonial rule in East Africa attempted to polarize these groups, increasing antagonism between them. German and 45, 325 Belgian administrators differentiated between them exclusively on the basis of cattle ownership: Anyone with fewer than 10 animals was Hutu, and anyone with more was Tutsi. Europeans mythologized the wealthier Tutsis as "black Caucasian" conquerors from Ethiopia who were a naturally superior, aristocratic race whose role was to rule the peasant Hutus. The colonists replaced all Hutu chiefs with Tutsi chiefs. These leaders carried out colonial policies that often imposed forced labor and heavy taxation on the Hutus. Education and other privileges were reserved almost exclusively for the Tutsis.

Ferocious violence between these two peoples marred Rwanda and Burundi intermittently after the states became independent in the early 1960s. After independence, the majority Hutus came to dominate the governments of both nations. In 1994, the death of Rwanda's Hutu president in a plane crash (in which Burundi's president also died) sparked civil war in Rwanda between the Hutu-dominated government and Tutsi rebels of the Rwandan Patriotic Front (RPF), based in Uganda. Hutu government paramilitary troops, still vengeful about decades of domination by Tutsis, systematically massacred Tutsi civilians with automatic weapons, grenades, machetes, and nail-studded clubs. Women, children, orphans, and hospital patients were not spared. An estimated 800,000 Tutsis and "moderate" Hutus (those who did not support the genocide) died. International media tracked these horrors, but outside nations did nothing to stop the fighting.

Well-organized and motivated Tutsi rebels seized control of most of the country and took power in the capital, Kigali, in 1994. That precipitated a flood of 2 million Hutu refugees mainly into neighboring Zaire, now the Democratic Republic of Congo (see page 340). Although the Tutsi-dominated government promised to work with the Hutus to build a new multiethnic democracy, Hutu insurgents, based mainly in eastern Zaire, continued to carry out attacks against Tutsi targets in Rwanda.

Rwanda now has a national unity policy aimed at reconciling Hutus and Tutsis; both occupy important government posts, and although the Tutsis still hold the political upper hand, they have encouraged the resettlement of Hutu refugees who fled the country during its troubles. In 2004, marking the 10-year anniversary of the genocide with a new policy meant to prevent its recurrence, Rwanda's government outlawed ethnicity. Now any Rwandan who speaks or writes of Hutus and Tutsis may be fined or imprisoned for practicing "divisionism." The country's past nevertheless continues to shape its future; for example, there is an ongoing defection of the faithful from Rwanda's majority Catholic Church to other religions, especially Islam but also smaller Christian denominations. Most of the country's worst massacres took place in Catholic churches, with the apparent knowledge and even participation of church leaders.

In Burundi, the Tutsi minority traditionally dominated politics, the army, and business. In 1993, there was a brief shift of power to the Hutu majority when Burundians elected the country's first Hutu president. Tutsis murdered him in October 1993, and a large-scale Hutu retaliatory slaughter of Tutsis ensued. An estimated 300,000 died in a decade of violence. Burundi's Tutsis and Hutus now have a power-sharing agreement in which the presidency rotates between them.

The peace has been maintained by the first-ever deployment of the African Union's **African Standby Force**. To help ensure the stability needed for economic and political progress throughout the region, the African Union established this continental military force. It takes orders from the AU's Peace and Security Council. The force will, in theory, deploy peacekeepers or peacemakers to ensure that the map of Africa, so filled with deadly conflicts in the past, has far fewer trouble spots in the future. The African Union has plans to create an African Court on Human and People's Rights and, like the European Union after which it is modeled, may also eventually seek a common currency.

Collective shame and embarrassment in the international community about the failure to stop the killing in Rwanda and Burundi led in subsequent years to numerous new antigenocide measures. These include a United Nations "early warning" system to prevent genocide and a new forum for crimes against humanity in the **International Criminal Court**, established in 2002 in The Hague, Netherlands.

West Central Africa

The subregion of West Central Africa is flanked by Cameroon and the Central African Republic to the north and the Democratic Republic of Congo to the south (**•Figure 9.24**). It contains seven countries: Cameroon, Gabon, Central African Republic, Congo Republic (often known by the shorthand Congo-Brazzaville), Democratic Republic of Congo (known by the shorthand Congo-Kinshasa or DRC), São Tomé and Príncipe, and Equatorial Guinea.

Colonialism and Modern Struggles

During the last quarter of the nineteenth century, the Congo Basin was virtually a personal possession of Belgium's King Léopold II, whose agents ransacked it ruthlessly for wild rubber, ivory, and other tropical products gathered by Africans. This lawless era inspired Joseph Conrad's famous novel *Heart of Darkness* (1902), in which the trader Kurtz, at the point of death, evokes the ravaged Congo region with the cry, "The horror! The horror!"

In 1908, the Belgian government formally annexed the greater part of the Congo Basin, creating the colony of Belgian Congo. In 1960, the Congo colony became the independent Republic of the Congo. In 1971, it took the name Zaire, meaning "river." Following the overthrow of Zaire's government in 1997, the country was renamed Democratic Republic of Congo.

In recent decades, the region's troubles have focused on its major power, the Democratic Republic of Congo, and have been inseparable from events in neighboring East Africa.

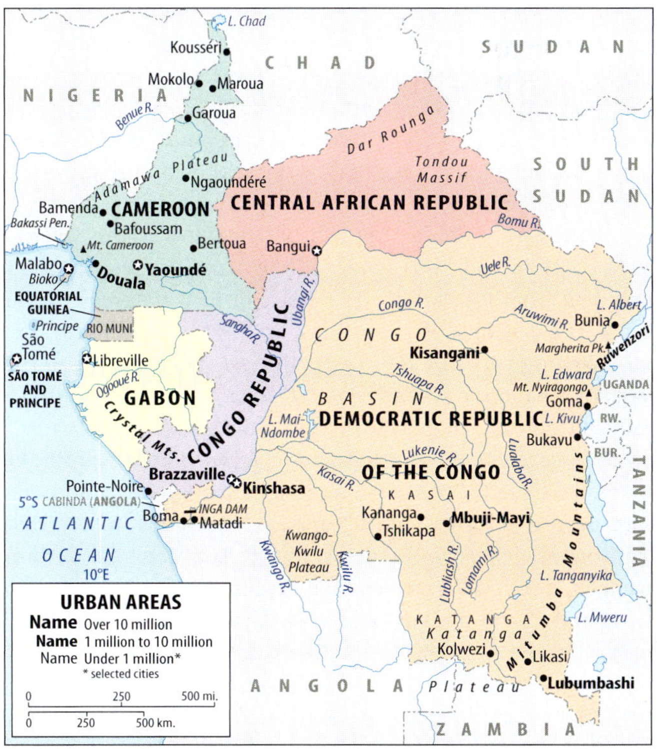

• Figure 9.24 Principal features of West Central Africa.

During his tenure in office from 1965 to 1997, then-Zaire's autocratic ruler, Mobutu Sese Seko, had neglected the country's development but acquired a massive fortune. Political opposition and public discontent with unemployment and inflation led to unrest in the early 1990s.

All-out war began after 1997, when Zairean army troops attempted to expel a force of indigenous Tutsis living in the country's far eastern region (for discussion of the region's Tutsis and Hutus, see page 338). The assault backfired when these Tutsis, supported by neighboring Tutsi-dominated governments in Rwanda and Burundi and by other sympathetic countries in the region, took up yet more arms and initiated fierce attacks on government forces.

The Tutsi counterattack quickly widened into an antigovernment revolt comprised of many ethnic factions and led by a non-Tutsi guerrilla named Laurent Kabila. Kabila's forces pushed westward virtually unopposed through the huge country and in May 1997 occupied Kinshasa with little loss of life. President Mobutu fled Zaire on the eve of Kabila's triumph. Kabila then declared himself president of the country he now called the Democratic Republic of Congo.

Like his predecessor Mobutu, he funneled much of the country's wealth and power into the hands of family and friends. He also obstructed international human rights investigations into the deaths and disappearances of thousands of Hutu refugees during the heady days of rebellion against Mobutu. And he failed to bring stability to the far east of his huge country, where an environment of chaos had given birth to his own bid for power.

In 1998, Hutu forces based in the eastern Congo-Kinshasa launched successive devastating attacks on Tutsi interests in Rwanda. Although the Tutsis of Rwanda had helped bring Kabila to power in Congo-Kinshasa, they now turned against him, forming the backbone of a rebel alliance bent on bringing down Kabila's young regime. The unrest led to something rather rare in the region, a broader fight dubbed **"Africa's First World War"** that involved nine countries and 20 rebel movements, and resulted in the deaths of more than 5 million people.

Although the fighting was sometime ferocious, it was not warfare that directly claimed most of the lives lost in this struggle between 1996 and 2001. Most of the casualties were civilians who died from starvation and disease or in widespread massacres directed against ethnic groups, usually Hutus killing Tutsis, but sometimes the reverse. The country appeared to be on the verge of dissolution when Laurent Kabila was assassinated by a bodyguard in 2001. He was succeeded by his son Joseph.

Conflict in Congo-Kinshasa has been enormously profitable for the warring factions, some foreign companies, and foreign governments (Uganda and Rwanda, for example, looted eastern Congo-Kinshasa of its gems, minerals, timber, agricultural produce, and wildlife, including elephant ivory from some of the country's national parks). Since 1998, most of the fighting has in fact been unabashedly about control over the Congo's vast mineral resources. Warlords and rebel proxies have been sustained by exploitation of minerals in the areas they controlled. Most of Congo's vast reserves of gold, diamonds, and copper are in the south center and west of the country, where the government is negotiating with the Chinese a vast minerals (coltan, tin, copper, diamonds, gold)-for-infrastructure deal (see page 334). In the east, where fighting to overthrow the government was centered, there is more gold. In this region, armed groups, now in concert with the Congo's army, control mines. There is a good chance that your cell phone contains coltan mined in this region and that it was not mined and exported "cleanly" as most diamonds are nowadays.

Although peace has been negotiated in this area, war could easily break out again. Now integrated into the Congolese army, ethnic Tutsi have been able to extend the area and the mineral wealth they control. Rival Rwandan Hutu militias, who control their own mineral wealth, have been carrying out reprisal attacks against Congolese civilians. Strife between Rwanda and the Congo is also enabled by expatriate sympathizers of the groups in Europe and the United States.

The Horn of Africa

The Galápagos Islands of Religion

In the extreme northeastern section of Sub-Saharan Africa, a great volcanic plateau rises steeply from the desert. This highland and adjacent areas occupy the greater part of the Horn of Africa, named for its projection from the continent into the Indian Ocean. This subregion includes the countries of Ethiopia, Eritrea, Somalia, and Djibouti (•**Figure 9.25**).

Of all these countries' peoples, Ethiopia's are the most ethnically and culturally diverse. About 45 percent, including the politically dominant **Amhara** peoples, practice **Ethiopian Orthodox Christianity**, an ancient branch of Coptic Christianity that came to Ethiopia in the fourth century from Egypt.

• **Figure 9.25** Principal features of the Horn of Africa.

Joe Hobbs

• **Figure 9.26** The churches of Lalibela, carved from volcanic rock in the twelfth and thirteenth centuries, are among Ethiopia's many Christian cultural treasures. For scale, note the men at left and right.

The entire area has had important cultural and historical links with Egypt, the Fertile Crescent, and Arabia. The Ethiopian monarchy based its origins and legitimacy on the union of the biblical King Solomon and the Queen of Sheba, who, tradition holds, gave birth to the first Ethiopian emperor, Menelik. Until a Marxist coup brought an end to the emperorship in the 1970s, Ethiopia's rulers were always Christian. Ethiopia has many outstanding Christian artistic and architectural treasures, including the 11 churches of Lalibela, carved from solid rock in the twelfth and thirteenth centuries (•Figure 9.26). Most of the rest of Ethiopia's people are either Muslims (who make up about 40% of the population) or members of Protestant, Evangelical, and Roman Catholic churches. There are still small numbers of **Falashas**, or **Ethiopian Jews**, in Ethiopia, a remnant of a very ancient and isolated Jewish population. The majority, about 100,000 people, have fled to Israel since the mid-1980s. Because this mountainous country has long served as an isolated refuge for such unique groups, it has been nicknamed the **Galápagos Islands of Religion**.

Perhaps the most unusual religion with roots in Ethiopia is that of Rasta, or the Rastafari movement of the Jah People (called **Rastafarianism** by non-Rasta). This faith, which originated in the 1930s in the West Indies island of Jamaica, has the central doctrine that the Ethiopian Emperor Haile Selaisse (1892–1975) was the earthly incarnation of Jah or Jehova (God). Ras Tafari was the Emperor's name before his coronation, and he was crowned on November 2, 1930, as the "King of Kings, Elect of God, and Conquering Lion of the Tribe of Judah (•Figure 9.27)."[6]

© Janine Wiedel Photolibrary/Alamy

• **Figure 9.27** The Lion of Judah.

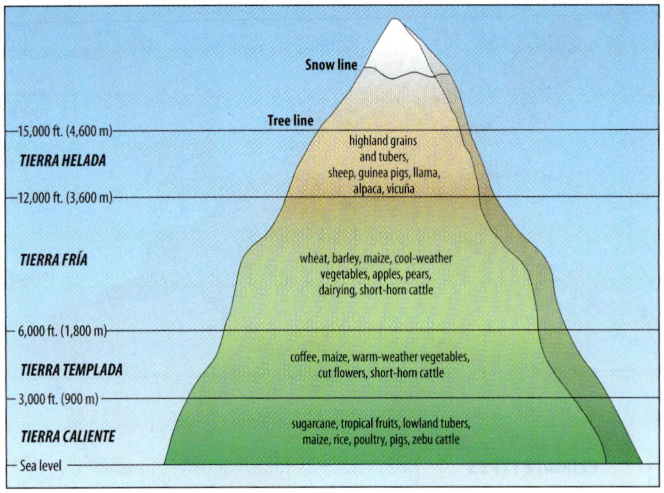

• **Figure 10.5** Altitudinal zonation in Latin America. Specific patterns of land use have evolved to take advantage of the region's diverse altitudinal zones. This graph shows elevations and general land use patterns, which may also be compared with land uses in Figure 10.6. Elevations shown here are for the equatorial region and are somewhat lower toward the poles for the respective land uses.

Source: Based on data from Clawson, 1997. Latin America and the Caribbean. Dubuque, Iowa: William. C. Brown Publishers.

Hemispheres. The largest segment, representing the world's largest continuous expanse of tropical rain forest, lies in the basin of the Amazon River system. There is more tropical rain forest in southeastern Brazil, eastern Panama, the western coastal plain of Colombia, on the Caribbean side of Central America and southern Mexico, and along the eastern (windward) shores of some Caribbean islands, particularly Hispaniola and Puerto Rico. Much of the original tropical rain forest vegetation of the islands has been lost to human activity, and even the vast Amazon forest is under considerable pressure (this problem, which is largely a reflection of Brazil's development, is described on pages 383–386).

On either side of the principal region of tropical rain forest climate, the tropical savanna climate extends to the vicinity of the Tropic of Capricorn in the Southern Hemisphere and, with more fragmentation, to the Tropic of Cancer in the Northern Hemisphere. Still farther poleward, in the eastern portion of South America, lies a large area of humid subtropical climate, with cool winters unknown in the tropical zones. Its Northern Hemisphere counterpart is north of the Mexican border in the southeastern United States. In South America, this climate is associated mainly with prairie grasses in the humid pampas of Argentina, Uruguay, and extreme southern Brazil. On the Pacific side of South America, a small strip of Mediterranean or dry-summer subtropical climate in central Chile is similar to that in southern California. This is an ideal climate for wine production, and Chilean wines are now well established on world markets.

The humid climates of Latin America have a fairly predictable spatial arrangement. But the region's dry climates and biomes—desert and steppe—are the product of local circumstances such as rain shadows or nearby cool ocean water. In northern Mexico, aridity is due partly to the rain shadow effect of high mountain ranges on either side of the Mexican plateau (a rain shadow is an area that is dry because it is sheltered from rain-bearing winds by a mountainous barrier). These dry climates are also partly associated with the global pattern of semipermanent zones of high pressure that create arid conditions in many areas of the world along the Tropics of Cancer and Capricorn.

In Argentina, the extremely high and continuous Andean mountain wall accounts for the aridity of large areas, especially the southern region called Patagonia. Here, the Andes block the path of the prevailing westerly winds, creating heavy orographic (mountain-induced) precipitation on the Chilean side of the border but leaving Patagonia in rain shadow. However, in the west coast tropics and subtropics of South America, the Atacama and Peruvian deserts cannot be explained so simply. Here, shifting winds that parallel the coast, cold offshore currents, and other complexities, as well as the Andes Mountains, combine to create the world's driest area. The Atacama Desert is drier even than the Sahara! The mountains restrict this area of desert to the coastal strip. Arid and semiarid conditions also prevail along the northernmost coastal regions of Colombia and Venezuela and in the region called the Sertão in northeastern Brazil.

Elevation and Land Use

With respect to human adaptations, one of the most significant features of Latin America's physical geography is a series of highland climates arranged into zones by elevation. This zonation results from the fact that air temperature decreases with elevation at a normal "lapse rate" of approximately 3.6°F (1.7°C) per 1,000 feet (c. 300 m). At least four major zones are commonly 28
recognized in Latin America (Figures 10.5 and •10.6): the *tierra*

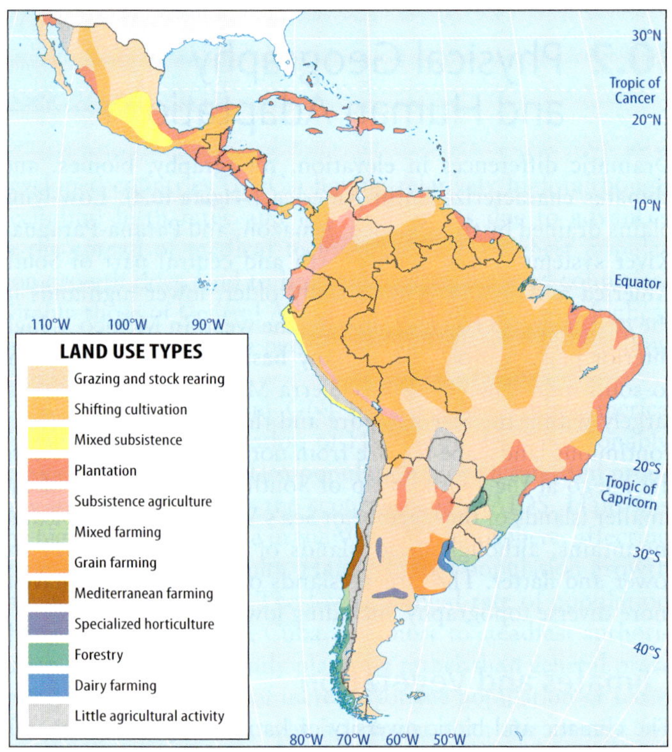

• **Figure 10.6** Land use in Latin America reflects a remarkable range of opportunities posed by varying latitudes and elevations.

Source: The Oxford School Atlas edited by Patrick Wiegand (OUP, 1997), copyright © Oxford University Press. Used by permission of Oxford University Press.

Joe Hobbs

• **Figure 10.7** Tending the coffee crop on a *finca* (plantation) in the *tierra templada* of southern Mexico. Note the coffee beans in the boy's right hand.

caliente **(hot country)**, the ***tierra templada*** **(cool country)**, the ***tierra fría*** **(cold country)**, and the ***tierra helada*** **(frost country)**. At the foot of the highlands, the *tierra caliente* is a zone embracing the tropical rain forest and tropical savanna climates. The zone reaches upward to approximately 3,000 feet (c. 900 m) above sea level at or near the equator and to slightly lower elevations in parts of Mexico and other areas near the margins of the tropics. In this hot, wet environment, the favored crops are rice, sugarcane, bananas, and cacao. Latin America's blacks are concentrated in many of the *tierra caliente* zones, a legacy of the slave trade, when they were forced to work the region's plantations.

The *tierra caliente* merges almost imperceptibly into the *tierra templada,* which flanks the rugged western mountain ranges and is the uppermost climate in the lower uplands and highlands to the east. Sugarcane, cacao, bananas, oranges, and other lowland products reach their uppermost limits in the *tierra templada,* but this zone is most famous as coffee habitat (•Figure 10.7). The upper limits of this zone—approximately 6,000 feet (1,800 m) above sea level—tend to be the upper limit of European-introduced plantation agriculture and modern commercial crops in Latin America. Densely inhabited sections occupy large areas in southeastern Brazil, Colombia, Central America, and Mexico. Although broadleaf evergreen (tropical rain forest) trees characterize the moister, hotter parts of this zone, coniferous evergreens replace them in parts of the zone's poleward margins. In places such as the highlands of Brazil and Venezuela, where there is less moisture, scrub forest or savanna grasses prevail.

Seven metropolises exceeding 2 million in population—São Paulo, Belo Horizonte, Brasília, Caracas, Medellín, Cali, and Guadalajara—are in the *tierra templada,* and two others—Mexico City and Bogotá—lie slightly above it. Four smaller cities that are national capitals and the largest cities in their respective countries are located in the *tierra templada*: Guatemala City, San Salvador, Tegucigalpa, and San José. Others, like Rio de Janeiro, which are situated at lower elevations, have close ties with predominantly residential or resort towns in these cooler temperature areas.

The *tierra fría* is comprised of high plateaus, basins, valleys, and mountain slopes within the great mountain chain that extends from northern Mexico to Cape Horn. By far the largest areas of this habitat are in the Andes, with significant areas also in Mexico (•Figure 10.8). Around the equator, the *tierra fría* begins at about 6,000 feet (c. 1,800 m) and extends upward to the **upper limit of agriculture** (represented by such hardy crops as potatoes and barley) and the **tree line** (the upper limit of natural tree growth) at about 12,000 feet (c. 3,600 m).

The *tierra fría* experiences frost and is often the habitat of a Native American economy with a strong subsistence component. In a classic process of marginalization, European colonization of Latin America drove some Native American settlements upslope and into the *tierra fría* zone, although some major populations (notably the Inca of Peru) had already 47

Joe Hobbs

• **Figure 10.8** Characteristic *tierra fría* landscape and agriculture in Andean South America.

Joe Hobbs

• Figure 10.9 The *páramo* at about 12,000 feet (c. 3,600 m) in the Andes of Colombia.

selected upland locations for their settlement before the arrival
362 of Columbus. These upland settlements are most extensive in Ecuador, Peru, and Bolivia and are also very evident in Colombia, Guatemala, and southern Mexico.

Another zone lies above the other three and consists of the alpine meadows, known as ***páramos***, along with still higher barren rocks and permanent fields of snow and ice (**•Figure 10.9**). This is the *tierra helada*. It generally lies between 12,000 feet (c. 3,600 m) and the lower edge of the snow line. It supports some grains and livestock (llama, alpaca, sheep) but is largely above the mountain flanks that are central to upland native settlement and agriculture. In the highlands of Latin America, the *tierra fría* tends to be a last retreat and a major home of the indigenous peoples, except in Guatemala and Mexico. Most of their settlements are rural villages based on subsistence agriculture. In some cases, notably in Bolivia and Peru, valuable minerals like tin and copper are located here, attracting large-scale mining enterprises into the *tierra fría* and the *tierra helada*.

The roof of South America is Mount Aconcagua in Argentina, rising to 22,841 feet (6962 m) in the Andes. Its slopes, like many of this higher Andean reaches, are covered in glacial ice. Andean glaciers are following the worldwide trend of melting. As the glaciers melt, they take consistent, dependable
76, 250, 428 water supplies with them. Some scientists predict that the Bolivian city of La Paz, with its million residents, will be the first large urban casualty of climate change. There are other natural threats as well.

Natural Hazards in Latin America

Adjoining a large section of the Pacific Ring of Fire and fronting two seasonal hurricane regions, Latin America is beset by natural hazards. A belt of land stretching from northern
Mexico all the way to Tierra del Fuego, the "Land of Fire" 21
at the southern tip of South America, has a violent history of earthquakes and volcanic eruptions. On the map of global tectonic plates (Figure 2.1), you can see the Nazca Plate and South American Plate confronting one another, with tremendous geological consequences including the rise of the Andes Mountain chain and those volcanoes and earthquakes. The largest earthquake ever recorded on earth was the titanic 9.5 quake in 1960, just off the coast of southern Chile.

Figure 2.1 also depicts an active tectonic area on the eastern side of the Americas: the Caribbean Plate is also in motion, causing earthquakes and a variety of associated tectonic features. You can see that one of this plate's boundaries passes through the island of Hispaniola, where the countries of Haiti and the Dominican Republic are located (**see Figure 10.1** and **Geography of Natural Hazards, next page**).

Hurricanes, giant low-pressure systems that feed on warm waters of the Atlantic and Pacific Oceans, have sometimes devastating impacts. A single storm can wreck the economy of an island nation, as Hurricane Ivan did by wiping out Grenada's nutmeg crop in 2004. The monstrous Hurricane Mitch (category 5, highest on the Saffir-Simpson scale) was an unusually late storm that made landfall in Honduras on October 28, 1998, and roamed for five days over that country and Nicaragua, El Salvador, Guatemala, and Mexico. Its ferocious winds and rainfalls of up to 2 feet (60 cm) brought floods, landslides, and storm damage—intensified by the region's widespread deforestation—that killed an estimated 15,000 people; inflicted enormous damage on property, crops, roads, power, and other infrastructure; and left disease epidemics in its wake. As if Mitch were not enough, the region had also suffered from drought and fires caused by El Niño in 1997 and 1998.

What is El Niño? It is best understood as a complete reversal of a normal climatic pattern. Every few years, things seem to go wrong with the winds and waters of the Pacific Ocean. Normally, the winds blow east to west, helping to promote the upwelling of cold, nutrient-rich water from the depths to the surface of the tropical eastern Pacific (**•Figure 10.10**). But sometimes, usually beginning in December, the winds reverse direction, blowing from west to east, suppressing the upwelling water and thus raising the surface temperature of the water by as much as 20°F (11°C). Because of its common occurrence in December, that event has been dubbed **El Niño**, meaning "the Baby" in reference to the Christ child, whose birth is celebrated in December. Less prosaically, meteorologists know it as **El Niño Southern Oscillation (ENSO)**.

El Niño conditions sometimes last a year or more, causing global climatic disruptions. These vary between events, but the typical El Niño pattern is illustrated in **•Figure 10.11**. Closest to the source of the condition, Peru experiences torrential rains. Unusually high rainfall also occurs in southern Brazil, the southeastern United States, western Polynesia, and East Africa. Drought or unusually high temperatures settle

GEOGRAPHY OF NATURAL HAZARDS | **Heartbreak in Haiti**

On January 12, 2010, an earthquake with a magnitude of 7.0 struck Hispaniola on a strike-slip fault zone just 16 miles (25 km) from Haiti's capital, Port-au-Prince. In a more developed country better prepared with earthquake-resistant infrastructure, such a quake might not have produced extensive damage. But in Haiti, by far the poorest country in the Western Hemisphere and one of the poorest on earth, this temblor was catastrophic. It killed an estimated 315,000 people and made nearly a million homeless (•**Figure 10.A**).

A unique combination of geologic, historic, and political factors made Port-au-Prince especially vulnerable to seismic catastrophe. Not only is the city close to the fault zone on which the earthquake struck, but much of Port-au-Prince lies atop soft sedimentary rock that amplified the seismic waves. Then there were the population issues. United States forces occupied Haiti for about two decades after World War I, building roads and other infrastructure that centralized urbanization in Haiti: Port-au-Prince became a magnet for settlement and development. Far surpassing the metric of a primate city, the capital was home to one in three Haitians when the earthquake hit. Most of the city's housing stock was made up of **informal settlements**, a term often used interchangeably with *slums* but meaning the occupants lack legal claims to the land and that the housing is not in compliance with planning and building regulations. Much of the housing is fashioned from recycled materials. Meeting the requirements of earthquake resistance is far beyond the financial means of the people of Port-au-Prince, where two-thirds of the labor force lacks formal employment. Much of Port-au-Prince is a sprawling slum, described this way by journalist Amy Wiletnz: "a heart-tugging city that functioned on a complicated, hypercharged fuel of chaos, exposed wiring, pig slop, smog, gingerbread turrets, hot cooking oil, rum, cockfights and bougainvillea."[1]

Xinhua /Landov

• **Figure 10.A** A magnitude 7.0 quake devastated Port-au-Prince, the Haitian capital, in December 2010.

The utter devastation of the 2010 earthquake gave Haitians the opportunity to refashion their urban geography. Assisted by geographers and other scientists from the University of Miami, a group of Haitian urban planners has come up with a plan to decentralize Haiti, redistributing large numbers of people to smaller Haitian cities at safer distances from fault zones. The plan will completely transform Haiti from a country dominated by the primacy of Port-au-Prince to one with a network of smaller urban "growth poles." These would be more suited to the economic assets Haiti could best develop: agriculture and tourism. Planners will locate schools, markets, and hospitals and health care centers in smaller towns and villages so that people will not need to leave these settlements and gravitate again to Port-au-Prince. Homes, stores, and offices will be constructed with earthquakes in mind and grow over time in modular fashion from basic shelters into permanent fixtures.

Unfortunately, as a metropolis ripe for seismic ruin, Port-au-Prince is probably not alone. Other vulnerable cities, which seismologists call "rubble in waiting," include Istanbul, Turkey; Karachi, Pakistan; Katmandu, Nepal; Tehran, Iran; and Lima, Peru. All of these cities continue to swell from rural to urban migration, and all share a surplus of informal settlements built from such dangerous materials as brick and mud-brick. University of Colorado seismologist Roger Bilham discussed the seismic hazards of "an unrecognized weapon of mass destruction: houses."[2]

in over northern Brazil, southeastern Canada and the northeastern United States, southern Alaska and western Canada, the Korean Peninsula and northeastern China, Indonesia and northern Australia, and southeastern Africa. When warmed, Pacific waters are far less rich in nutrients, so typically fish catches plummet and food chains are disrupted, leading to massive wildlife losses. The most severe recent El Niño events occurred in 1982–1983 and in 1997–1998, but it is not unusual for three or four moderate events to occur every decade.

10.3 Cultural and Historical Geographies

It is perhaps unfortunate that this region came to be known as Latin America because there were no "Latins" among its inhabitants before the end of the fifteenth century. In 1492, when the first Europeans arrived (perhaps after seafaring Polynesians reached the west coast of South America; see page 291), the region was home to an estimated 50 to 100 million **Native**

All photos courtesy of Joe Hobbs

• **Figure 10.20** In addition to the Native Americans of Latin America, the leading racial types are **(a)** Europeans, **(b)** blacks, **(c)** mestizos, and **(d)** mulattoes.

Rica have significant white European ethnic groups today (•Figure 10.20a). These are mainly families of immigrants from Europe who arrived relatively recently, during the late nineteenth and early twentieth centuries. Scattered districts in other countries are also predominantly European.

Black Latin Americans of relatively unmixed African descent live in the greatest numbers on the Caribbean islands and along the Atlantic coastal lowlands in Middle and South America (Figure 10.20b). These are the areas to which African slaves were brought during the colonial period, mainly to work on sugar plantations. An estimated 3 to 4 million were sold in Brazil alone. Slavery was gradually abolished during the nineteenth century, although not until the 1880s in Brazil and Cuba. By that time, slavery had generated large fortunes for many owners of plantations and slave ships. The African peoples it introduced to the region have made cultural contributions that today provide a varied, vibrant, and important part of Latin American civilization.

Latin America has escaped many of the racial tensions and violence that have battered much of the world. This may be because a majority of Latin Americans are of mixed heritage, the most common being a mixture of Spanish and Native American that has resulted in a heterogeneous group known as **mestizos** (Figure 10.20c). A person of mixed African (black) and European (white) ancestry, sometimes with Native American blood as well, is known as **mulatto** (Figure 10.20d); in some locales, the term used is, like the linguistic term, **Creole**. It must be noted, however, that in most places, socioeconomic discrepancies are strongly associated with ethnicities in the region. A thin stratum of people of mainly European origin often dominate government and business sectors, with mestizos and mulattoes making up the bulk of the middle and lower classes. Native Americans almost always have the lowest standing, and in some countries such as Guatemala, systematic exclusion and violence has been directed against them from time to time.

Mestizos have not lost all traits of their Amerindian forebears. Although it can be hazardous to generalize about ethnicities, the mestizos' inheritance from their Native American roots includes the collective exploitation of land, the accumulation of lifelong personal bonds, the devotion to a whole series of supernatural beings (who have Christian names but non-Western origins), the ritualized remembrance of deceased relatives, and reverence for political leaders.[5] As will be seen later in the chapter, reverence for political leaders cannot be taken for granted in this region today.

Despite enormous impacts, some Native American cultures have proven remarkably resilient, as we will see throughout the rest of the chapter. Some, like the indigenous peoples of Nicaragua's Miskito coast, are actively pushing for independence (they call themselves the Communitarian Nation of Mosquitia). Many are technologically savvy, using the Internet and social media to elicit understanding and political and economic support in their struggles. Some are championed by Western celebrities and activists, but others, including some of the groups pushing back against oil development in the Ecuadorian Amazon, have managed to gain a footing without such assistance.

Some governments actively protect the lives and ways of life of Amerindians. New constitutions expanded indigenous rights in Colombia in 1991, Peru in 1993, Bolivia in 1994, and Ecuador in 1997. In these countries, Native Americans now have more rights over land, and the governments recognize communal rights over some resources. Brazilian law forbids mineral extraction by outsiders on aboriginal lands, but illegal diamond and gold mining has led to many bloody skirmishes between Native Americans and prospectors in several Indian reserves in the country. On the negative side of this protection, as the Native Americans see it, many resources of Brazilian and other national Indian reserves are placed under government control, and often indigenous peoples are forbidden from setting up their own businesses.

10.4 Economic Geography

Latin America is generally a region of less developed countries (LDCs). An estimated one-third of its residents live in poverty. However, both poverty and unemployment have diminished in

recent years for the region as a whole, and the middle class is growing. In most countries, this is thanks to an abundance of raw materials; Mexico is an exception in that it relies more on exports of manufactured goods rather than commodities. The commodities boom is tied mainly to China's dynamic growth. As in Africa and elsewhere, Chinese interests are scouting out every possible prospect in Latin America that could help feed
333 that nation's voracious appetite for minerals, energy, and food. Latin America's trade with China was 16 times bigger in 2009 than it was in 1999, and China now stands as Latin America's leading trade partner.

In some Latin American countries, this raises the same fears as in Africa: that China is seeking to develop a classic mercantile relationship with Latin America, taking raw materials (both agricultural and industrial) and profitably selling its own manu-
45 factured goods. But for now, the influx of Chinese capital is welcomed. It helped Latin America to weather the Great Recession and emerge with economic growth rates generally well above those of industrialized countries like the United States.

Although Latin America's overall situation is improving, the region suffers from notoriously large discrepancies between the haves and the have-nots. The glitter of the great metropolises like Mexico City and Rio de Janeiro, with their forests of new skyscrapers, may distract attention from poverty, but portions of these cities are massive, squalid slums, known as ***favelas*** in Brazil and ***barrios*** in Spanish-speaking countries (•**Figure 10.21**). Such **shantytowns**, which are practically universal in the world's LDCs, are full of unemployed, underemployed, and sometimes undernourished people. "*Ni ni*" these people are called throughout much of the region: neither educated nor employed. Their sites and situations are noteworthy: remarkably, many *favelas* immediately border some of the most costly real estate in Rio de Janeiro and other cities; others are in the least desirable, most hazard-prone areas of the rural landscape, making headlines when mudslides slam through them, taking scores of lives. Many people prefer their plight in the *favelas* to the conditions of the depressed rural landscapes from which they came. Glimpses of life in these slums may be seen in the Brazilian film drama *City of God*. Brazil's preparations to host the 2014 Soccer World Cup and the 2016 Summer Olympics included major works to improve conditions in the *favelas*.

Latin American economies have born many of the traditional hallmarks of LDCs. Trying to overcome their economic problems and constrain popular discontent, for example, many Latin American governments have borrowed heavily from the international banking community. In the early 2000s, unpaid loans had reached staggering proportions, and the Latin American debtor nations were having great difficulty in mustering even the annual interest payments, let alone generating surplus capital to pay toward the principal of these troubling loans—a
dilemma all too characteristic of the world's LDCs. Like other 47
LDCs, Latin America also has historically been over-reliant on non-value-added goods like cash crops and minerals. As we will see, free-trade agreements form the foundations of a recent push to move away from commercial agriculture and raw materials and toward manufactured exports.

Commercial Agriculture

Although the total number of Latin Americans employed in farming has not declined much in recent decades, the value of

Joe Hobbs

•**Figure 10.21** Many of Latin America's poorest people live in ramshackle housing. This is the shantytown of Belen in Iquitos, Peru. The great Amazon River is just out of view to the left.

INSIGHTS | Fair Trade Fruits

Bananas are the most popular fruit in the United States, with more than 8 billion sold annually—or about 26 bananas per person per year. Now it is possible to be a discriminating banana buyer. At many North American and European grocery stores, one may find bananas, pineapples, other fruits, chocolates, and coffees bearing "Fair Trade Certified" stickers (•Figure 10.B). These are higher in price than the standard products, but consumers who buy them are generally motivated by a sense of social responsibility. Consumers keen to reduce their environmental footprint are also willing to pay a premium for fair trade products that are organically produced.

In standard trade transactions, only a tiny fraction—2 cents per pound of bananas, for example—of the final price paid in the supermarket actually goes to the grower. Intermediaries, including importers, distributors, and "ripeners" (who spray ethylene gas to turn the bananas yellow), all get a cut. Fair trade buyers deal directly with the farmer cooperatives they helped organize, avoiding brokers and intermediaries and guaranteeing higher prices—typically, 18 cents per pound for bananas—for the farmers. The fair trade organizations also help the farmers establish schools and health clinics.

Before socially conscious consumerism became popular in the United States, the **fair trade movement** developed in Europe as people began taking notice of the impacts of collapsing coffee prices. Whereas farmers could earn $3.18 per pound for their coffee in 1999, a glut in the world coffee supply, caused largely by surging production in Vietnam, sent prices down to 47 cents per pound by 2001—the lowest in a century, adjusted for inflation. Europeans learned that African and Latin American coffee farmers, never wealthy to begin with, were now destitute, so poor that their young children were compelled to work in the fields to help make ends meet. Green Party politicians and others stepped forward with the fair trade solution, seeking to guarantee what they considered a living wage for the producers of these commodities.

Joe Hobbs

• **Figure 10.B** Fair trade certification of coffee sold in a U.S. health food store reflects changes in consumers' awareness and an improvement in the producers' livelihoods.

Coffee is the world's second largest traded commodity (after oil), so fair trade coffee could affect many people. Now more than a million coffee growers in Tanzania, Rwanda, Ecuador, Nicaragua, and other countries belong to the cooperatives that sell their products through fair trade channels instead of to commercial producers. Fair trade products are becoming more mainstream in the United States, especially as even large retailers like Walmart realize that social and environmental justice can be good for their bottom line.

agriculture in national economies has experienced a marked percentage decline as economies have become more diversified. But in many countries, more than half of all export revenue is still derived from farm products, and some countries rely mainly on one or two agricultural commodities. As in other LDCs, overreliance on a narrow range of exports makes the Latin American countries economically vulnerable to changes in market conditions, competition from other sources, and
45 changing consumer appetites. Reliance on coffee and bananas has whipsawed the economic fortunes of many countries, especially those in Central America such as Honduras and Guatemala that came to be known disparagingly as "banana republics." In colonial fashion, the American-owned United Fruit Company owned vast tracts of land for the production of bananas and other agricultural commodities in these countries and a had powerful influence on local politics. In Guatemala, President Árbenz developed land reform policies that were seen as a threat to the United Fruit Company. He was subsequently ousted in 1954 in a CIA-led coup (it has been noted that CIA Director Dulles was a significant United Fruit Company shareholder). Today's fair trade movements represent a reversal of the exploitative relationships that marked the United Fruit Company era (see Insights, this page).

Types of Farms

Farms in Latin America can be divided into two major classes by size and system of production. Large estates with a strong commercial orientation are known as ***latifundia*** (sing., *latifundio*). These estates, whether called ***haciendas***, plantations, or some other name, are owned by families or corporations (•Figure 10.22). Some have been in the hands of the same family for centuries. The desire to own land as a form of wealth and a symbol of prestige and power has always been a strong characteristic of Latin American societies. Spanish and Portuguese sovereigns granted huge tracts to members of the military nobles who led the way in exploration and conquest. In sporadic land reform programs, some of this land has been reallocated to small farmers by government action. But in many countries, a very large share of the land is still in the hands of a small, wealthy, landowning class.

Minifundia (sing., *minifundio*) are smaller holdings with a strong subsistence component. The people who farm them generally lack the money to purchase large and fertile properties; they farm marginal plots, often on a sharecropping basis. Individual landowners are often burdened by indebtedness, the threat of foreclosure of farm loans, and the fragmented

Joe Hobbs

• **Figure 10.22** Henequen (sisal) plants on a plantation in Mexico's Yucatán region brought great wealth to the ancestors of the hacienda's current owners.

nature of farms, which become smaller as they are subdivided through inheritance and the reversal of promised governmental programs of land distribution. Such farms produce food for family use and also for the local market (•**Figure 10.23**). The crops most commonly raised, especially in Middle America, are maize (corn), beans, and squash, although many other crops are locally important, especially in the different climatic zones of the highland regions.

Minerals and Mining

Latin America is a large-scale producer of a small number of key minerals, notably petroleum, iron ore, bauxite, copper, tin, silver, lead, zinc, and sulfur. Only a handful of Latin American nations gain large revenues from exporting these minerals. These countries have been able to develop significant infrastructure, including new highways, power stations, water systems, schools, and hospitals, in the process boosting employment.

Most of the extracted ores from Latin America are shipped to China and other overseas consumers in raw or concentrated form. The region has scattered iron and steel plants, as well as smelters of nonferrous ores, which process metals for use in Latin American industries or for export. Deposits of high-grade iron ore in Venezuela and central Brazil are the largest known in the Western Hemisphere and are among the largest in the world; the two countries are Latin America's main producers and exporters of ore. In Latin America, Brazil is the main producer of iron and steel by far, with Mexico second. Most of the region's production of bauxite—the major source for aluminum—comes from Jamaica, Brazil, Suriname, and Guyana. The deposits in these countries are generally located near the sea and are of critical importance to the industrial economies of the United States, Canada, and China. Venezuela has huge bauxite deposits and the sources of energy needed to process the ore; the country is a major aluminum exporter.

Chile is the largest copper producer in Latin America and also ranks number one in the world. Most of Latin America's known reserves of tin are in Peru, Bolivia, and Brazil, and these countries produce most of the region's output. The silver of Mexico, Peru, and Bolivia—sought since the very beginning of Spanish colonization—is found mainly in mountains or rough plateau country. Mexico and Peru are the largest Latin American producers of silver, lead, and zinc. Latin America's petroleum has long been extracted in the Caribbean Sea and Gulf of Mexico area, particularly the central and southern Gulf coast of Mexico and northern Venezuela. Other oilfields in Latin America are widely scattered, with the main ones located along and near the Atlantic coast in Brazil, Argentina (in Patagonia), and Colombia, and in lowlands along the eastern flanks of the Andes in every country from Trinidad and Tobago to Chile. Natural gas is extracted in many areas that produce oil, but Latin American production is not yet of major world consequence. Mexico, Venezuela, Argentina, and Trinidad and Tobago are the largest gas producers in Latin America. Venezuela is also the only Latin American nation in the Organization of Petroleum Exporting Countries (OPEC) and was one of its founding members in 1960.

Strong growth characterized Latin American economies after 2003, particularly because of China's superheated economic growth. Brazilian iron ore, Chilean and Peruvian copper, and Venezuelan oil have been in ever-increasing demand to feed China's appetite for commodities. The flip side is that if China's 234
boom were to go bust, Latin American economies would fall too. These countries need to diversify to protect themselves.

Joe Hobbs

• **Figure 10.23** Highland Maya market, Guatemala.

Free-Trade Agreements

Latin America's heavy reliance on unprocessed minerals has periodically dealt crushing economic blows. Many Latin American countries are trying to reduce their dependence on

• **Figure 10.24** Latin American countries belong to a variety of economic associations.
Source: Based on The Economist, April 11, 1998, p. 125.

raw materials and boost their exports of value-added manufactured products, thus improving their financial situations. To do so, they have formed or joined free-trade agreements (FTAs), particularly with their giant neighbor to the north (•**Figure 10.24**) (see Regional Perspective, page 371).

Four nations in South America—Brazil, Argentina, Uruguay, and Paraguay—belong to the **Southern Cone Common Market**, known as **Mercosur**, which was established in 1991. In value of products traded, this is the third largest trade group in the world, after NAFTA and the European Union. Mercosur reduced tariff and other long-standing barriers that had existed between the countries, promoting a huge increase in trade. Its members have worked to coordinate economic policies and a better negotiating position with the United States and other powers. The more than 250 million people of this union can live and work in any of the member countries and have the same rights as citizens of those countries. This stipulation brought immediate relief to an estimated 3 million illegal workers.

Patterned after NAFTA, another U.S.-brokered trade organization was ratified by the U.S. Congress in 2005. This was the **Dominican Republic–Central America Free Trade Agreement (DR-CAFTA)**, composed of the United States, Costa Rica, El Salvador, Guatemala, Honduras, Nicaragua, and the Dominican Republic. Using NAFTA as its foundation and DR-CAFTA as a building block, the United States is taking the lead in establishing a hemisphere-wide trade organization called the **Free Trade Area of the Americas (FTAA)**. Like the political organization known as the **Organization of American States (OAS)**, it would include all of the countries of North, Middle, and South America, with the exception of Cuba. Originally envisioned for establishment in 2005, this agreement is proving very difficult to achieve because some of its loudest critics in Latin America—Brazil in particular—insist that the United States is including too many restrictions in its vision of "free" trade. Mercosur members, including Brazil, want to see the United States drop the huge subsidies of many agricultural products and steel that keep American farmers and factory workers in business but that shut out Brazilian products. In return, Mercosur would endorse patent, copyright, and other **intellectual property rights (IPR)** protection of U.S. industries. With the FTAA stalled, the United States has chosen to negotiate one-on-one trade agreements with members of the proposed FTAA, beginning with Chile in 2004 and Peru in 2007.

There are other regional trade associations, including the Central American Common Market, the Andean Community, and the Caribbean Community (CARICOM) (see Figure 10.24).

Sending Money Home

More and more Latin American families, particularly from Mexico, Central America, and the Caribbean, have come to rely on having at least one member work abroad to help out the family economy. Among the largest sources of income for Latin American economies are the **remittances**, or earned savings, sent home by people working abroad, especially in the United States. In 2011, Latin Americans working in the United States sent home more than $61 billion, about $3 billion less than in 2008 before the Great Recession began, but significantly more than in 2009 and 2010. The workers typically send these payments home in periodic electronic transfers of $200 to $300 each. Mexico is the world's third largest remittance market (after China and India), with $24 billion infused yearly into the country from Mexican laborers in the United States.

The greatest numbers of these estimated 35 million immigrant workers, of whom perhaps a third are illegal or undocumented, are in the U.S. states of California, New York, 394 Texas, Florida, and Illinois, but all 50 states have at least some. An estimated 75 percent of them send monies home, in many cases to grandparents caring for the working parents' children; families often remain divided for many years while the father or both parents work abroad. Collectively, these remittances are so valuable—far surpassing the amount supplied to Latin America in foreign aid and foreign investment—that home countries are searching for ways to tax the income or otherwise find ways to funnel more of it into national development. Some critics say that Mexico and other Latin American countries have become so dependent on remittances that they neglect to invest their own resources in national development. Honduras is one such country: remittances amount to 20 percent of the country's GDP.

One recent and important twist in the traditional one-way flow of remittances is that some experienced workers are

REGIONAL PERSPECTIVE | The North American Free Trade Agreement (NAFTA)

The most economically significant FTA is the **North American Free Trade Agreement (NAFTA)**, of which Canada, the United States, and Mexico are members (**•Figure 10.C**). In terms of both economic value and geographic area, it is the largest trading bloc in the world. Citizens of these countries make up about half of the 942 million people of the hemisphere. The essential goal in establishing this free-trade agreement, like most, was to reduce duties, tariffs, and other barriers to trade among the member countries, theoretically strengthening all the economies. The United States had a strong if perhaps understated NAFTA goal of reducing Mexican immigration into the United States by boosting Mexico's prosperity.

NAFTA was already controversial before it went into effect in 1994, and it remains so today. Among U.S. citizens, there were fears that factories would close down and jobs would flow south to Mexico. American environmentalists feared that the U.S. environment would suffer because Mexican factories and other polluters, unable to afford the cleanup of their industries, would set new and lower standards that U.S. companies could then match. Mexican officials feared that a surge in their consumption of American-made products would so increase their trade deficit with the United States that Mexico's monetary reserves would be exhausted.

Some U.S. factories did close, in part because production was cheaper south of the border—notably in the ***maquiladora*** factories (discussed on page 377)—but NAFTA was not the exclusive cause of the shutdowns. Lower—cost manufacturing for American companies shifted to a host of countries outside NAFTA and the Western Hemisphere. Trying to maximize benefits for its three members in the global marketplace, NAFTA does require all its member countries to abide by **rules of origin** specifying that half or more of the components of any manufactured good must originate in Canada, the United States, or Mexico. This is designed to prevent a fourth country from using Mexican labor in the assembly of its own component parts. Critics of NAFTA say the agreement should have demanded that an even larger proportion of the component parts originate in the member countries to boost their economies.

NAFTA did create more manufacturing jobs in Mexico but in the process contributed to a growing disparity between the wealthier north and the poorer south. The higher-paying jobs became concentrated in a belt in the north. In the more agricultural south of the country, where few of these factories were established, Mexican farmers lost incomes or livelihoods because of falling prices for corn, rice, beans, and pork. Ironically, this pushed a new stream of poor immigrants into the United States. NAFTA gradually eliminated most Mexican tariffs on agricultural imports from the United States. The large, heavily subsidized American agricultural businesses can raise grain or livestock at prices below production cost. A pig, for example, can be raised and sold in Iowa for one-fifth of what it would cost to raise and sell it in Mexico. Cheap American agricultural products flooded the Mexican market, making it impossible for Mexican farmers to compete and driving many out of business; one-third of Mexico's 18,000 swine producers went out of business soon after NAFTA went into effect. On the positive side, Mexican companies have been able to turn the cheaper U.S. pork into sandwich meats sold back at a profit to American consumers.

Corn occupies a cherished place in Mexico's culture and nutrition, and increasing integration of the Mexican and U.S. economies led to a corn crisis there in 2006. The United States began diverting more of its corn crop into the production of ethanol, a biofuel. Subsequent shortages of corn on the food market drove prices higher in the United States. Those increases were passed on in corn exported to Mexico and matched by domestic corn producers in Mexico. The young government of President Calderon found itself besieged by protesters demanding that Mexico's corn producers be shielded from U.S. imports (then accounting for one-fourth of Mexico's corn) and that the government intervene to keep this most basic of all foodstuffs affordable to the poor. At the same time, genetically modified corns began appearing in southern Mexican fields, and environmentalists and other anti-GM activists chimed in their demands for "food sovereignty" with the protection of
Mexico's corn heritage. These events swirling 86
around corn came to be known as Mexico's **"tortilla wars."**

In sum, in the United States and Canada, NAFTA has been beneficial to farmers but not to factory workers, while the reverse is true in Mexico. Both the United States and Canada have trade deficits with Mexico, importing more from that country than exporting to it. The hoped-for reduction of emigration to the United States did materialize, but for the "wrong" reason: the economic downturn in the United States made emigration from Mexico less attractive. Meanwhile, NAFTA appears to be losing some steam, with trade among its three members declining about 10 percent between 2001 and 2011.

• Figure 10.C The trinational NAFTA flag symbolizes the integrated economic ambitions of its members.
Source: NAFTA

Joe Hobbs

• **Figure 10.25** Warm, turquoise waters of Caribbean and other Latin American beaches are an irresistible draw for tourists from El Norte. This is the "Maya Riviera" of the Yucatán Peninsula's east coast.

returning to their home countries to establish companies that invest in the United States. These so called **multi-Latina companies** are aiding the revival of certain downtrodden sectors of the U.S. economy, including cement and steel mills.

Tourism in Latin America

Tourism has become a major regional economic asset to Latin America, generating critical foreign exchange; only oil exports are more valuable. Countries in which tourist dollars amount to more than 50 percent of the nation's foreign exchange include Antigua and Barbuda, the Bahamas, and Barbados. During what many North Americans see as their too-long winters, they are bombarded by television and print advertisements of vacation in paradise: Jamaica and hosts of other destinations of the Caribbean basin and western Mexico. A wide variety of vacation experiences are available, from the hedonistic resort complexes that tourists seldom leave (one resort on Jamaica actually is called Hedonism), to the insular sea voyage that offers day trips ashore in exotic destinations, to whitewater and rain forest adventures of the ecotourist variety (•Figure 10.25).

The region benefits from the proximity of its year-round warmth to the wealthy and mobile and seasonally chilled U.S. and Canadian populations. Tourism revenues reflect that **distance-decay relationship**: the highest tourism receipts in Latin America flow to Mexico, the nearest neighbor to the wealthy countries, but tend to fall off for more far-flung destinations.

10.5 Geopolitical Issues

The central geopolitical reality of Latin America is that it is in America's backyard. For better or worse, the United States has staked its geostrategic claim to the region. Much of the United States itself is comprised of land wrested from Mexico in conflicts of the early nineteenth century. In 1823, U.S. intentions for Latin America were formulated in a policy known as the **Monroe Doctrine**, after U.S. President James Monroe announced that the United States would prevent European countries from undertaking any new colonizing activities in the hemisphere. Many subsequent interventions, wars, investments, and other U.S. activities may be seen in light of this seminal policy. Even more consequential was the 1904 **Roosevelt Corollary** to the Monroe Doctrine, whereby the United States declared that it had the right to supervise the internal affairs of Latin American countries to ensure U.S. national security.

Ever since the Monroe Doctrine was proclaimed, the United States has had a particularly firm hand in Central America. Washington has dealt unapologetically and sometimes harshly with perceived threats to the United States from this region. Examples include U.S. actions to support governments against nationalist insurgencies, or insurgents against leftist governments: El Salvador and Nicaragua in the 1980s, Cuba in the 1960s, and Nicaragua in the 1950s were the most significant instances. Perhaps the most-cited intervention took place in 1954, when the U.S. Central Intelligence Agency ousted the Guatemalan leader in a coup. Among other things that earned American denunciation as communistic, President Árbenz pushed for the redistribution of land in a country where 2 percent of the population owned
70 percent of the arable land. 61

The Panama Canal

Built between 1904 and 1914 by American contractors, the Panama Canal is a legacy of American hegemony in Central America. The Canal was intended mainly to serve U.S. commercial and strategic interests. For decades, it was sovereign U.S. territory, carefully monitored by an American military presence, and the United States kept more than half of the transit fees charged to ships using it.

The Panama Canal today is one of the world's most critical chokepoints: 14,000 ships a year, traveling 80 shipping routes and representing about 5 percent of the world's cargo volume (and 16% of the United States'), use the 50-mile (80 km)
shortcut from the Atlantic Ocean to the Pacific. A 14-day sea 177
voyage between New York and San Francisco via the canal cuts 7,800 miles (12,530 km) and 20 days from the trip (•Figure 10.26; see also Figure 10.31). The United States is a major user; about one-seventh of all U.S. foreign trade is carried through the Canal.

In the late 1970s, the administration of U.S. President Jimmy Carter accepted the Panamanians' complaint that U.S. control of the canal was an outmoded vestige of colonialism. Carter negotiated the transfer of authority over the canal to Panama in a treaty that went into effect in 1999. Now the revenues from ships transiting the canal are exclusively Panama's (more information on Panama's ownership of the canal begins on page 378).

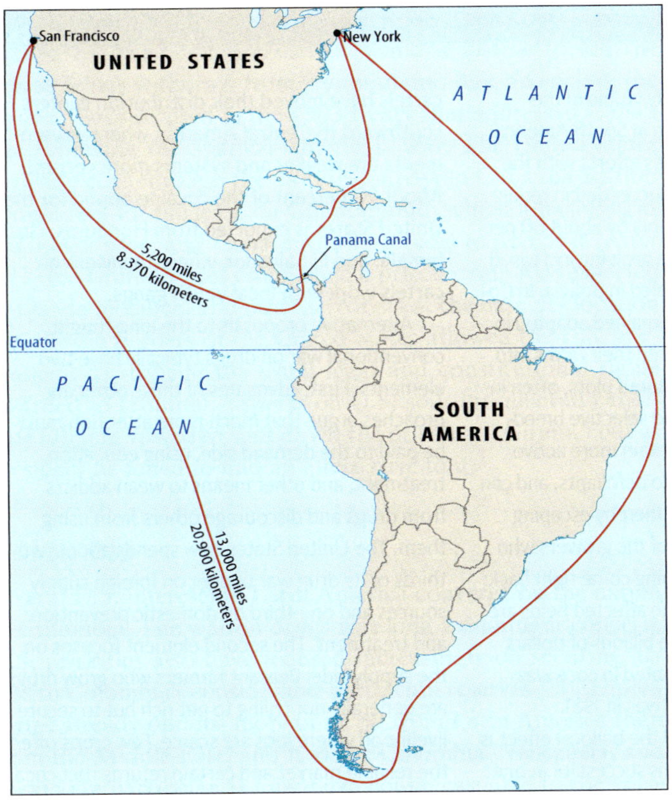

• **Figure 10.26** Distance, time, and money are saved by use of the Panama Canal.

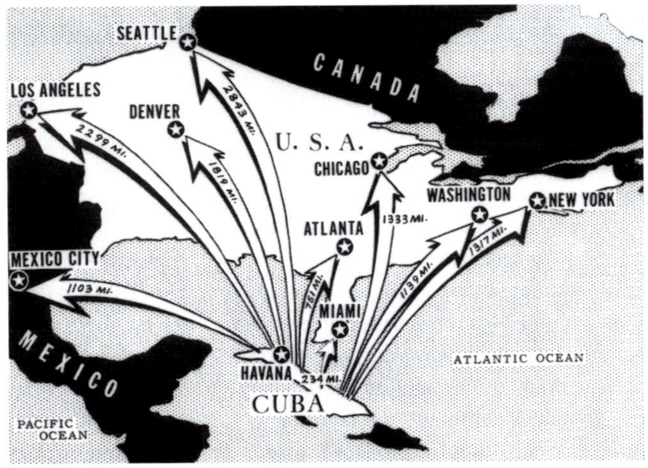

• **Figure 10.27** The 1962 Cuban Missile Crisis brought the United States and the Union of Soviet Socialist Republics close to the brink of nuclear holocaust. This 1962 map shows the range that Soviet missiles would have if stationed in Cuba.

From a strategic and economic defense standpoint, however, the United States is still deeply involved with the canal. The U.S. Navy is concerned that its two access points (Balboa in the Pacific and Cristóbal in the Atlantic) are now controlled by a company with strong backing from China, the canal's second largest user. The United States has already effectively invoked the Roosevelt Corollary since returning control of the canal; in 1989, U.S. troops invaded Panama and ousted its ruler, Manuel Noriega, whom the Americans accused of facilitating the illegal trade of drugs into the United States. The U.S. war on drugs has been another central geopolitical issue of the region since the early 1970s (see Geography of Drug Trafficking, next page).

Cuba

The most consequential test of U.S. commitment to its interests in Latin America came with the **Cuban Missile Crisis** of 1962. Cuba is as close as 90 miles (144 km) from the United States. The frosty relations between the United States and Cuba's Marxist regime, led after 1959 by a firebrand revolutionary named Fidel Castro, were emblematic of the Cold War. Those tensions came dangerously close to war and even nuclear holocaust in 1962, when the Soviet Union began to build facilities in Cuba that would have accommodated nuclear missiles capable of striking Washington, D.C., New York, and other cities in the eastern United States (•Figure 10.27). As it so often does today, remote sensing played a vital intelligence role then. Photographs taken from U-2 aircraft provided undeniable evidence of missile launch infrastructure on the ground in Cuba.

Although the missile threat was defused under intense pressure from the U.S. government, led at the time by President John F. Kennedy, Cuba has remained a storm center of political affairs in the Western Hemisphere. After several attempts to assassinate Castro—even with the unlikely ploy of an exploding cigar—and a disastrous attempt by Kennedy's administration to overthrow him by military force (in an incident known as the **Bay of Pigs invasion**), the U.S. administration settled on a course of economic and political isolation of Cuba.

Despite the resumption of diplomatic and economic relations with most of the remaining Communist countries in the world—China most notable among them—the United States has remained steadfast in imposing sanctions on Cuba. This anomaly to some extent reflects domestic political realities in the United States, where a large population of **Cuban Americans** (including 800,000 in Florida alone) who fled Castro's Cuba tend to vote for the politicians who keep up the most pressure against Castro. The Cuban Americans are against any relaxation of the virtual ban on American tourism to Cuba, which is in place because the **Trading with the Enemies Act** forbids Americans from spending more than $300 in Cuba. The failing health of Fidel Castro and the discovery of significant oil reserves off Cuba's shores led to unprecedented discussions among U.S. lawmakers about reopening American doors to Cuba. One fear was that U.S. businesses would lose important opportunities if the doors remained closed. Recent legislation allows U.S. states to sell food and medicine to Cuba on a cash-only basis, and about 40 states do so.

United States, through a pipeline that would cross Chile to a Pacific port. Native American Bolivian peasants, backed by labor unions, student groups, and opposition politicians, complained that royalties for Bolivia from the gas sales would be so low that the common people would never realize any benefit from them. They cited Bolivia's history as a colony stripped of its silver and tin to enrich the Spaniards and local aristocrats, and insisted that again only foreigners and Bolivia's elite would gain from the gas project. These opponents wanted the gas to remain in Bolivia, where it could form the basis for new local industries. (They also resented the pipeline's passing through Chile, a focus of Bolivian anger ever since Chile, following an eighteenth-century war, seized lands that cut Bolivia off from the sea.) Their votes were instrumental in bringing Evo Morales, himself an Aymara, to office. Morales championed their cause by announcing that his government would nationalize (impose government control over) all of Bolivia's critical energy and other resources and use them for the benefit of the people. The so-called "ideology of fury" expressed by the Aymara in Bolivia rattled nerves among the elites in other countries of Latin America with large indigenous populations, including Ecuador, Peru, Paraguay, Guatemala, and even Mexico.

The following section explores more issues among the peoples and lands of the region, first in Middle America and then in South America.

10.6 Regional Issues and Landscapes

Middle America

The region most geographers prefer to call Middle America, also known as Mesoamerica, includes Mexico (its largest country in area and population); the much smaller Central American countries of Guatemala, El Salvador, Belize, Honduras, Nicaragua, Costa Rica, and Panama; and the numerous island countries in and near the Caribbean Sea, the largest being Cuba, Haiti, the Dominican Republic, and Jamaica. The region is a vibrant patchwork of physical features, ethnic groups, cultures, political systems, population densities, and livelihoods. Most of the political units are independent, but a few are still dependencies of outside nations.

Mexico: Higher and Further

The federal republic of Mexico—officially, *Los Estados Unidos Mexicanos* (United Mexican States)—is the largest, most complex, and most influential country in Middle America (•**Figure 10.28**). The country's great population of 115 million in 2011 makes Mexico by far the world's largest nation in which Spanish is the main language. Compared to most of the world's countries, Mexico is a spatial giant, but its large size tends to

• **Figure 10.28** Principal features of Mexico.

go unrecognized because it lies in the shadow of the United States. Triangular in shape, Mexico's area of 756,000 square miles (c. 2 million sq km) is nearly eight times that of the United Kingdom, and its elongated territory would stretch from the state of Washington to Florida if superimposed on the United States (see Figure 10.2). Its capital, Mexico City, stands now as the world's second largest city in population (after Tokyo-Yokohama), with 27 million inhabitants. Mexico is one of the most urbanized countries of Middle America, with about 78 percent of its people living in cities. In these and many other attributes, Mexico in the Latin American context seems to live up to its motto: "Higher and Further."

Mexico has been described as one of the globe's nine "pivotal countries," a country whose economic or political collapse could cause international refugee migration, war, pollution, disease epidemics, or other international security con-
237, 240 sequences. The country's stability is enormously important, particularly to its giant neighbor to the north. Mexico's natural and human capital is much greater than many outsiders realize. Mainly because of the wealth of its tropical forests in the south, Mexico ranks as the world's third most important megadiversity country, after Brazil and Indonesia. Mexico ranks highly among the world's producers of such commodities as metals, oil, gas, and sulfur, with oil the country's most important export. The Mexican oil industry is a government monopoly (carried on by a government corporation, Petróleos Mexicanos, or PEMEX), and the United States is its largest foreign customer, taking in 90 percent of the country's oil exports. Mexican oil represented 11 percent of oil imported to the United States in 2011, making Mexico the second largest foreign supplier after Canada.

Mexico's oil will run out—some analysts say by 2030. The country is trying to diversify and grow its economy, the second largest in Latin America (after Brazil), away from reliance on commodities like oil. Mexico has made substantial gains in its development of manufacturing and component assembly
371 industries, especially since the initiation of NAFTA. The manufacturing sector is now Mexico's leading source of revenue, followed by oil and remittances from abroad. High-technology exports grew from 8 percent of total exports in 1990 to 22 percent in 2011. About one-third of all manufacturing employees in Mexico work in the thousands of factories located in metropolitan Mexico City.

The large *maquiladoras* of Mexico, churning out automobiles, textiles, electronics, and computers, are the symbols of the new Latin American free-trade-oriented economies. These have become an exception to the previous spatial pattern of Latin American manufacturing enterprises. Larger operations were almost always located in the larger cities, including the greatest producing metropolises of Mexico City, São Paulo, and Buenos Aires, representing the region's three greatest manufacturing countries, Mexico, Brazil, and Argentina, respectively. Consider the geographical concept of *situation:* these *maquiladoras* are placed where location serves them best: as close as possible to the great market of the United States—the principal destination for most of their output—or deep in areas like the country's southern Yucatán region where labor costs are lowest.

Exports from Mexico to the United States have more than tripled since NAFTA went into effect in 1994. This bilateral trade is in Mexico's favor, with imports from Mexico exceeding exports to Mexico by 20 percent. Mexico is the third largest provider of imports to the United States, after China and Canada. For its part, the United States exports more goods to Mexico than to any other country but Canada (13 percent of U.S. exports go to Mexico).

The *maquiladoras* have surged in growth and diversification since NAFTA went into effect; about 2,700 of these factories, employing nearly 1 million Mexicans, are now operating in the U.S. border region alone. The *maquiladora* is designed to boost exports of manufactured goods by using tax breaks, cheap labor, and other incentives to attract foreign industries. The factories produce jeans, cars, computers, and other goods, mainly from components that are imported duty-free because they are assembled for export only (**•Figure 10.29**). There has also been growth in industries supporting the *maquiladoras,* such as small-scale retail and transportation services. Among

AP Photo/Damian Dovargane

• Figure 10.29 The *maquiladora* is globalization in a nutshell. This woman is assembling electronic devices for Samsung, a Korean firm, in a *maquiladora* in the Mexican border town of Tijuana for sale in the United States.

the most recent developments are **energy *maquiladoras*.** These are power generation plants—built by American firms just inside Mexico, where pollution standards are much lower than in the United States—that provide electricity to the U.S. grid. Their power source, natural gas, comes from Texas, and
415, Mexico supplies their coolant waters.
418 In the global marketplace, Mexico's *maquiladoras* are evolving. After 2000, Mexico lost some of the competitive edge represented by its inexpensive labor. Hundreds of *maquiladoras* closed, and hundreds of thousands of their workers lost jobs. Why? Mexico's clothing, textile, and other low-tech industries could not match Chinese rock-bottom labor costs. Mexico itself became flooded with Chinese imports, even of pottery imitating Mexico's traditional wares.

While conceding the loss of low-tech, low-skill industrial advantage to China (and also nearby El Salvador, where labor is also cheaper), Mexican manufacturers hung on to Mexico's edge in exports of flat-screen televisions, computers, aircraft parts, and other high-tech and heavier goods that are more expensive to ship from China. The *maquiladoras* began a slow and steady recovery and became more attractive as labor
269 wages climbed in China. They have also begun to merge into specialized clusters, for example, with Tijuana concentrating on flat-screen televisions and Ciudad Juárez on car parts. It is remarkable that, at least so far, the *maquiladoras* in these border towns have not been affected by the drug-related violence that has claimed more than 40,000 lives in Mexico.

There are both inseparable ties and divisive issues between Mexico and the United States. Huge disparities in wealth and power come between them. Mexicans have not forgotten that Mexico lost more than half its national territory to the *gringos* in the **Mexican War** (known to Mexicans as the **American Invasion**) in 1847. Mexicans have enormous national and cultural pride and often feel misunderstood and underappreciated by their northern neighbors. In the next chapter, we explore immigration and other issues in relations between the United States and Mexico.

Central America

The Panama Canal

Between Mexico and South America, the North American continent tapers southward through an isthmus containing seven less developed countries known collectively as Central America. Five of these countries—Guatemala, Honduras, Nicaragua, Costa Rica, and Panama—have seacoasts on both the Pacific and the Atlantic Oceans (via the Caribbean Sea; •Figure 10.30). The other two have coasts on one body of water only: El Salvador on the Pacific and Belize on the Caribbean. If the seven countries were one, they would have an area only about one-fourth that of Mexico (see Table 10.1), with a total population about 40 percent that of Mexico.

At the southern end of Central America, the narrow Isthmus of Panama separates the Pacific and the Atlantic oceans by a mere 50 miles (80 km). Even looking at a map or globe from a distance, you can easily appreciate that this is the best place to

• **Figure 10.30** Principal features of Central America.

locate transcontinental shipment routes. In 1855, the Panama Canal Railway became the first transcontinental railway, and it remains in use today. Colombia is considering whether it should build a second, Chinese-funded railway across its portion of the Isthmus.

The Panama Canal, as discussed earlier (page 372), is of great importance to both Panama and the world (•Figure 10.31). The idea of a shortcut between the Atlantic and Pacific Oceans

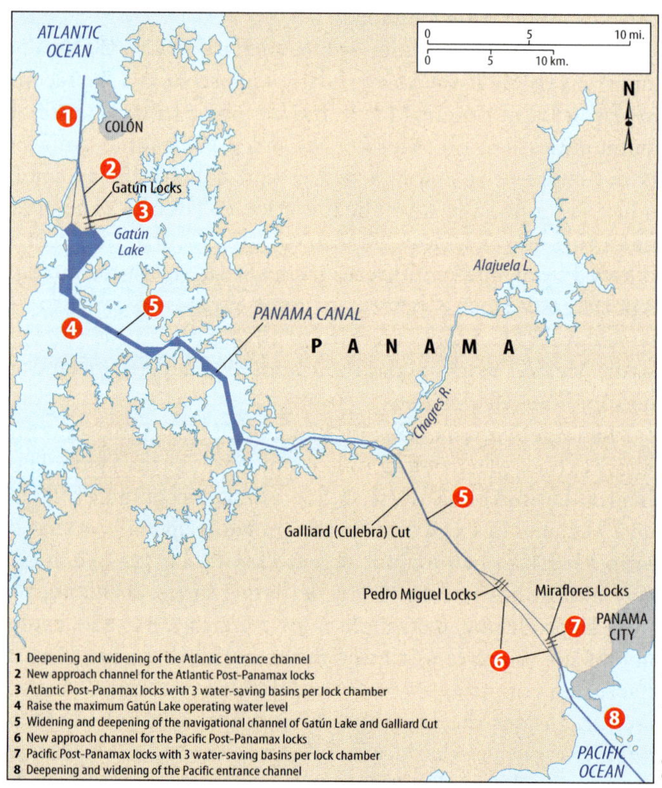

• **Figure 10.31** Profile and plan of the Panama Canal.

had been around for a long time before the Panamanian site was chosen. In the early 1800s, Nicaragua was the first choice because of natural landscape features that appeared better able to accommodate a crossing without massive construction (Nicaragua is still considering building this canal, which, unlike even the upgraded Panama Canal, would accommodate the world's megaships). In 1846, however, Colombia (which then controlled Panama) and the United States signed a treaty supporting a project to cut a canal across the much narrower Isthmus of Panama.

Interest in the project picked up substantially in 1849 with the discovery of gold in California. At that time, prospectors and speculators had to sail to the port of Colón on the east side of Panama, make the 50 miles of land crossing in any way possible, and then reboard a ship in Balboa and sail to San Francisco. After 1855, the Panama Canal Railway (funded by New York business interests) provided a reliable mode of transit from port to port. This train, however, required a break of bulk of goods and passengers a canal
266 would not.

In 1878, Colombia granted permission to a French firm to build a canal across the isthmus near the railroad line. Ferdinand de Lesseps, who had seen the Suez Canal to completion in 1869, designed the project. The French quickly bought control of the railroad line and began digging in 1882, with a plan to build a sea-level transit requiring no **locks** (the devices that allow a ship to move from one water level to another). This venture went bankrupt in 1889 and was followed by another French attempt, which also failed. Construction difficulties and a high death toll due to yellow fever and other diseases were to blame. By the beginning of the twentieth century, the French had left. The United States had just participated in the 1898 Spanish-American War and had more geopolitical interest than ever before in this prospective waterway.

In 1903, Panama broke away from Colombia, and within weeks, the United States recognized the new government. Shortly thereafter, the two governments struck an agreement to build the canal, which ultimately required three sets of locks. Construction began in 1904 and, at a cost of $380 million and many lives lost to yellow fever, bubonic plague, and malaria, was completed in 1914. The first crossing was made under a banner that read: "THE LAND DIVIDED; THE WORLD UNITED."

Work on the canal has never ended (Figure 10.31). To meet anticipated demand, especially from surging exports of China's manufactured goods, the canal recently underwent the largest expansion ever. Panama's citizens approved the expensive project in 2006 with the expectation that it would quadruple income from transit fees. The construction project aimed to overcome problems delaying and prohibiting the passage of many ships. To double capacity and accommodate larger cargo vessels up to the size known in the shipping trade as "post-Panamax," new channels with longer and deeper locks were added at each end of the canal. These and other measures doubled the canal's capacity. Thanks to growing trade, expanding ports, and residential and commercial development, Panama's economy has boomed in recent years; it has invited comparison with that of Dubai, another place that has capitalized on its strategic crossroads location. 168

South America

Venezuela's Petroleum Politics

Oil resources in the coastal area around and under the large Caribbean inlet called Lake Maracaibo (•Figure 10.32) have been a huge blessing for Venezuela, the world's 13th largest oil producer in 2011. Oil makes up about 95 percent of the country's exports by value. The country was one of the founding members of the Organization of Petroleum Exporting Countries (OPEC) in 1960 and is still a member. Oil revenues 176 have fueled urbanization and industrialization and generated wealth. Yet Venezuela is rather like Nigeria or Angola, an oil-rich but oil-cursed country where the vast majority of people remain poor. 328

The most prominent member in the lineup of left-leaning Latin American leaders elected in recent years was Venezuelan President Hugo Chávez. Invoking the revered name of Simón Bolívar (1783–1830), the Venezuelan-born independence leader who spearheaded a regional quest for liberation from Spain, he proclaimed to be undertaking a **"Bolivarian Revolution."** (Chávez had his hero exhumed and examined in 2010, ostensibly to determine how Bolívar died.) Chávez had his own ideas about liberation: he rewrote the constitution, put cronies on the supreme court, cracked down on the media, and jailed political foes. The Chávez agenda appealed to the poorest classes by promising a redistribution of wealth like that championed by Cuba's Fidel Castro (Chávez's strongest ally in the region). Chávez's political philosophy was not communism, but what he called "twenty-first-century socialism" (and others dubbed "Chavismo"), with the armed forces and

• **Figure 10.32** Principal features of Venezuela.

20 percent of global carbon emissions. The challenge is to use mechanisms like the United Nations' REDD (Reducing Emissions from Deforestation and Forest Degradation) Program to persuasively make the case that conservation pays; the ecological
35, 253 services provided by the forests are enormously valuable. The governments of Brazil and other Amazon Basin countries generally acknowledge the importance of the area's natural state but also argue that the area is so vast that portions of it can be reasonably developed to help advance their economies.

Since the 1970s, Brazil in particular has aggressively pursued development in five major areas of the Amazon: exploitation of iron ore and other minerals, resettlement of "excess" populations from Brazil's crowded southeast, farming and ranching, timber exploitation, and hydroelectric development. Each of these serves as an agent of deforestation. All are ultimately linked to Brazil's expanded road network in Amazonia. To begin its development of the Amazon, in the 1970s Brazil began construction of the **Trans-Amazon Highway** and its feeder system (•Figure 10.36). The main line of the interoceanic highway runs roughly east to west from the Atlantic toward its ultimate, long-awaited destination: the Peruvian coast, finally reached in 2011. This road construction was part of an international project called the **Regional Initiative for the Infrastructure Integration of South America**. The final link crossing the Andes is intended to give Brazilian exporters an easier outlet to Asian markets. Principal feeder lines include the controversial BR-364 through the Brazilian states of Acre and Rondônia; BR-163, running north–south through the heart of the region; and BR-319, now under construction between the free-trade zone of Manaus and Rondônia. BR-319 is of particular concern to environmentalists because it runs through pristine forest to the "arc of deforestation" that was deforested precisely because roads like BR-364 gave access to the forest. This arc

• **Figure 10.36** Deforestation and road development in the Brazilian Amazon.
Source: Adapted with permission from the April 22, 2004 issue of The Christian Science Monitor (www.CSMonitor.com). © 2004 The Christian Science Monitor.

of deforestation, on the southern edge of the Amazon Basin, is visible as the mustard color in Figure 10.36.

Here is how that arc was created: the first wave of settlers into Amazonia consisted mainly of these hardscrabble farmers and ranchers. Brazilian authorities advertised the Amazon as a "land without people for people without land." Brazil began transferring its landless poor from the overcrowded southeast of the country, up along the feeder roads into the wilderness of the Amazon where, with government cash subsidies in hand, they began a destructive cycle of land use. The process is a classic example of marginalization as described on page 47, with poverty fueling environmental destruction.

Road construction in Amazonia initiates a sequence of events, illustrated well by Highway BR-364 (seen in •Figure 10.37). First, the main road is cut as a swath through the forest. Landless peasants, lured by cheap or free land, cut and burn the rain forest along the main road and the smaller lanes linked to it. Typical of farmers' experiences with tropical slash-and-burn agriculture elsewhere, they may be able to wrest only three to five years of corn, rice, or other crops from the soil before it is exhausted. They then move on to clear and cultivate 219
new lands—the typical pattern of shifting cultivation.

Tropical hardwoods are harvested and sold as well. About 40 percent of the wood cut from the Amazon is shipped overseas. The largest single destination is the United States (consuming about a third of the total), followed by China, but collectively, the European Union nations take in 40 percent. In 2007, Brazil announced that it would open vast new tracts of the Amazon for large-scale logging but promised there would be strict monitoring to ensure that the resource was not overharvested and that none of the 70 percent of the forest owned by the government would end up in private hands. Brazil also shifted some of the burden back to the consuming countries, insisting that they should buy timber only from these zones.

Brazil and other Amazon Basin countries have passed legislation to slow or halt illegal logging and other unlawful uses that contribute to the rain forest's destruction. Authorities in Brazil have slowed the illegal logging of valuable mahogany trees, but the problem has boomed in the Peruvian Amazon. Nearly 80 percent of the estimated 1.6 million cubic feet (45,000 cu m) of mahogany exported each year from Peru to the United States—where it becomes furniture, acoustic guitars, home decks, and coffins—is illegal. Despite U.S. support for laws prohibiting the illegal logging of mahogany, enforcement has been lax, and legal and illegal mahogany is often imported to the United States in the same shipments.

If Amazonian lands used for farming and timber were abandoned and left to regrow, they would return to mature tropical rain forest in a matter of decades. In Amazonia, however, the typical pattern is that large cattle operations move in to use the areas conveniently cleared for them by the farmers and loggers. As about 60 million cattle (one-third of Brazil's total cattle population) feed and tread over these Amazonian lands, the soils are exposed to excessive cycles of wetting and drying and to relentless bombardment by ultraviolet radiation. This turns the soil into the bricklike substance known as laterite,

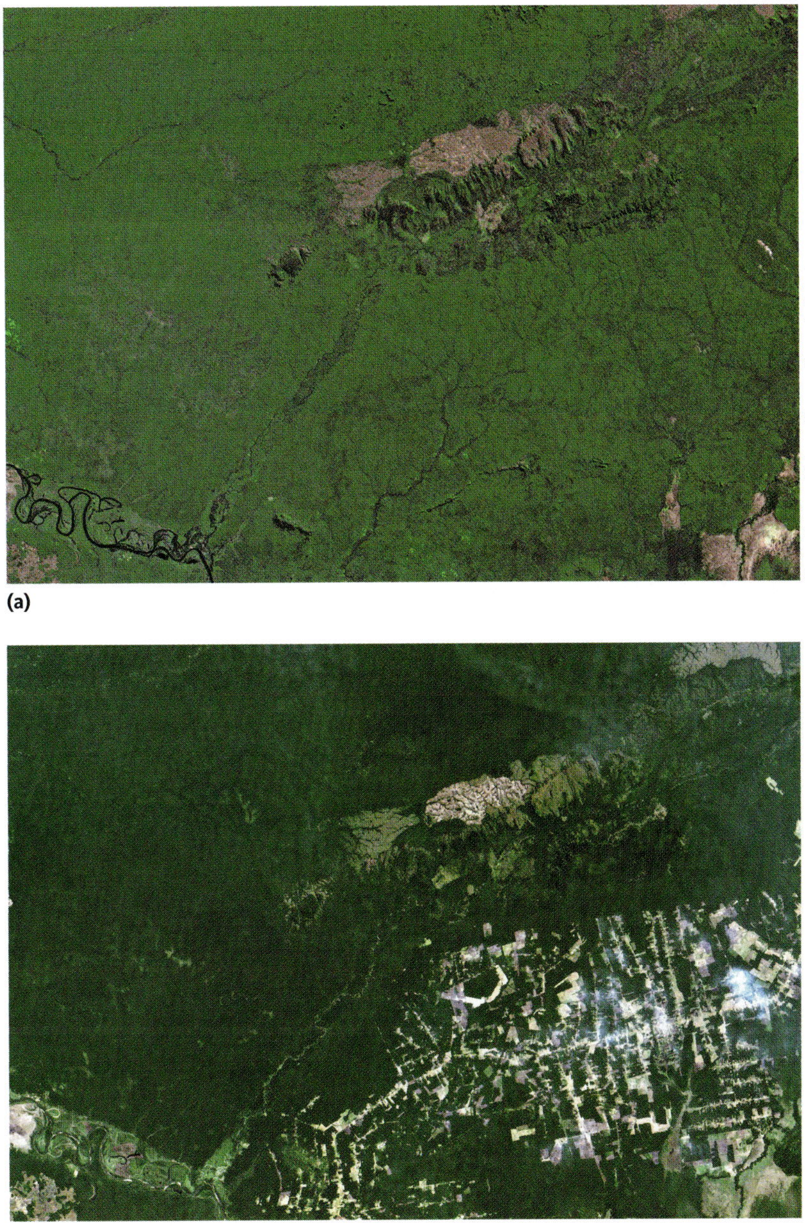

(a) Planet Observer/Universal Images Group via Getty Images

(b) Planet Observer/Universal Images Group via Getty Images

• **Figure 10.37** True color satellite image showing the first stage of deforestation in Amazonia: clearing land for cultivation along Highway BR-364 in western Rondonia State, Brazil.

219, 345 on which it is all but impossible for forest regrowth to occur. When ranchers have exhausted the land in this fashion, they move their cattle on to the next plots cleared by farmers. This **shifting ranching** follows the shifting cultivation; it is part of a cycle that is exceptionally destructive of the rain forest.

The Amazon is gradually becoming urbanized in the process. Between 1990 and 2010, the population of Brazil's Amazon region grew by 50 percent to 20 million, the majority of these living in cities and towns. There are now five cities in Amazonia with more than 300,000 residents—a threshold that is attracting Brazil's national retail chains. American-style shopping malls have begun sprouting in Amazonia.

These processes and the other uses of the forest, especially the boom in commercial soybean cultivation, have resulted since the 1960s in the removal of about 20 percent, or 1.6 million square miles, of the tropical rain forest cover of the Amazon Basin. In recent years, the annual rate of deforestation in Brazil has slowed somewhat to about 4,600 square miles (12,000 sq km), or about the size of Connecticut. National law in Brazil stipulates that 80 percent of every tract in the upper Amazon, and 50 percent in more developed regions, must remain forested, but there is little enforcement. About 20 percent of the Amazon Basin's forest, or 1.6 million square miles, has been razed. The World Wildlife Fund projects that at current rates of deforestation, 55 percent of the Amazon forest will be gone by 2030. Brazilian authorities have announced a plan to reduce deforestation by 72 percent by 2017. But there is concern among ecologists that the Brazilian forest has reached a

"tipping point," from where the vegetation will shift into a savanna biome with less biodiversity. Forest dieback has been accelerated recently by unprecedented droughts. These unusual dry spells are related to El Niño and to another weather phenomenon called the Atlantic dipole, which warms Atlantic waters. There were widespread forest fires in 2005 and again
358 in 2010. A positive feedback loop is at work: with less foliage there is less moisture in the water cycle, making the local climate hotter and drier.

Climate change threatens not only flora and fauna but also indigenous cultures relying on hunting, gathering, and simple agriculture to survive. Previously, their adaptation to environmental stress was to move, but that has become nearly impossible as other Amazonian interests stake claims to the forest. Development in the Amazon has also increasingly provoked conflicts with the region's indigenous populations. There is particular concern now about how fossil fuel production will affect the Kichwa, Achuar, Schuar, and other Native Americans. After 2004, 70 percent of Peru's lowland rain forest areas, many inhabited by Native Americans, were zoned for oil and gas development, and indigenous advocates worry that exploration will introduce new diseases to the native populations. There are plans to build 800 miles (1,280 km) of pipelines that would take natural gas from lowland rain forest areas to coastal refineries and ports. Another big natural gas project is planned for a part of the Brazilian Amazon inhabited by indigenous people. In Ecuador, where oil already makes up half the country's exports by value, there are plans to double production, again mainly in areas where there are significant Native American populations.

In almost all cases, the Native Americans in the areas affected believe that little, if any, benefit will come to them from these developments. They are becoming activists, using a combination of threats against the oil interests and pleas for more benefits from the projects. Some of the native groups have powerful advocate allies abroad and have come to rely on the Internet to make their cause known; an example is the Kichwa people of Sarayaku (http://www.sarayaku.com), who live in one of the Ecuadorian oil blocks slated for development.

The Kichwa were buoyed by the 2006 election of Ecuadorian President Rafael Correa, who as a candidate had expressed his support for their cause. Correa appealed to the international community to use debt-for-nature swaps or other means to finance the protection of sensitive cultural and environmental sites
61 in the rain forest. In exchange for the funding, Ecuador would prohibit oil drilling in these areas. Correa set forth particular challenges; for example, if the international community will give

Joe Hobbs

• Figure 10.38 Are ecotourism and cultural tourism helping to protect indigenous people and natural environments in the Amazon? This community of Bora Native Americans in Peru's Amazon region promotes itself as a "genuine Indian village." Its residents wear native costume for visitors.

Ecuador $3.6 billion, exactly half the projected earnings from extracting oil in the Yasuni National Park, Ecuador will leave the ground there untouched. European powers were interested in this cause until their sovereign debt crisis made such generosity difficult. Some Native American communities are develop- 88
ing their own ecotourist and cultural tourist enterprises, hoping both to demonstrate to governments that there are alternative economic uses for the rain forest areas and to gain valuable support and publicity from international visitors (**•Figure 10.38**).

Amazon rain forest destruction has become a *cause célèbre* for environmentalists around the world, who denounce the development efforts of Brazil and the other countries. Most of the region's governments and people see it differently, denouncing foreigners as hypocrites who have cut down their own forests and who are intent on stunting development of the rain forest countries. There are alternative development schemes, such as the commercial extraction of valuable exports like rubber and brazil nuts from intact tropical forests. These take place in areas like Brazil's **extraction reserves**, but they are very limited in economic impact relative to the conventional development enterprises in Amazonia, and their long-term prospects for growth are uncertain. "Sustainable development" of the Amazon is elusive but, if achieved, would have enormous benefits for people and environments within and far beyond the forest biome.

The book continues, and ends, with the United States and Canada.

summary

- Latin America lies south of the United States and includes Mexico, the Central American countries, the island nations of the Caribbean Sea, and all the countries of South America. It is called Latin America because of the post-fifteenth-century dominance of Latin-based languages and the Roman Catholic Church imposed by European colonizers. The term *Middle America* refers to Mexico, Central America, and the Caribbean.
- Latin American landscapes and livelihoods are correlated in altitudinal zones, which include *tierra caliente, tierra templada, tierra fría,* and *tierra helada.*

- Large areas of humid climates and biome types are in the lowlands of Latin America, with tropical rain forests occupying major river basins, especially in Brazil and Venezuela. Grasslands are found poleward of the tropics, and there is a zone of Mediterranean climate on the western coast of Chile.
- Dry climates and biomes, including desert and steppe, exist because of rain shadow and other factors. There are major desert regions on the west coast of South America and northern Mexico.
- Latin America's population is about 8.5 percent of the world's total, and its growth rate is above the world average. Most of the people live on the rim of the landmasses, with major population bands in the mountain valleys and, in some cases, on high plateaus or on the flanks of the Andes Mountains in western South America.
- Before the arrival of the Europeans in the late 1400s, Latin America had been home to highly advanced indigenous civilizations, notably the Aztecs, Maya, and Inca. There were many other Native American groups that did not develop urban livelihoods. All these cultures declined precipitously after Europeans arrived. Although some of the losses are associated with war and with mining and other land uses, the biggest killer was European diseases, especially smallpox.
- The region's major ethnic groups today are Native Americans, mestizos, mulattoes, blacks, and Europeans. Political and economic systems still tend to favor people of European extraction, but non-Europeans have found increasing power through democratic political systems. Many reject the free-trade and capitalist systems that they believe favor the ethnic Europeans.
- Latin America has been urbanizing rapidly, with now more than two-thirds of its population living in cities. There continue to be steady rural-to-urban migration and a steady stream of legal and illegal immigrants northward into the United States. The remittances of those workers are an important resource for Latin American economies.
- *Latifundia* and *minifundia* are two major agricultural systems. There is a strong plantation economy, especially in Central America, the Caribbean, and coastal Brazil, with sugar and fruits dominant. The fair trade movement raises prices for consumers of coffee, bananas, and other produce but helps ensure better wages for Latin American farmers. Ownership of farmland is concentrated in the hands of relatively few people, creating imbalances that land reform efforts have so far not redressed.
- Minerals have played a major role in the history, economic development, and trade networks of Latin America. Petroleum is of growing importance, especially in Mexico and Venezuela. The global commodities boom related especially to China's growth has helped Latin American economies in recent years.
- The Latin American countries have formed or joined several free-trade agreements. The largest in volume and value is the 1994 North American Free Trade Agreement (NAFTA). One of its goals is to promote economic growth in Latin America so that migration to the United States becomes less attractive to unemployed Latin Americans.
- Tropical Latin America's proximity to temperate and affluent North America has helped make tourism one of the region's most important industries.
- U.S. interests in Latin America dominate the region's geopolitical themes and issues. The Monroe Doctrine and its Roosevelt Corollary were designed to justify U.S. activities and interventions in the region. Historical efforts have been directed at suppressing Communist or leftist interests in the region and at safeguarding passage through the Panama Canal. Modern interests focus on promoting trade, fighting drug trafficking, and guaranteeing secure access to oil in Venezuela and Colombia.
- Mexico is the largest, most populated, and economically most developed of the Middle American nations. It is the largest Spanish-speaking country in the world.
- The North American Free Trade Agreement (NAFTA) has played a major role in economic change in Mexico. Much of Mexico's economy is based on production in the *maquiladora* factories of the U.S. border region, where NAFTA has benefited the country. NAFTA has been less beneficial to the mainly agricultural south, where farmers have a difficult time competing with subsidized U.S. farmers.
- Central America is made up of Guatemala, Belize, El Salvador, Honduras, Nicaragua, Costa Rica, and Panama. Panama's history over the past century has been shaped by U.S. interests in the Panama Canal, which is now being enlarged.
- In Venezuela, major settlement is concentrated in the uplands. Discovery of oil a century ago in the coastal regions around Lake Maracaibo prompted rapid industrial growth. Venezuelan President Hugo Chávez has led a self-proclaimed "Bolivarian Revolution," ostensibly aimed at using the country's oil wealth to enhance the prospects for the poor.
- Until recently, much of Colombia's territory was controlled by two rebel groups. With U.S. economic and military assistance, the government is fighting back and has wrested much control from the rebels.
- Brazil is the giant of South America. It is larger than the 48 lower United States but has more than 100 million fewer people. Settlement clusters along the coast, but Brazil is promoting settlement in the interior savanna and rain forest. Brazil's economy is the largest in South America, and Brazil has the region's widest disparities between wealthy and poor people.
- Brazil has a large realm of tropical rain forest centered in the Amazon River system, which drains almost 3 million square miles (7.8 million sq km).
- Periods of economic change in Brazil have been marked by booms and busts in sugarcane, gold, diamonds, rubber, and coffee. Soybeans, especially for the Chinese market, are increasingly important in Brazil's agricultural export economy. Beef and orange juice concentrate are also prominent exports. Sugarcane is grown mainly for Brazil's ethanol fuels, which are in increasing demand as fossil fuel costs rise. Iron ore and steel are vital nonagricultural exports.

- The Amazon has a greater volume of water than any other river in the world. The Amazon Basin extends into nine countries, and about half of it lies in Brazil. Development in the Amazon region is controversial, in part because of the rich biodiversity in its rain forest. The Trans-Amazon Highway and its feeder roads are opening Amazonian wilderness to development. Shifting cultivation is often followed by shifting ranching in a very destructive land use cycle.
- Oil in both Ecuador and Peru has begun to be extracted in the tropical lowlands draining into the Amazon River. Its development is controversial because many indigenous peoples live on lands slated for oil development, and they are generally opposed to the industry.

Key Terms + Concepts

alternative development (p. 374)
American Invasion (p. 378)
Amerindians (p. 360)
Aztec-Tanoan language family (p. 362)
Aztecs (p. 361)
balloon effect (p. 374)
barrio (p. 367)
Bay of Pigs invasion (p. 373)
"Bolivarian Revolution" (p. 379)
Chibcha (p. 362)
Creole (p. 366)
creole languages (p. 363)
crop substitution (p. 374)
Cuban Americans (p. 373)
Cuban Missile Crisis (p. 373)
decertification (p. 373)
distance-decay relationship (p. 372)
Dominican Republic–Central America Free Trade Agreement (DR-CAFTA) (p. 370)
El Niño (p. 358)
El Niño Southern Oscillation (ENSO; El Niño) (p. 358)
energy *maquiladoras* (p. 378)
ethanol (p. 382)
extraction reserve (p. 386)
fair trade movement (p. 368)
favela (p. 367)
Free Trade Area of the Americas (FTAA) (p. 370)
hacienda (p. 368)
hinterland (p. 353)
Hokan-Siouan language family (p. 362)
hurricane (p. 358)
Inca (p. 362)
Indians (p. 360)
intellectual property rights (IPR) (p. 370)
informal settlements (p. 359)
latifundia (p. 368)
lock (p. 379)
maquiladora (p. 371)
María Lionza (p. 365)
Maya (p. 360)
Mayan language family (p. 363)
megalanguage (p. 363)
Mercosur (p. 370)
mestizo (p. 366)
Mexican War (p. 378)
minifundia (p. 368)
Monroe Doctrine (p. 372)
mulatto (p. 366)
multi-Latina companies (p. 372)
narcoterrorist organization (p. 375)
Native Americans (p. 359)
Nazca (p. 362)
North American Free Trade Agreement (NAFTA) (p. 371)
Organization of American States (OAS) (p. 370)
Oto-Manguean language family (p. 362)
pampa (p. 350)
páramos (p. 358)
Penutian language family (p. 363)
Plan Colombia (p. 374)
Plan Patriot (p. 375)
pre-Columbian times (p. 365)
Quechua-Aymaran language family (p. 363)
Regional Initiative for the Infrastructure Integration of South America (p. 384)
remittances (p. 370)
"River Sea" (p. 383)
Roosevelt Corollary (p. 372)
rules of origin (p. 371)
Santería (p. 365)
shantytown (p. 367)
shifting ranching (p. 385)
Southern Cone Common Market (Mercosur) (p. 370)
Teotihuacános (p. 361)
tierra caliente (hot country) (p. 356)
tierra fría (cold country) (p. 357)
tierra helada (frost country) (p. 357)
tierra templada (cool country) (p. 357)
tortilla wars (p. 371)
Totonac language family (p. 362)
Trading with the Enemies Act (p. 373)
Trans-Amazon Highway (p. 384)
tree line (p. 357)
upper limit of agriculture (p. 357)
war on drugs (p. 374)
Washington Consensus (p. 375)

Review Questions

1. Where are the main areas of population concentration in Latin America?
2. What and where are the principal climate and biome types of Latin America?
3. What are the four elevation zones? What human uses are generally associated with them?
4. Why was Port-au-Prince especially vulnerable to earthquake damage, and what plans are there to remake Haiti's urban geography in the wake of the 2010 earthquake?
5. What causes El Niño? What effects are usually associated with it in particular regions?
6. What and where were the major Native American civilizations prior to the arrival of the Europeans?
7. What are the major pre-Columbian and modern languages of Latin America, and where are they spoken?
8. What are the principal faiths in Latin America?
9. What are the principal cash crops and minerals exported from Latin America?

10. What are the region's most important existing and planned free-trade agreements?
11. What are remittances? What role do they play in Latin American economies?
12. What country outside the region is most significant in the geopolitics of Latin America, and how has it exercised its influence in the region?
13. What drives the economies of towns on the Mexico–U.S. border?
14. What are energy *maquiladoras*? What resources do they use, where is their product exported, and why are they located where they are?
15. What considerations went into the construction of the Panama Canal? What steps are being taken to improve the canal?
16. What is the balloon effect, and what is its relationship to the "war on drugs"?
17. In what ways is Brazil the giant of South America?
18. What is the Trans-Amazon Highway system? What role does it play in the region's development and deforestation?
19. What cycle of land use is particularly destructive to the Amazon rain forest?
20. Why are Native Americans of the Amazon Basin generally opposed to fossil fuel development there?
21. What were the hallmarks of Brazil's economic history? What exports are particularly strong today, and why?
22. How committed is Brazil to ethanol, and what interests do other countries have in this fuel source?

Notes

1. Amy Wilentz, "The Dechoukaj This Time," *New York Times*, February 7, 2010, p. WK-12.
2. Quoted in Andrew C. Revkin, "In Quake-Threatened Cities, Quick Growth Invites Disaster," *New York Times*, February 25, 2010, p. A-1.
3. Quoted in Alfred M. Tozzer, "Landa's *Relación de las Cosas de Yucatán:* A Translation," in *Papers of the Peabody Museum of American Archaeology and Ethnology,* vol. 18 (Cambridge, MA: Harvard University Press, 1941), p. 78.
4. Bernard Comrie, Stephen Matthews, and Maria Polinsky, *The Atlas of Languages* (New York: Facts on File, 2003), p. 151.
5. Simon Collier, Harold Blakemore, and Thomas E. Skidmore, eds., *The Cambridge Encyclopedia of Latin America and the Caribbean* (Cambridge: Cambridge University Press, 1985), p. 159.
6. Mike Ceaser, "An Oil for Food Policy, or Buying Off U.S. Influence?" *The Christian Science Monitor,* August 25, 2005, p. 7.
7. David S. Cloud, "Like Old Times: U.S. Warns Latin Americans Against Leftists," *New York Times*, August 19, 2005, p. A3.

Online Resources

CourseMate: Make the most of your study time by accessing everything you need to succeed in one place. Read your textbook, take notes, review flashcards, watch videos, complete activities, take practice quizzes, and more—online with CourseMate. Log in at **www.cengagebrain.com**.

GEOGRAPHY OF NATURAL HAZARDS (continued)

atmospheric pressure cut paths of destruction, sometimes causing large losses of life and property. Storms in May 2011 caused many deaths in Mississippi and Missouri. A tornado ranking EF5 (the strongest possible) on the Enhanced Fujita scale simply leveled a long swath of Joplin, Missouri (•**Figure 11.B**).

Successive days or even weeks of thunderstorm activity sometimes produce localized or widespread flooding. In the summer of 1993, weeks of heavy rains over portions of the Missouri and Mississippi River watersheds produced a flood of historic proportions. Flooding and wind damage are typical results of the hurricanes that batter the East Coast and Gulf of Mexico regions between June and October. The most catastrophic in U.S. history in terms of lives lost was a hurricane in 1900 that killed more than 8,000 people, mainly in storm surges, in Galveston, Texas. In storm damages, the most costly prior to 2005 was Hurricane Andrew, which came ashore south of Miami, Florida, in 1992, causing $27 billion in property losses. Those losses were dwarfed by the impacts of Hurricane Katrina, which roared onto the coasts of Louisiana, Mississippi, and Alabama in August 2005 (see discussion on pages 437–439). Yet another precipitation-related natural hazard is the **blizzard**, a combination punch of heavy snowfall and high winds that strikes the United States—typically in the Midwest and Northeast—an average of 11 times each year.

Joe Hobbs

• **Figure 11.B** The monstrous EF 5 tornado that struck Joplin, Missouri, on May 22, 2011, killed 124 people and destroyed this high school, a hospital, and many homes. This was among the first of a string of deadly twisters that struck "Tornado Alley" that year.

A more pernicious natural hazard, a kind of disaster in slow motion, is the drought that sometimes rakes the country, especially in the West. The worst was the eight-year drought that produced the **Dust Bowl** of the 1930s, ruining crops and livelihoods and causing mass migrations of broken farming families to leave Oklahoma and adjacent areas, mainly to settle in the West. This story is told superbly in John Steinbeck's novel *The Grapes of Wrath*. Recently, a drought lasting longer than the Dust Bowl years afflicted the western and especially southwestern United States. Climatological records suggest that such lower precipitation may be the norm for this region, which may previously have enjoyed decades of abnormally high precipitation. Human populations in the region boomed in those years, perhaps far beyond levels that can be sustained by local water supplies.

pressure on the forests, and with their rebound also came increasing numbers of moose, beaver, and other wildlife.

Immediately east of the Appalachians, between Georgia and the vicinity of New York City, lies the rolling to hilly Piedmont ("foothills"), an area with a large population and good soils for farming. The Piedmont is fairly narrow, and its eastern edge is marked by the **fall line**, a natural boundary where rapids and waterfalls commonly mark the head of navigation on many rivers and have long supplied hydropower. Cities such as Columbia, South Carolina; Richmond, Virginia; Washington, D.C.; and Philadelphia, Pennsylvania, are located along the fall line.

East of the fall line is the Atlantic Coastal Plain, which is continuous with the Gulf Coastal Plain to the south. Forming a large crescent from eastern Texas up to the New Jersey shore, the coastal plains are characterized by sandy and largely infertile soils (although river valleys are more agriculturally productive), a generally flat topography, and many large river mouths and bays indenting the coast. There are many important wetland areas along the coast and inland, such as the Everglades in Florida and the Great Dismal Swamp in Virginia.

North of the Gulf Coastal Plain lies a small region of highlands composed of the Ozark Plateau and the Ouachita Mountains, sometimes collectively referred to as the Interior Highlands. Divided by the Arkansas River Valley, this is the largest area (but not with the highest elevations) of uplands between the Appalachians to the east and the Rockies to the west.

The interior of North America is dominated by a vast, roughly triangular-shaped plain, reaching from Texas to Ohio, to northern Alberta. Millions of years ago, this land was submerged beneath a shallow sea. Today, most of this area is in the watershed of the Mississippi River and its many tributaries, including the Ohio and Missouri Rivers. The Interior Plains, as the area is collectively called, is hillier and more rolling in the east, north of the Ohio River, but becomes more level farther west, even as average elevation increases markedly from east to west. The eastern sections (the prairie), once dominated by long

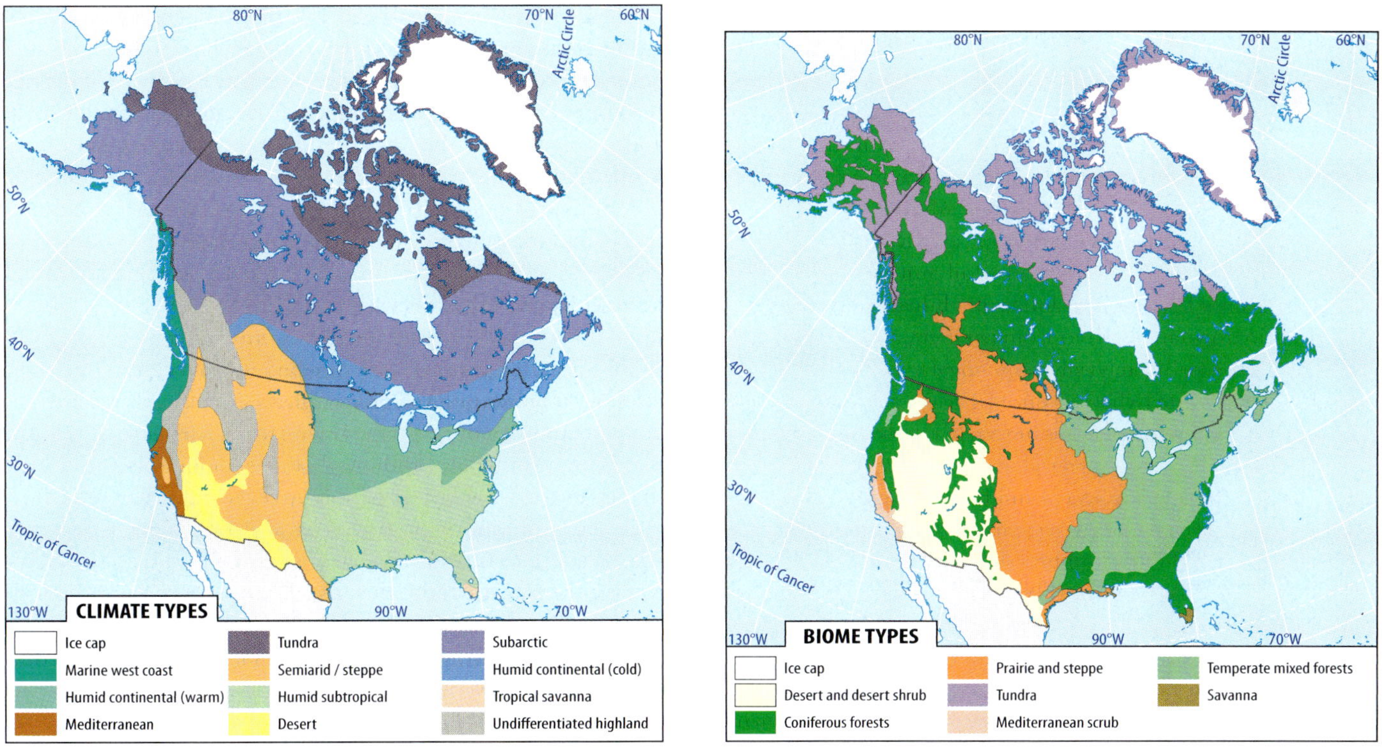

• **Figure 11.5** Climates (left) and biomes (right) of the United States, Canada, and Greenland.
Sources: (left) Based on Rand McNally's Classroom Atlas, 2003. (right) Based on World Wildlife Fund ecoregions data, 1999.

grasses with scattered wooded areas, have mostly been turned into built-up and agricultural land; only about 10 percent of the original ecosystem is intact. The western third of the region is called the Great Plains, a semiarid steppe dominated by short grasses that is mostly flat but higher in elevation than many of the peaks of the Appalachian Mountains (see •**Figure 11.7**). The Great Plains, once referred to as the Great American Desert before its agricultural potential was realized, has some of the world's finest soils and most productive farmlands but is only lightly populated. A more rugged area that includes the Black Hills (site of Mount Rushmore; •**Figure 11.8**) and the colorful terrain known as the Badlands disrupts the otherwise generally level terrain of the western plains in the western Dakotas and eastern Montana.

The Great Plains end abruptly to the west alongside the Rocky Mountains, a long chain of mountain ranges stretching

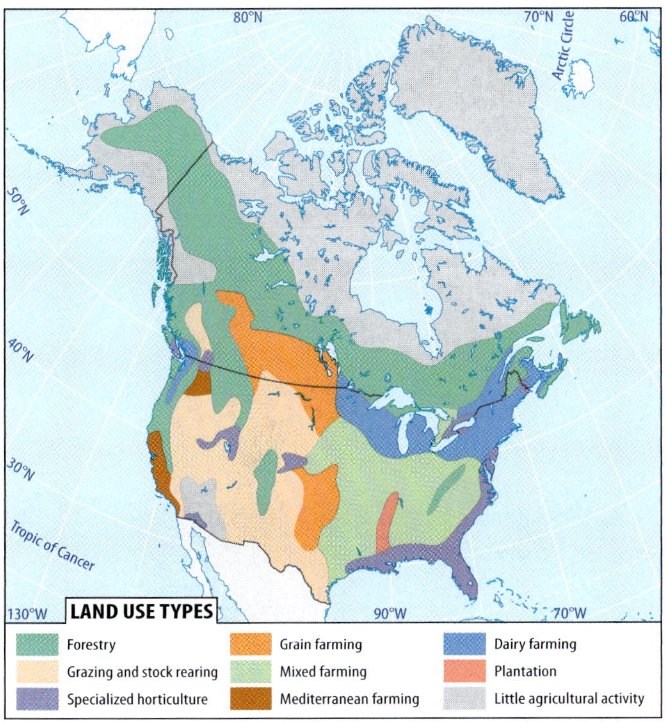

• **Figure 11.6** Land use in Canada, the United States, and Greenland.
Source: The Oxford School Atlas edited by Patrick Wiegand (OUP, 1997), copyright © Oxford University Press. Used by permission of Oxford University Press.

Joe Hobbs

• **Figure 11.7** The Appalachian Mountains in western New York state. This highly scenic portion of the Appalachians is called the Allegheny Mountains.

• **Figure 11.8** The Black Hills are sacred to many Native American groups. While nearby Mt. Rushmore was being sculpted, Indian chiefs asked the sculptor Korczak Ziolkowski to carve an image of the Oglala Lakota warrior Crazy Horse in the Black Hills. Work began in 1948. When completed, this will be the largest sculpture in the world, exceeding the Mother Russia statue depicted on page 129.

from British Columbia to New Mexico (•**Figure 11.9**). The tallest mountain in the Rockies is Mount Elbert in Colorado at 14,433 feet (4,399 m). High and rugged, with few easy passes, the Rockies are the source of many major rivers, including the Rio Grande, Colorado, Missouri, Fraser, Saskatchewan, and Columbia. Recreation, especially skiing and visiting parks such as Yellowstone National Park in Wyoming and Banff National Park in Alberta, is a major draw to the Rockies.

• **Figure 11.9** The Maroon Bells Wilderness in the Colorado Rockies.

West of the Rockies is a geologically complex region of alternating basins and lower mountain ranges. The Columbia Plateau lies in the northern part of this region; south of there, it includes the Great Basin to the west and the Colorado Plateau to the east and is often called the "basin and range country." Arid in the north and true desert in the south, as tall mountains block rainfall to the east (in the Rockies) and west (in the Sierra Nevada), this is the driest area of North America. Canyons and dry valleys carve many plateaus, in many cases presenting stunningly beautiful landscapes. The Grand Canyon (5,000 ft [1,500 m] deep) and the lowest point on the continent, Death Valley (282 ft [86 m] below sea level), are located in this region. Ranching, mining, and energy extraction (oil, gas, and coal) are the economic mainstays of this region. Despite the dryness, agriculture is also important here, with major dams impounding river water that is diverted toward irrigation projects and urban areas.

A series of mountain ranges parallel the Pacific Coast from Alaska south to California. These ranges—including the Alaska Range, the Coast Mountains, the Cascades, the Sierra Nevada, and the Coast Ranges—are all tall, rugged, and glaciated in places. The tallest mountains in North America (Denali, or Mount McKinley, 20,320 ft [6,194 m]), in Canada (Mount Logan, 19,551 ft [5,959 m]), and in the contiguous United States (Mount Whitney, 14,505 ft [4,421 m]) are located in these ranges. The Pacific Coast is prone to earthquake activity, and several active volcanoes are found in Alaska and the Pacific Northwest. Fishing and logging are important to the economy, as is agriculture in two very productive lowlands, the Willamette Valley of Oregon and the Central Valley of California (•**Figure 11.10**).

Stretching across northern Alaska and into the Canadian territories lies the Arctic Coastal Plain, a flat, marshy area that has few resources except for vast quantities of oil lying offshore in the Beaufort Sea.

• **Figure 11.10** California's Central Valley, one of the breadbaskets of the United States. The rolling hills bordering the western side of the valley (at left) belong to California's Coastal Range. The two aquaducts carry water from Northern California to Southern California.

Climates and Land Uses

The United States has more climatic types than any other country in the world, and even Canada is quite varied. The wide range of economic opportunities afforded by this climatic variety helps to make this world region so wealthy.

Canada's and Alaska's tundra climate, with its long, cold winters and brief, cool summers, promotes vegetation of mosses, lichens, sedges, hardy grasses, and low bushes. The subarctic climate zone, also in Canada and Alaska, has long, cold winters and short, mild summers, with a natural vegetation of conifer-
121 ous (boreal) forest resembling the Russian taiga. In 2003, an unusual coalition of energy and forestry companies, environmental groups, and Native American communities joined an agreement to preserve at least 50 percent of Canada's subarctic forests and to develop the other half in sustainable ways. Population is very sparse in the subarctic and tundra climate areas. Scattered peoples engage in trapping, hunting, fishing, mining, logging, and military activities; many live largely on welfare. The severe winter in vast sections of the tundra and subarctic zones requires unusual adaptations—like having to wait until the spring thaw to bury loved ones who died during the winter.

The humid continental climate with short summers ("humid continental cold" in Figure 11.5) is characterized by long, cold winters and short, warm summers. Agriculturally, emphasis is on dairy farming except in the Far West, where spring wheat production dominates. The humid continental long-summer climate ("humid continental warm") has cold winters and long, hot summers; agriculturally, this belt, which includes the agricultural riches of the Midwest, emphasizes dairy farming in the east and corn, soybeans, cattle, and hogs in the midwestern (interior) portion.

In the humid subtropical climate zone, winters are short and cool—with some cold snaps—and summers are long and hot. Agriculture varies, with local production of cattle, poultry, soybeans, tobacco, cotton, rice, peanuts, fruits, and vegetables.

Within each of these three humid climate regions, the pattern of natural vegetation is complex, with areas of coniferous evergreen softwoods, broadleaf deciduous hardwoods, mixed hardwoods and softwoods, and prairie grasses. Far southern Florida has a small area of tropical savanna climate, which is a major climate in adjacent Latin America. The state of Hawaii has a tropical rain forest climate, but most of the original as-
289 sociated vegetation has been cleared (•**Figure 11.11**). Along the Pacific shore of the United States and Canada, the narrow strip of marine west coast climate is associated with the barrier effect of high mountains near the sea, ocean waters that are warm in winter and cool in summer, and winds prevailing from the west. The mild, moist conditions have favored a magnificent growth of giant conifers (most notably the redwood and the Douglas fir) that provides the basis for the lumber industry, and the now untouched giants, the sequoias (see Figure 2.6i). Lush pastures support dairy farming, as they do in the corresponding climatic region in northwestern Europe.

The Mediterranean, or dry-summer subtropical, climate region of central and southern California, where almost all

Joe Hobbs

• **Figure 11.11** Na Pali coast, island of Kauai, Hawaii.

rain comes in winter, is associated with irrigated production of cattle feeds, vineyards, vegetables, fruits, cotton, and numerous other crops (see Figure 11.10). These products, together with associated livestock, dairy, and poultry, make California the leading U.S. state in total agricultural output.

In the semiarid steppe climate region, occupying an immense area between the Pacific coast of the United States and the landward margins of the humid East and extending north into Canada, temperatures range from continental in the north to subtropical in the south. The natural vegetation of short grass, bunch grass, shrubs, and stunted trees supplies forage for cattle ranching, which is the predominant form of agriculture. More moist areas are used for wheat (both winter wheat and spring wheat). Other crops are grown in scattered irrigated areas, often associated with major rivers such as the Columbia, Snake, Arkansas, and Rio Grande.

The desert climate of the U.S. Southwest is associated with scattered irrigated areas and settlements emphasizing mining, recreation, and retirement. The high, rugged mountains of the Rockies and the Sierra Nevada have undifferentiated highland climates varying with latitude, elevation, and exposure to moisture-bearing winds and to the sun (•**Figure 11.12**).

Joe Hobbs

• **Figure 11.12** The Sierra Nevada range and Mono Lake (top, just left of center).

World. After the Spaniards and their successors introduced the horse, the Lakota and some other groups gave up what had been a village and farming way of life, migrating west to become horse-mounted hunters of the bison (buffalo) and other game on the Great Plains.

In what is now California and Oregon, the Penutian language family included the Miwok and Klamath-Modoc languages, among others. Most of these peoples were not farmers but had very productive hunting and gathering economies that took advantage of the region's diverse and abundant resources. These natural bounties supported relatively dense human populations.

Farther to the north, but still on the Pacific coast, the **Mosan language family** included branches of Chemakuan, Salish, and Wakashan. The Northwest Coast peoples who spoke these languages also had rather dense populations that thrived on an abundance of sea mammals, fish, and land game. They generally developed complex societies in which the custom of the potlatch played a prominent role. **Potlatch** was a system in which rank and status were determined by the quantity and quality of material goods one could give away—an interesting variant and inversion of the "keeping up with the Joneses" tradition of later Anglo North America.

The **Algic language family** members were spread across all 10 Canadian provinces and 20 U.S. states, mainly in the northern tier. Some of its better-known languages include Arapaho, Blackfoot, Cheyenne, and Cree. The tribes that spoke the Algic languages relied on the region's vast forests for game animals, fuel, shelter, and tools, complementing their hunting and gathering with fishing and farming; some groups were seminomadic and lived in villages.

To the northwest of this linguistic region, in west central Canada and southeastern Alaska, the **Na-Dene language family** was prominent. This group includes the **Athabascan subfamily** with its languages, including Koyukon; Apache; and in a remote outlier, the Navajo of the Southwest (the largest tribe in the United States, with more than 270,000 members; see Figure 1.5, page 7). Most of these peoples lived either in the subarctic region of coniferous forests and wetlands or in the tundra, where agriculture was impossible. They hunted caribou and other game, and their populations were limited.

Across the rest of Alaska, throughout northern Canada, and in Greenland, the **Eskimo-Aleut language family** dominated. The peoples who carried these tongues into North America were among the last to migrate from Asia, probably between 2500 and 1000 B.C.E., and they did so in boats, as the former land bridge was inundated by then. The so-called **Eskimos**—an Algonquian term meaning "raw meat eaters"—prefer to call themselves **Inuit**, meaning "the people." The Inuit and the **Aleuts** were indeed characterized by a carnivorous diet, as their livelihood was based on hunting sea mammals, caribou (known as reindeer in the Old World), and fish.

What all these diverse groups had in common was an exceptionally well-developed set of adaptations for living in the local environment. Native American economies spanned a remarkably wide continuum, from simple foraging and hunting to complex agricultural systems. These were not static adaptations but changed and generally improved over time—for example, through greater mastery of finer tools. As the Lakota and others demonstrated by giving up village agriculture in favor of a nomadic life based on horses, these cultures were capable of dramatic adjustments to changing opportunities.

It was once supposed that the Native Americans always lived within the limitations imposed on them by local ecosystems, but it is now known that they transformed landscapes in significant ways by fire and other means and may even have hunted populations of large animals to extinction; this is the Pleistocene overkill hypothesis (see page 40). One trait apparently shared by most if not all the Native American groups was their deep reverence for the natural world, as is clearly seen in Chief Seattle's words below. Animals, forests, geological features, and the people themselves were tied together intimately in animistic belief systems that emphasized the kinship 228, 302, 322
between these diverse elements of life on earth.

> So that they will respect the land, tell your children that the Earth is rich with the lives of our kin. Teach your children that we have taught our children that the earth is our mother. Whatever befalls the Earth befalls the sons of Earth. If men spit upon the ground, they spit upon themselves. This we know: the Earth does not belong to man; man belongs to the Earth. This we know. All things are connected like the blood which unites one family. All things are connected.
>
> —Chief Seattle of the Suquamis

European Impacts on Native Cultures

There are two very different narratives of what took place in North America in the centuries following 1492. For European newcomers, these were times of settlement, development, taming the frontier, and "civilizing the savages." For the Native Americans, these were times of depopulation and cultural demolition. Popular T-shirts and bumper stickers on Indian reservations in the western United States these days read "NATIVE AMERICANS: FIGHTING TERRORISM SINCE 1492."

How many Native Americans lived in what is now Canada, the United States, and Greenland when Columbus made landfall in 1492 is uncertain. The most common estimate is 1 million, with about three-quarters of that population in what is now the United States and almost all the rest in Canada. As in Latin America and Australia, the European contact initiated years of population losses among Native Americans through disease, famine, and warfare. In the United States, Native 292, 364
American populations crashed to an all-time low of fewer than 250,000 between 1890 and 1910, but they are now back up to 1.8 million. Their growth has been particularly strong in the Great Plains, where there are now more Native Americans (and bison) than there were in the late 1870s.

In Canada, the current population of Native Americans is about 1 million. Collectively, these peoples refer to themselves as the **First Nations** in acknowledgment of their pre-Columbian cultures and claims to the land. In Canada, this

• **Figure 11.15** Modern Native American reservations and other lands in the United States and Canada.
Source: From Carl Waldman, Atlas of the North American Indian (Facts On File, 1985). Reprinted with permission of the Publisher.

definition excludes the Inuit, who were relative latecomers but who are referred to as among the **First Peoples**. The indigenous Greenlanders, known as the *Kalaallit*, are related linguistically and ethnically to the Inuit of Canada, Alaska, and Siberia. The official name of their island is Kalaallit Nunaat, meaning "Land of the Greenlanders."

In the United States, most of the Native American populations are in the west; in Canada, they are in the north and west. The two countries have dealt in different ways with their indigenous populations and claims to land and other rights, with government–Native American relations historically better in Canada.

In the United States, an estimated one-third of all Native Americans today live on the **reservations** that the government established as remnants of their former tribal territories or sometimes set up thousands of miles from native homelands, forcing the people to relocate there (•Figure 11.15). "The Res," as Native Americans call it, is among the poorest communities of the United States, often plagued by high rates of incarceration, alcoholism, drug abuse, depression, broken families, teen suicide, and unemployment. The Lakota's Pine Ridge Reservation in South Dakota (•Figure 11.16), for example, has an annual per capita income of just $7,770 (compared to the national U.S. average of $27,330). However, it must also be noted that even these lamentable conditions represent a considerable improvement. Native Americans have growing political and economic clout, with poverty and unemployment rates dropping and education levels rising.

Gambling revenues have contributed to economic and other improvements in many Indian communities. The semiautonomous status of their territories allows casinos to circumvent

Joe Hobbs

• **Figure 11.16** Lakota at the Wounded Knee Massacre Site, Pine Ridge, South Dakota. Their reservation is prominent in the poverty map on page 413.

Joe Hobbs

• **Figure 11.17** The Seneca Allegany Casino in Salamanca, western New York State, is in the Allegany Indian Reservation. Most of the people living here are from the Seneca, Iroquois, and Cayuga tribes.

many restrictions that prohibit casinos elsewhere (•Figure 11.17a). About one-third of the roughly 563 federally recognized Native American tribes in the United States have developed casinos since Congress legalized gambling on their lands in 1998. This **gaming industry** has been a mixed blessing for the Native Americans, with a wide variety of successes and failures in different places. In some, gambling revenues have supported community education and student scholarships, health care, and cultural centers, but in others they have lined a few pockets and failed to benefit many members of the community. Some Native Americans complain that several of the groups that have established casinos are not legitimate, historical tribes to begin with. Non-native people often resent casinos because they view them as a windfall of special treatment for individuals who have not earned it. In all cases, casinos have introduced significant changes.

In Hawaii, as in North America, incursions by Europeans —Captain Cook's visits were the first in 1778 and 1779— brought huge changes to the native people. Polynesians may have populated the island group as early as 300 C.E., making a living by growing bananas, coconuts, breadfruit, and
291 taro and by fishing. Hawaiian society was based on a ranked caste system, and taboos governed many aspects of daily life and relationships with the spiritual world. The islands were a united indigenous kingdom in 1893, when a group of American businessmen, backed by the U.S. military, overthrew the monarchy, and took control (for more information on Polynesian geography and culture, see Chapter 8). The United States formally annexed Hawaii five years later. Diseases, including smallpox and influenza, introduced accidentally by outsiders, nearly wiped out the indigenous population: there were an estimated 1 million ethnic Hawaiians in the late eighteenth century but just 22,000 in 1919. Today, the **Kanaka Maoli**, as the native Hawaiians call themselves, number 250,000. There has been much ethnic mixing among this population. The native *Hawaiian* language a member of the Malayo-Polynesian subfamily of the Austronesian family, is undergoing a major revival following a long decline. Hawaiian became an official state language in 1978—the only indigenous language in the United States to have this distinction. Major education efforts ensued. As recently as 1983, only 50 children spoke the language, but now more than 2,000 do.

The situation for indigenous peoples in Canada is unique,
even on a worldwide scale. In 1999, Canada ceded nearly one- 301
quarter of its total land area to the Inuit peoples, who called their land **Nunavut** ("our land" in Inuktitut; see Figure 11.15). These 733,600 square miles (1.9 million sq km) were carved from Canada's Northwest Territories and generally lie north of 60°N, surrounding the northwest part of Hudson Bay and extending well within the Arctic Circle. The total population is approximately 30,000, scattered through about 30 communities. About nine-tenths of the capital required to keep the territory viable comes from Ottawa, the federal capital of Canada, making the per capita cost of maintaining Nunavut higher than for any other political unit in Canada. The Canadian government has allocated more than $1 billion to be distributed in the form of services, supplies, and subsidies until 2016 to the Inuit population, representing about $38,000 a year per person. Large subsidies also go to Native American populations in the Northwest Territories.

All across Canada, settlements of indigenous claims to ancestral lands have led to federal government financial support of the communities, a measure of self-government for the indigenous peoples, and royalty income to them from mines and fossil fuel industries. Indigenous peoples who have relocated to Canada's cities are showing major improvements in education and employment. Still, the average income of Canada's indigenous peoples is about a third lower than the population as a whole. Canada's record with its indigenous peoples is not unblemished; in the 1950s, for example, the government forcibly relocated a group of Ungava Inuit from Hudson Bay 1,200 miles (1,920 km) north to inhospitable Ellesmere Island. The move, which caused these people enormous suffering, was meant to help confirm Canadian sovereignty over the island.

European Settlers and Settlements

European settlement of what are now Canada and the United States took place in a series of waves propelled by religious persecution in Europe, colonization of new lands by European powers, and then the expansionist efforts of newly independent Canada and the United States. Canada's core region of Québec and Ontario developed with the French entry into the interior of North America. The French founded Québec City in 1608, at the point where the Saint Lawrence estuary leading to the Atlantic narrows sharply. Their fortifications on a bold eminence allowed control of the river there.

From Québec, fur traders, missionaries, and soldier-explorers soon discovered an extensive network of river and lake routes with connecting portages reaching as far as the Great Plains and, via the Mississippi River, to the Gulf of Mexico. Montréal, founded later on an island in the Saint Lawrence River, became the fur trade's forward post for the interior wilderness. Between Québec City and Montréal, a thin line of settlement grew along the Saint Lawrence, forming the agricultural base for the colony. Population grew slowly in this northerly outpost where the winter was harsh and only French Catholics were welcome.

French Canadians numbered only about 60,000 when Britain conquered the region in 1759 (France was expelled from Canada in 1763). They did not join the English-speaking colonists of the Atlantic Seaboard in the American Revolution and by their refusal laid the basis for the division of the continent of North America into two separate countries (see page 423). Until the late twentieth century, French Canadians increased rapidly in numbers. About 20 percent of Canada's people today are French in language and culture. They are concentrated in the lowlands along the Saint Lawrence River, where they form the majority population in the province of Québec. The French Canadians have clung tenaciously to their distinctive language and culture, with some attempting twice since 1980 to gain
424 independence for the province of Québec.

A few British settlers came to Québec after the conquest, and this immigration increased rapidly during and after the American Revolution of 1775–1783. These migrations set the foundations for the British segment of Canada's core. Many of the early English-speaking immigrants were Loyalist refugees from the newly independent United States, whose rebellion French Canada had refused to join. They were soon joined by more English settlers emigrating directly from Europe. These newcomers formed a large minority in French Québec, including a British commercial elite in Montréal that persists to this day, but they also settled west of the French on the Ontario Peninsula. By 1791, the British government decided to separate the two cultural areas into different political divisions, known at that time as Lower Canada (Québec) and Upper Canada (Ontario), taking the directional terms from their positions on the Saint Lawrence River.

Farther west, the prairie region of Canada was settled much later, in the decades after 1890 (see •**Figure 11.18**). Settlers found some good land in an area with long and harsh winters, short and cool summers, and marginal precipitation. They could grow only hardy crops for subsistence. Later came specialization in spring wheat for export, and the prairie region became the Canadian part of the North American Spring Wheat Belt.

The settlers of this region were mainly English-speaking people from eastern Canada and the United States, but they included notable minorities of French Canadians, Germans, Ukrainians, and others. The ethnic makeup of the area still

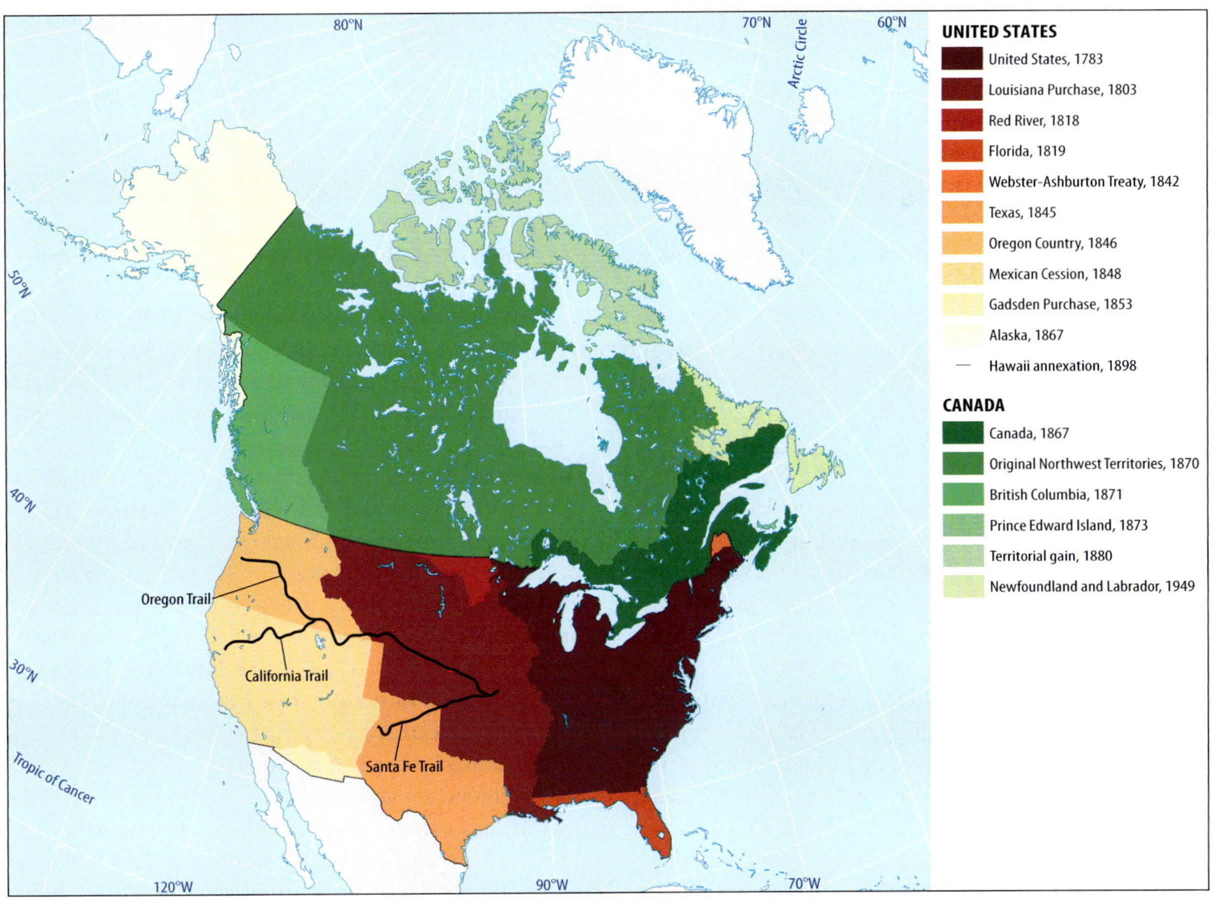

• **Figure 11.18** Territorial acquisitions of the United States and Canada.
Source: Based on John Haywood, ed., Atlas of World History (Barnes and Noble Books, 1997).

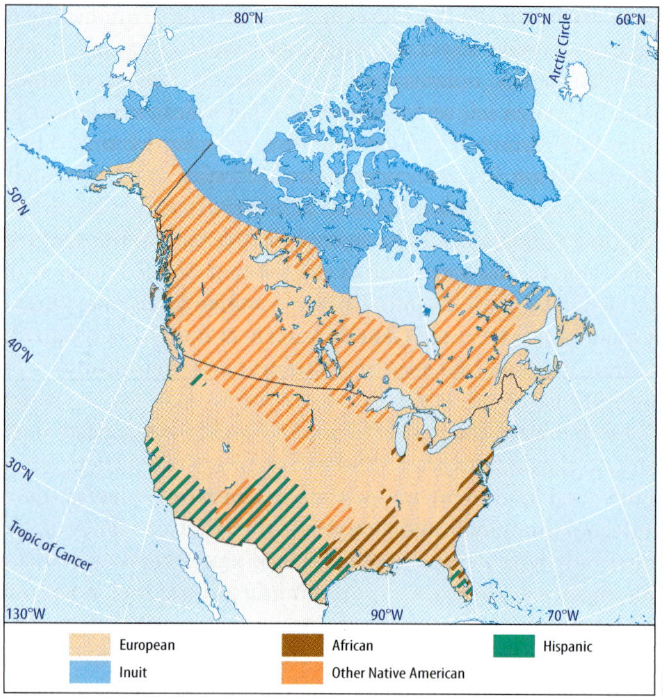

• **Figure 11.19** Major ethnic groups in the United States and Canada.

reflects them. Canada's ethnic mix today includes about 6 million people of English origin, 3 million Irish, and 4 million Scottish (see •**Figure 11.19** and **Table 11.3**).

Other ethnic groups would come to join Canadian society, which by law in the 1988 **Multiculturalism Act** was recognized as a multicultural society. These laws acknowledged that Canada's Native American and non-European groups would probably never become assimilated in Canadian society and that the country would probably be more stable—avoiding the kind of racial tensions that wracked the United States—if traditional identities were recognized and encouraged. Most Canadians embrace this legislation, but some argue that Muslim immigrants in particular are challenging Canada's freedoms; mirroring an issue in Britain, there has been a debate over Muslim women wearing the face-cloaking
71 veil known as the *niqab*.

The first European settlers of what is now the United States were Spanish, Dutch, English, Swedish, and French, with the majority being English. Most of their settlements were along the Eastern Seaboard and were linked to Europe by the long but generally reliable sea passage across the Atlantic. Northeastern urban development began early. In the early 1600s, English settlers founded Boston, and Dutch settlers established New Amsterdam (which the English annexed and renamed New York in 1664). The English founded Philadelphia in the late 1600s and Baltimore in the early 1700s. All four were major urban centers by the time of the American Revolution.

Development of the interior began with exploration along waterways: the British up the Chesapeake Bay and its tributaries, the Dutch up the Hudson in the early 1600s, and the French up the Saint Lawrence and Mississippi rivers. Westward expansion proceeded very slowly; up to about 1800, there was little European settlement beyond the Appalachian region of what is now the eastern United States. The floodgates of expansion then opened, and wave after wave of immigration took place.

Table 11.3 Ethnic Groups of Canada and Their Origins

Major Ethnic Groups	Dates
Native peoples	pre-1600
French	1609–1755
Loyalists from the United States	1776–mid-1780s
English and Scots	mid-1600s on
Germans and Scandinavians	1830–50, c. 1900, 1950s
Irish	1840s
African Americans	1850s–70s
Mennonites and Hutterites	1870s–80s
Chinese	1855, 1880s, 1990s
Jews	1890–1914
Japanese	1890–1914
East Indians	1890–1914, 1970s on
Ukrainians	1890–1914, 1940s–50s
Italians	1890–1914, 1950s–60s
Poles	1945–50
Portuguese	1950s–70s
Greeks	1955–75
Hungarians	1956–57
West Indians	1950s, 1967 on
Latin Americans	1970s on
Vietnamese	1970s on
Filipinos	1990s

Sources: E. Herberg, *Ethnic Groups in Canada: Adaptations and Transitions*. Copyright © 1989 Nelson Canada, Scarborough, Ontario; and Census of Canada, 2001.

There were several phases of territorial acquisitions in what Americans considered their **Manifest Destiny**—the conviction that Americans were destined to acquire new lands all the way to the Pacific, "from sea to shining sea" as the song goes. (see Figure 11.18). First, they established themselves along the Ohio and eastern Mississippi watersheds following the American Revolution in 1783. Then, in the **Louisiana Purchase** of 1803, the United States bought from France (for 3 cents per acre!) about 23 percent of what are now the lower 48 states. This included much of the western watershed of the Mississippi and the lands of the Missouri River watershed. The United States then picked up major lands in the south and west. It purchased Florida from Spain in 1819 and took Texas, the Southwest, and California as spoils of war with Mexico in 1845 and
1848. It established the Oregon Territory in 1846. In the 1853 378
Gadsden Purchase, the United States bought a small strip of

present-day Arizona and New Mexico from Mexico. Finally, it bought Alaska from Russia for 2 cents per acre in 1867. The young nation thus acquired a vast land empire much like that
127 built earlier by Russia, although by different means.

The **Homestead Act** of 1862 made many of these new regions almost irresistible magnets for settlement, as it allowed a pioneer family to claim up to 160 acres of farming land for a fee of about $10. Two major trails, the ancestors of today's interstate highway system, were developed in the 1840s to channel the newcomers into the frontier zones (see Figure 11.18). From a trailhead in Independence (near Kansas City), Missouri, the **Oregon Trail** took a northerly track across what are now Nebraska, Wyoming, and Idaho to the fertile Willamette Valley of the Oregon Territory, with a southern branch (the **California Trail**) leading to what is now Sacramento, California. The southern route from Independence, known as the **Santa Fe Trail**, snaked across Kansas and New Mexico. Army posts were set up along the way to facilitate and protect the westbound migrants. They also served as military garrisons in the forceful subjugation of Native American peoples, whose stories are told in broken treaties, massacres, and decimation of political leadership.

Accessed in part by the Oregon and Santa Fe trails, the U.S. Great Plains region was settled rapidly between 1860 and 1890. A major impetus for settlement of the far west was the **California Gold Rush** of 1849. No Panama Canal existed at that time, so most of the prospectors, known as **Forty-Niners**, made their way either tortuously along the Oregon and California trails, across the Isthmus of Panama by land, or by sea for a 17,000-mile (27,200 km) voyage around South
372 America, which could take up to seven months.

Most of the homesteaders, Forty-Niners, and other settlers of the American frontier were of English, German, Scandinavian, and other western and northern European ethnicities. The United States ultimately developed into a nation in which about 99 percent of the people were either immigrants or of immigrant stock from a vast array of ethnic backgrounds. The larger groups of today include more than 45 million of German origin, over 20 million of Irish origin, and over 20 million of English origin. Although some kept their own ethnic identities—still found in ethnic neighborhoods, especially in the East, or in small rural communities in the Midwest—most were blended into the melting pot that produced what might be called the **mainstream European American culture**.[2] The ethnographer David Levinson describes that culture as one based on institutions such as the English language and British educational principles brought by the British colonists, combined with core American values that developed on U.S. soil. These include English as the national language, religious tolerance, individualism, majority rule, equality, a free-market economy, and the idealization of progress.

Added later to the mix were immigrants from Eastern and Southern Europe, represented today by more than 11 million Americans of Italian origin, over 6 million of Polish origin, and over 5 million Jews. Many of these immigrants settled initially in the larger cities of the U.S. Northeast, where earlier immigrants of mainly northern and western European extraction often discriminated and directed violence against them (•Figure 11.20). It took several generations for these newer immigrants to become part of the mainstream ethnic fabric of America. Other large groups have had more difficult careers with the white majority—which is projected to become a minority soon after 2050—or are relatively new contributors to the American ethnic mix.

Joe Hobbs

• **Figure 11.20** One poor immigrant group after another – Irish, Italian, Polish – stayed in the tenements of Manhattan's Lower East Side before moving up the economic ladder.

Ethnic Minorities in the United States and Canada

Former U.S. Secretary of State Condoleezza Rice aptly described slavery as "America's birth defect."[3] The plantation economy that supported the American South was built on the exploitation of more than 500,000 black slaves brought to
America from Africa between 1619 and 1807. People of mainly 322
African ancestry, known as **blacks** and **African Americans**, occasionally as **Afro-Americans**, and formerly as **Negroes**, number about 39 million, or 13 percent of the U.S. population. The group also includes many people of mixed black and other ethnic extraction and blacks who came from regions other than Africa. Blacks were acknowledged as the largest minority in the United States until 2000, when census figures revealed that the number of people identifying themselves as Hispanic slightly exceeded the number of African Americans.

Unlike the French Canadians, African Americans are not largely confined to one part of the country. Some regions, especially in the eastern half of the United States, have a very strong black cultural presence (Figure 11.19), and African Americans are majorities in a number of large cities. In the so-called **Great Migration** between 1915 and 1970, about 6 million blacks moved out of the former Confederacy states of the Old South, settling especially in the inner cities of the Northeast and Midwest. They tended to have more education than the blacks who stayed in the South, and had higher employment numbers and a more stable family life than Northern-born blacks. These are the general characteristics of the "migrant advantage" shared by Irish, Italian, and other immigrants in U.S. historical geography.

Race relations have played an important role in the country's political and social history. Conflicting attitudes toward black slavery were among the issues that nearly split the United States into two countries in the American Civil War. The **Thirteenth Amendment** to the Constitution ended slavery in 1865, but the economically and politically subordinate position of blacks has been slow to change. As recently as the 1960s, there were legal struggles to end official **segregation** between the races in parts of the United States—for example, in the state of Alabama, where there were separate sections for whites and blacks on buses and where the races were expected to live in different neighborhoods, attend different schools, and never intermarry.

Today, although a segment of the African American population has moved into the middle class and above, a large portion resides in impoverished and troubled **ghettos** of central cities, where there are problems in housing, jobs, and crime (•**Figure 11.21a**). About 40 percent of all people incarcerated in U.S. prisons are black. The legal battles to eliminate segregation have ended, but "hearts and minds" have been slower to change, and African Americans continue to struggle for equality. The 2008 election of Barack Obama as the 44th President of the United States marked a milestone in the nation's history. As with the growing trend of interracial marriages, the inauguration of President Obama raised widespread hopes that remaining racial barriers in the United States would weaken and fall.

In the 1840s, when the United States won from Mexico the great swath of the American Southwest, West, and Texas, it also became home to many Spanish speakers of European or mestizo descent. The growth rate of people of Spanish or mixed Spanish background—the Hispanic American (or **Latino**) population—has in recent years been more rapid (about 3% per year) than that of any other group, both domestically and through the major migration stream flowing actively from Latin America (Figure 11.21b). About half of the growth is from natural increase and another half from immigration, both legal and illegal. About 50 million people, or 16 percent of the U.S. population, are Hispanic. The majority is of Mexican origin and is concentrated in California, Texas, New Mexico, Arizona, and Colorado. Other important Hispanic groups include people of Puerto Rican origin, most heavily represented in New York City and in the Northeast and increasingly in Florida, and people of Cuban origin, who live mainly in southern Florida. Previously, Hispanic immigrants settled overwhelmingly in the "gateway" states of the southwestern United States, but today, many leapfrog these states and put down roots all across the country.

Minorities make up about one-third of the U.S. population, and they are rapidly gaining momentum toward that date in the 2040s when non-Hispanic whites will drop below 50 percent of the population. Demographically and politically, a gap is emerging in the United States between a young, ethnically diverse population and an older, mainly white one. There are four states with majority populations of ethnic minorities: Hawaii (77%), New Mexico (60%), California (60%),

Paul Fusco/Magnum Photos

(a)

Joe Hobbs

(b)

Joe Hobbs

(c)

• **Figure 11.21** Ethnic urban landscapes of the United States: **(a)** a black-run business in Los Angeles; **(b)** a Latino neighborhood in Chicago; **(c)** Chinatown in San Francisco.

and Texas (55%). Twenty-one percent of the nation's minority populations live in California, and 12 percent live in Texas.

Asian Americans now make up about 4 percent of the U.S. population. Immigrants from China were especially important in the construction and service sectors of the West Coast economies of both Canada and the United States between 1849 and 1882. Chinese laundries and the Chinese "coolies" who did the grueling labors of building western railways became the stuff of stereotype and legend, and there were sore points in Chinese-Anglo relations. Chinese immigrants brought an indelible cultural presence in the form of "Chinatown" landscapes in major cities, particularly in the nineteenth and early twentieth centuries, and especially in the western regions of both countries (Figure 11.21c). Chinese food has become even more widespread than Mexican food across the continent.

Asian Americans have long been steadily moving out of the central city and into white and mixed neighborhoods; they include among their ranks not only the Chinese but also large populations of ethnic Japanese, Korean, Vietnamese, and Filipinos, among others. The closest approach to Asian minority dominance of a state is in multiethnic Hawaii, where Asian Americans, especially of Japanese descent, make up more than 40 percent of the population. Race relations between the Anglo majority and Asian groups in both the United States and Canada have generally been good in recent decades. The last major episode of tension was during World War II, when large numbers of civilian Japanese Americans, regarded as potential collaborators with the enemy, were rounded up and interned in prison camps in the West.

Asian Americans in the United States today have the distinction of being the most highly educated of all the ethnic and religious groups except for Mormons and Jews. Among their ranks, the ethnic Japanese, Koreans, Chinese, East Indians, and Filipinos have generally prospered economically, but the ethnic Vietnamese, Lao, and Cambodians are poorer. The largest Asian populations in the United States are the Chinese and Filipino, each over 1 million. The largest in Canada are ethnic Chinese (about 1 million, more than double their numbers in 1984), East Indian (mainly of the Sikh religion), and Filipino. Recent large Asian migrant flows into Canada include Koreans and Vietnamese. Asian minorities have intermarried with other ethnic groups in both countries, and their mixed offspring call themselves **Hapas**, a Hawaiian word meaning "half" that people of mixed Pacific Island and other ethnicities also use to describe themselves.

These brief sketches only scratch the surface of the ethnic mosaics of the United States and Canada. Nearly 200 non-Native American ethnolinguistic groups are represented in these countries, which are rather rare examples of nations that have generally welcomed immigrants into their social fabric. The "melting pots" of Canadian and American societies are truly remarkable. You can travel all around the world and never see anything like their great wealth of ethnic diversity.

Nonindigenous Languages and Faiths

The non–Native American religions and languages prevailing in Canada and the United States reflect both early colonial influences and more recent waves of immigration. English and French are Canada's official languages, with French usage confined largely to the traditionally French southeast (•Figure 11.22). The United States does not have an official language (despite repeated efforts to pass a constitutional amendment that would establish it as English), but 96 percent of U.S. residents speak English. The language map well illustrates the growth of Hispanic populations in the Southwest and southern Florida. **"Spanglish,"** a hybrid tongue of Spanish and English, is moving from Hispanic neighborhoods of southern California into the mainstream media and culture of the United States. Spanglish uses "code switching," substituting words or phrases from one tongue when speaking in another (for example, "*Vamos a la store para comprar milk*," meaning "Let's go to the store to buy some milk") and also coining new words, like *perro caliente* for "hot dog."[4] In the larger cities of both the United States and Canada, there is a dizzying array of languages associated with ethnic neighborhoods.

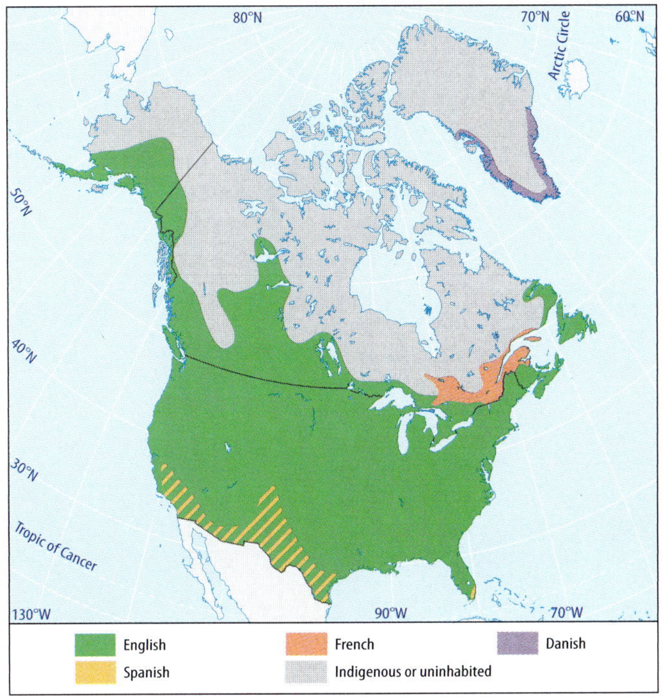

• **Figure 11.22** Nonindigenous languages of the United States, Canada, and Greenland.

Source: Based on Bernard Comrie, et al., The Atlas of Languages, Revised Edition (New York: Facts on File, 2003).

Laws guarantee religious freedoms in both the United States and Canada, and both countries have a rich fabric of faith (•Figure 11.23). Christianity prevails in both countries, with numerous Christian churches represented. The largest single denomination in both countries is Roman Catholic (26% of all Americans and 43% of Canadians), brought to the United States by Irish, Italian, Polish, Filipino, Hispanic, and other immigrants, and to Canada mainly by the French. Hispanics are replenishing the ranks of American Catholics who have left the church in recent decades and are also introducing their own forms of charismatic worship into the country. The next largest group in the United States is Baptist (about 17%), then

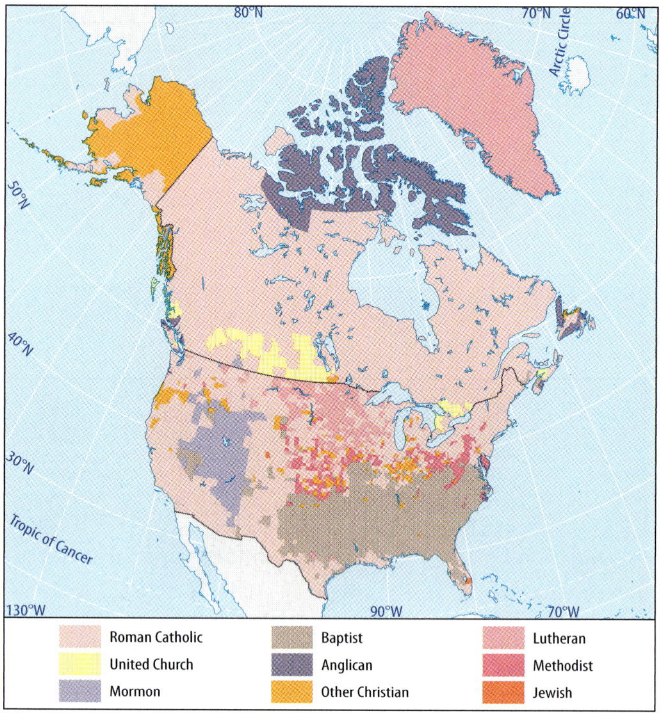

• **Figure 11.23** Religions of the United States, Canada, and Greenland.
Source: Based on Edwin Scott Gaustad, Philip L. Barlow, and Richard W. Dishno, New Historical Atlas of Religion in America (Oxford University Press, 2001).

Methodist (about 7%), and Lutheran (about 5%). The other Abrahamic faiths are well represented by about 5 million Jews and 1 million Muslims. Islam is the fastest-growing religion in the United States, as it is in the world.

11.4 Economic Geography

The United States and Canada are very wealthy nations. Per capita GNI PPP figures for the countries are $45,640 and $37,280, respectively, compared to the world's highest, Luxembourg, at $59,590. Both countries belong to clubs representing the world's strongest economic powers: the Group of 8 (G-8) and the Group of 20 (G-20). Together, the United States and Canada generate an impressive 21 percent of the global gross national product.

The United States has the world's largest economy and is by far the world's largest producer and consumer of goods and services. It consumes far more than it produces. It is also the world's leading example of "consumption overpopulation" (see page 58). With just about 5 percent of the world's population, the United States has a third of the world's wealth and each year consumes, for example, more than 18 percent of the world's energy, 27 percent of the world's commercially harvested wood, and 40 percent of the world's illegal drugs. Even with China's surging consumerism, the average American consumes 53 times more goods and services than the average Chinese.

Generally, in the globalized economy, the United States may be seen as having built an economic structure of consumption. By contrast, China and most other major economies have built economies of production and export. The United States borrows money from other countries (it has the world's largest federal foreign debt—much of it to China—as well as the world's largest national debt), and it sells off assets to finance its consumption. The nation has a trade deficit (for which China alone accounts for about one-third), exporting $1.5 trillion worth of goods while importing $2.3 trillion annually. Once the world's largest exporter of goods, the United States lost that position to Germany in 2003 and became third (after China) in 2007. The United States' share of global gross domestic production fell from 31 percent to 23 percent between 2000 and 2011. In the same period, the four leading emerging economies of Brazil, Russia, India, and China—the so-called BRICs—saw their share of global gross domestic production rise by 3 percent to a combined 11 percent. 133, 233, 380

This discussion of American affluence may lead many of you to wonder who is being written about. In relative terms, the United States is extremely well-to-do on the global scale. But also in relative terms, the "average" American has been going nowhere economically for a number of years. Adjusted for inflation, the median income of the American family (with half making more, half making less) in 2011 was $49,445, roughly where it was in 1996, and is more than 7 percent below its 1999 peak. The gains of the boom years in the 2000s have been wiped out. A typical man's earnings are less than they were in 1978, again adjusted for inflation. Although a typical woman makes 77 percent of what the typical man does, her income has been rising more than his. There was a clear correlation between education and income, leading many people during the Great Recession to go to or go back to college.

When I travel abroad, people often do not believe me when I say there are poor people in America. But there are. They were more than 46 *million*, or 15 percent of the population, in 2011. About the same number of people lack health insurance. Twenty-two percent of American children are poor. In many of the countries I visit—Egypt, Vietnam, or Madagascar, for example—the $22,314 that made up the poverty line for a family of four in the United States in 2011 sounds like a great fortune, but of course it does not go far in many places in America (*where* you live is very important, isn't it?) (•**Figure 11.24**).

Surprisingly perhaps, the majority of the urban poor (55%) lives in the suburbs rather than the inner cities. The biggest asset in Americans' net worth is their homes, which have declined in value in recent years. Many are "underwater," meaning that they are worth less than the money owed on them. In the heady days before the Great Recession, homebuyers were lured by seductive credit terms into buying homes that were far beyond their means. 421

In the wake of the Great Recession, unemployment was highest in the West (with Nevada and California in the number one and number two spots, respectively) and in the South. This trend represents a reversal from the pattern of previous decades, and it is not clear whether it is temporary or will be long-lasting. Nationwide, as much as 9 percent of the eligible workforce was unemployed in 2011, up from 4.4 percent in 2007, on the eve of the Great Recession. It is still not bad when compared to Zimbabwe's 95 percent unemployment rate! That

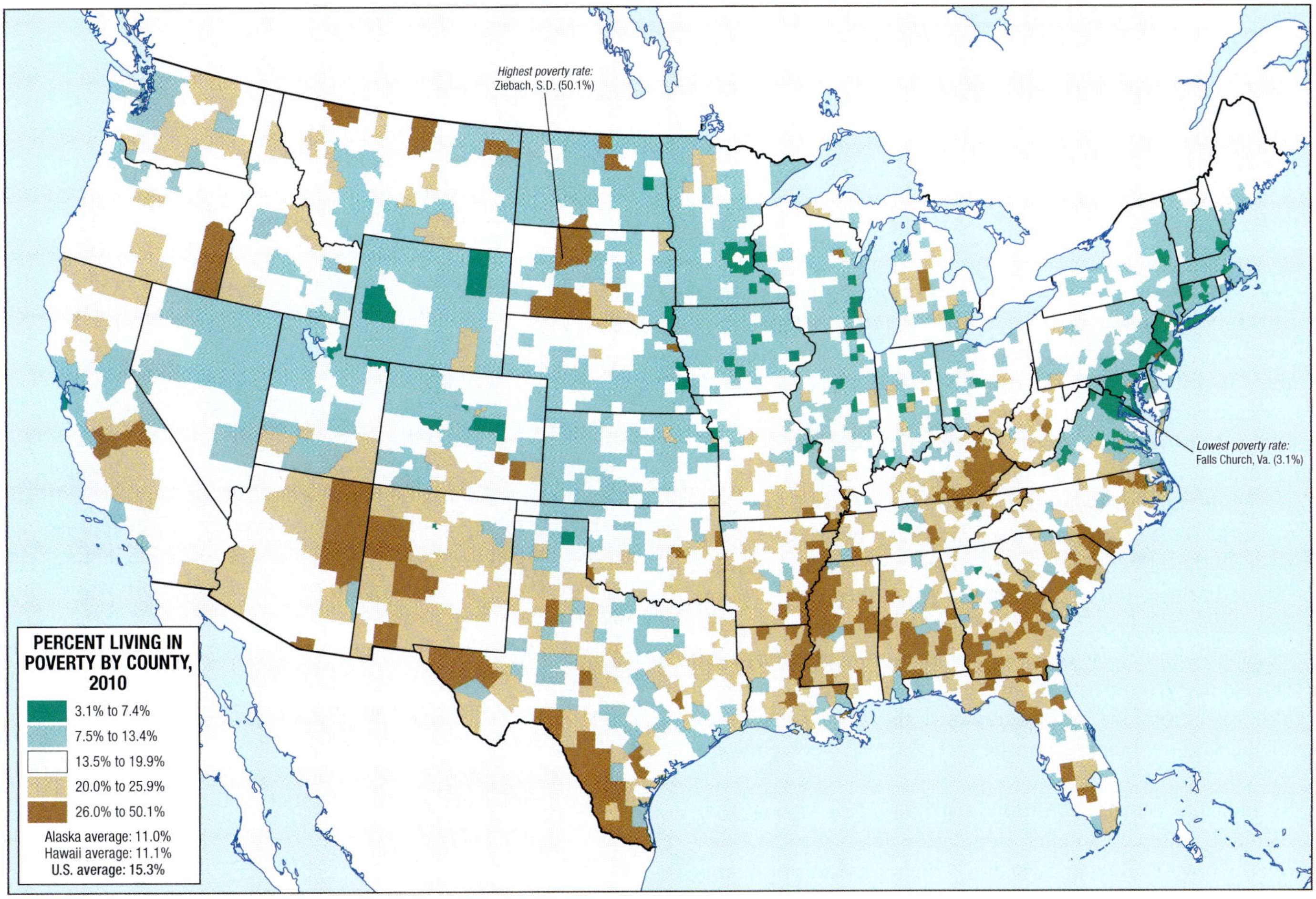

• **Figure 11.24** Poverty in America. If we want to tackle poverty, we can use this map to show us where to work.

is the highest in the world, followed by Turkmenistan with a 70 percent unemployment rate. The European microstate of
65 Monaco registered a zero percent unemployment rate.

There are many ways to measure poverty in the United States. Again, there is a clear correlation between education and affluence. The United States can boast about being on the fortunate side of the digital divide globally, but there is a digital divide within the country that contributes to a cycle of poverty. Only 4 out of 10 households with incomes of less than $25,000 had Internet access at home in 2010. Being cut off from the Net is not trivial: it translates into lower-quality health services, education, and career opportunities. About 30 percent of Americans lack any kind of access to the Internet. Many will come on board initially with smartphones, but it is almost impossible to fill out job applications or type up a résumé on a smartphone.

Technically, the United States made its way out of the Great Recession but recovery was slow. Falling tax revenues meant huge cuts in city, state, and federal budgets. Canada weathered the recession much better than the United States did, by 2011 recouping the number of jobs lost in 2008–09. In 2011, Canada had a jobless rate of 7 percent.

Sources of the Region's Affluence

This region, particularly the United States, is blessed with very large endowments of some of the world's most important natural assets. A number of geographic, political, and other circumstances have complemented this resource wealth in the development of these economically powerful countries. Both countries are among the five largest in the world. A wide range of environmental settings has both allowed and stimulated full use of human ingenuity in economic development. The combined population of both countries is large, yet neither country is "people overpopulated." A large popu- 58
lation represents both a big pool of prospective labor and talent to promote economic growth, and a vast market for the goods and services produced. Both countries developed mechanized economies early and under favorable conditions. Innovations continue to replace human labor with machine labor, generally increasing efficiency and productivity (while also leading to unemployment in traditional manufacturing sectors).

Peace and stability within and between these countries have provided a good climate for investment and economic development. Since the Civil War ended in 1865, the United States has managed to keep all of its major conflicts off its own shores. Generally, cooperative relations between the United States and Canada also help. Both have an overall sense of internal unity despite episodes of poor race relations in the United States and the separatist movement in Québec. Both countries have a track record of continuity in political, economic, and cultural institutions. Such stable conditions are conducive to economic growth.

The distribution of national wealth in both countries, however, is imbalanced. In America especially, there is a profound gap between the rich and poor. The top 20 percent accounted for 50 percent of the nation's wealth, while the bottom 40 percent accounted for 12 percent of the wealth, in 2010. Another measure was that the wealthiest 1 percent of Americans took in more than 20 percent of the country's total income, while the poorest 10 percent took in less than 2 percent of the total. This represents the greatest gap between rich and poor of all the MDCs and has led to growing social discord in the United States. Large segments of the United States' population lack a "safety net" in the form of health insurance and other social services. In 2011, a movement called **Occupy Wall Street** swelled from protests directed against that epicenter of wealth to nationwide protests against the "1 percent"—the monied class—by members of the "99 percent"—everyone else (•**Figure 11.25**).

Canada has a more equitable distribution of income than the United States, a smaller population (about 10%) living below the poverty line, and generally higher indexes of quality of life. Long dominated by liberal politics, Canada has a more socialistic approach to its population than the United States does (for example, with taxpayer-funded universal healthcare) and tends to be more in line with Europe than with the United
90 States on most social and political issues. Taken as a whole, however, Canada and the United States make up a region where most people enjoy the "good life."

An Abundance of Resources

The United States and Canada are the epitome of "neo-Europes," lands that European immigrants chose to settle largely because of their resemblance to the European environments and their potential for production of wheat, cattle, and other vital products. The United States has more high-quality 45 arable land than any other country in the world. A much smaller proportion of Canada is arable, but it has more farmable land than all but a handful of countries. Such abundance has helped these two countries become the largest food-exporting region of the world.

Canada and the United States also have abundant forest resources. Canada's forests are greater in area than those of the United States, but not as varied. Wood has been an essential material in the development of both countries, prompting the less developed countries of the world today, such as Brazil and Malaysia, to cry foul when American environmentalists denounce tropical deforestation. Even now, the United States cuts 35, 386 more wood for lumber each year, produces more wood pulp, and produces more paper than any other country in the world. The United States both exports and imports large quantities of wood, with the exports going largely to Japan. U.S. domestic industries can meet only 60 percent of their lumber needs, and their imports come mainly from Canada.

Canada is the world's largest exporter of wood (•**Figure 11.26**). Sales of Canadian timber to the United States have long been a sore point between the two countries as the United States accused the Canadians of timber subsidies (pricing this commodity artificially low). Canada is delighted to have found 420 a new customer in China, having convinced China that wood is superior to concrete in construction of homes and schools in earthquake-prone areas. The United States had bought 60 percent of Canada's timber exports, but by 2011 that figure fell to 40 percent as China bought up 20 percent of Canada's exports.

The United States and Canada are also very rich in mineral resources. These played a major role in the earlier development of the region's economy, and mineral output is still huge.

Reuters /JESSICA RINALDI/LANDOV

• **Figure 11.25** The Occupy Wall Street movement that erupted in 2011 expressed widespread dismay about the growing gap between the haves and have-nots in the United States.

Richard Kolar/Animals Animals

• **Figure 11.26** Canadian timber en route to East Asia.

Canada is a major exporter of oil and natural gas, mainly to the United States, which in 2012 obtained 21 percent of its imported oil from Canada—more than from any other country. Although still a major mineral producer, the United States is also a major importer, especially of oil and industrial raw materials. This situation reflects the partial depletion of some American resources and the enormous demands and buying capacity of the U.S. economy. Recent years have seen impressive growth in proven reserves of fossil fuels in Canada and the United States. Exploitation of new reserves of oil in North Dakota and elsewhere within its borders made the United States at least temporarily reverse course in its trend of dependence on foreign oil imports. In 2005, the United States imported 60 percent of its oil; in 2011 that figure was 47 percent.

A spectacular development in fossil fuel resources in recent years has been the exploitation of natural gas (and, to a lesser extent, oil) lying in shale deposits. When energy prices were low, it was uneconomical to coax this energy from the ground, but as prices rose, so did the incentive to tap these hydrocarbons. The resources are obtained through an energy and water-intensive process known as **hydraulic fracturing**, or **fracking**. Water, sand, and chemicals are pumped underground at high pressure through a well into shale deposits thousands of feet deep. This causes fissures and layers in the rock to crack. The sand particles keep the fissures open, allowing natural gas to flow up the well. At the surface, the gases are separated into dry gases and natural gas liquids. One of the most important products is ethylene, used in making many plastics. Fracking is a special challenge and even a dilemma in drought-prone areas like the U.S. West. In using enormous amounts of water, the process draws down on aquifer waters that are also needed for irrigation and other human uses.

Some Canadian provinces and American states are particularly well endowed with these resources, Alberta, British Columbia, North Dakota, Texas, Arkansas, Oklahoma, Virginia, West Virginia, and Pennsylvania among them (•Figure 11.27). Some of these states are in the Rust Belt, and the shale gas industry has the potential to revitalize their long-depressed econ-
419 omies. There are tens of thousands of new jobs, not only for those who work directly to get the energy out of the ground, but also in manufacturing the equipment and parts for the rigs and supplying services like food and lodging.

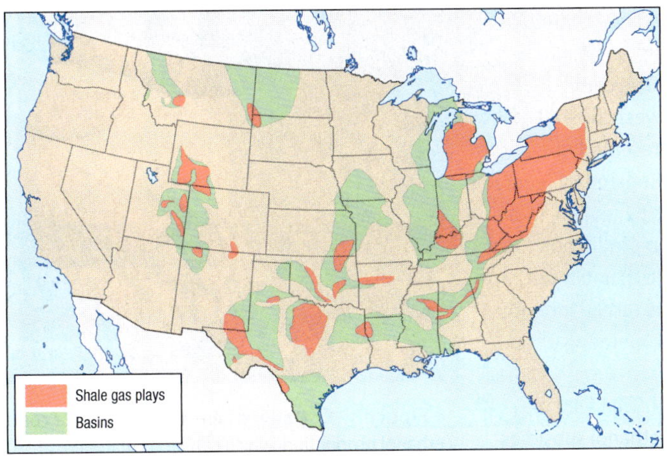

• **Figure 11.27** Principal U.S. shale gas reserves.

Is all of this news too good to be true? Yes: There are known and suspected downsides to fracking. Gas can leak into water reservoirs. You may have seen YouTube videos of someone turning on the kitchen water tap and igniting their "flammable water."

A number of other countries, including China, Argentina, Mexico, and South Africa, have deposits of shale gas, but the United States is far ahead in the technical capacity to exploit them.

Abundant energy resources have been essential to the region's economies. Coal was the largest source of energy during the nineteenth-century industrialization of the United States. Today, the country leads the world in estimated coal reserves (about 25% of the total). In coal production, China and the United States are first and second, with China producing 35 percent and the United States 18 percent. Canada is also an important coal exporter, with reserves and production that are small on a world scale but large in terms of Canada's needs.

Environmental concerns about the impacts of this most polluting of fossil fuels raise long-term questions about what the United States can do with its generous coal endowment. As it stands, the United States gets more than half of its electricity from coal-fired plants, with natural gas supplying 22 percent and nuclear power 20 percent. Drawing on large uranium resources in each country, nuclear power is an important secondary source of electricity for both (see Geography of Energy, page 416).

Petroleum fueled much of the development of the United States in the twentieth century. For decades, the country was the world's greatest producer, having such large reserves that the question of long-term supplies received little thought. The United States is still exploiting its 21 billion barrels, an estimated 1.6 percent of the world's proven oil reserves (a figure that may increase following the recent find of a new oilfield deep under the Gulf of Mexico). But its huge economy demands 7 billion barrels a year—as much as the four next largest consumers, China, Japan, India, and Brazil, combined—far more oil than the country can produce. The United States does possess reserves of petroleum that can be produced from oil shale, but only at great economic and environmental cost. This is why the United States imports so much oil. In 2011 it imported 11 percent more than it did during the 1973 energy crisis, when President Carter became the first American leader to insist that the U.S. must reduce its dependence on foreign oil. The top five suppliers of imports are, in descending order, Canada, Mexico, Saudi Arabia, Nigeria and Venezuela. Note that the top two suppliers are America's nearest neighbors, and that only one Middle Eastern country (Saudi Arabia) is among these five.

With demand and prices so high, the United States has been pressed to develop its own domestic petroleum supplies. It is going after increasingly technically challenging and environmentally risky reserves, including the Macondo Prospect in
the United States' territorial waters of the Gulf of Mexico. 417

INSIGHTS | Trade Barriers and Subsidies

Many of the world's LDCs are convinced that the best way to develop is not to borrow money from the richer countries but to trade with them. After all, with their larger populations and lower labor costs, the LDCs should easily be able to churn out products at much more favorable prices than those sold by the MDCs. Buyers around the world should be snapping up those affordable products, pumping desperately needed money back into the developing countries. This exchange does not take place as freely as it could, however, because of two tools the MDCs wield as much as possible to protect
45 their own producers: subsidies and tariffs.

Subsidies are expenses paid by a government to an economic enterprise to keep that enterprise profitable, even if the market would not. Tariffs are essentially taxes that a government imposes on foreign goods, basically for the same reason, to ensure that domestic products make it to market, even though they are not produced at competitive costs. Alone or in combination, subsidies and tariffs can make it all but impossible for a less developed country to join the world economy as a trading partner.

There are often struggles between the MDCs and LDCs over the issues of subsidies and tariffs, especially in meetings of the World Trade Organization (WTO). At these meetings, the LDCs are represented by a bloc that calls itself the **G-21 (Group of 21)** as a "play on numbers" against the more affluent G-20.

Often bitter debate between these groups focuses on subsidies that the United States, the European Union, Japan, and other MDCs have used for several decades to prop up certain sectors of their economies, especially agriculture. The subsidies were created to help farmers and the countries get through especially difficult times, including the **Great Depression** of the 1930s in the United States, when the livelihoods of so many farmers were threatened, and post-World War II Europe and Japan, when the specter of malnutrition loomed. Over the years, the critical need for subsidies passed, but the sectors that received them were loath to part with them. Federal supporters of subsidies found political capital in maintaining them. Today, farming subsidies are something of a sacred cow in many of the MDCs, and any effort to reduce or remove them would have domestic consequences—especially at the ballot box. Altogether the MDCs pay out $1 billion a day in subsidies that support the farmers who make up less than 1 percent of their population. In the United States, subsidies are concentrated on growers of the big commodity crops such as wheat, rice, corn, soybeans, and cotton (these tend to be large, corporate-owned farms). The government views farmers of the "specialty crops" such as fruits and vegetables as self-sustainable (these tend to be smaller, family-owned farms). Under market pressure from cheaper imports, such as Chinese garlic, these farmers are now clamoring for subsidies too.

Cotton is a fine example of how the subsidies and related tariffs work domestically and what their impact is in the LDCs. The United States spends over $4 billion each year to subsidize its cotton farmers. The country produces far more cotton than it consumes, and most of the prodigious output is exported. The high level of production made possible by the subsidies drives the global price of cotton down to around 50 cents per pound. This is less than the African production cost of around 65 cents per pound. The African cotton producer not only cannot compete in the world market but also loses money doing so. Cotton is, or should be, big business for Africa. If the U.S. subsidies were reduced or removed and world cotton prices rose, the potential economic gains for several African countries would be enormous. The United States, however, has been slow to reform cotton policies except in ways that clearly serve its own interest. In recent legislation, tariffs were reduced on some imported African textiles—but only those made from American cotton. The European Union, Japan, South Korea, and other countries are just as
strident in protecting some of their products. 332

The G-21 countries, speaking for two-thirds of the world's farmers, demand that the MDCs reduce or eliminate such subsidies and tariffs or that the LDCs be compensated for the revenue lost to them. They threaten to begin erecting tariffs of their own (a proposal the MDCs say is bad for the developing economies) if the MDCs do not yield. In a familiar negotiating sequence, the MDCs give up little or no ground in such impasses, causing the G-21 delegates to walk away from trade talks in frustration.

Would removal of subsidies hurt the small American family farmer, whose cause is championed by entertainers Willie Nelson and Neil Young? That would depend on how subsidy cuts were targeted. Sixty-five percent of the subsidies go the top 10 percent of America's agricultural producers, the large corporate farming industries. It is possible that the unkindest cuts could be avoided with sacrifices imposed instead on the bigger businesses, thereby perhaps helping some of the world's poorest countries escape the poverty trap on their own.

Manufacturing now only accounts for about 10 percent of the U.S. economy (down from as much as 30% in the 1950s), having fallen to the forces of globalization that moved traditional industries like textiles to countries with lower manufacturing costs, such as China and, more recently, Vietnam. Manufactured goods make up two-thirds of U.S. exports by value, however, and the nation still has the global lead in many high-tech products. It is also important to keep in mind that manufacturing has not declined as much as the figures might suggest; instead, its relative contribution has been displaced by growth in other sectors of the economy, like finance, insurance, and real estate.

Ironically, perhaps the best prospect for more growth in U.S. manufacturing is investment from China as that country looks to produce goods for the American market that are cheaper to produce in the United States than to export from Asia. Wages for Chinese workers are rising, already beginning to push manufacturing jobs back to the United States; this process is known as insourcing (or re-shoring, that is, the reverse of outsourcing or offshoring; see page 235). For now, though, the United States hopes to maintain its global leadership in design and innovation, profiting from a knowledge economy even while other countries actually make products. As one American economist put it, "We want people who can design iPods, not make them."[5] But many worry that design and innovation will also move offshore, leaving the United States with no competitive edge.

Cities often serve as economic barometers. In the first part of the twentieth century, wealth and jobs clustered in what are now Rust Belt cities like Cleveland and Detroit, where there were large concentrations of heavy industry and capital. Now the economic growth is concentrated in "smart cities" where the knowledge economy is driven by higher education levels and a larger supply of highly skilled workers.

The United States is the world leader in the production of software. In the 1990s, there was such a speculative frenzy
70, 72, 88, 94, 269, 271, 434 about IT that a classic **economic bubble** developed. Tens of millions of Americans started investing in stocks for the first time, bidding up the values of companies—even companies that produced no tangible products—to lofty heights. There was talk of a "new economy" in which the old paradigms of corporate investment and growth no longer applied. However, the **tech bubble** burst in 2000. Clever or lucky investors withdrew their investments in time, but others lost fortunes that had existed on paper. An economic recession set in. The economy gradually recovered. The tech sector was approached more cautiously. High-tech industries increasingly built what came to be known as a **platform economy**, with companies like Dell operating headquarters in the United States but outsourc-
235 ing virtually all work on its products to other countries.

A far larger bubble burst in 2008. The implosion began in the United States housing sector, where through so-called "subprime loans" millions of under-qualified American bor-
394 rowers bought homes they could not afford. Banks foreclosed on the unpaid properties, housing prices slumped, and a full-fledged crisis developed in the banking industry. This **credit crisis** in the United States spread quickly worldwide, sparking a sell-off in global stock exchanges, a widespread economic recession, and plummeting prices for oil and the other commodities that had supported booming economies. The Great Recession had begun.

U.S.-Canadian Economic Relations

Each of these countries is a vital trading partner of the other, although Canada is much more dependent on the United States than vice versa. In 2011, Canada supplied 14 percent of all U.S. imports by value and took in 19 percent of all U.S. exports, making Canada the leading country in total trade with the United States. That same year, the United States supplied 50 percent of Canada's imports and took in 75 percent of its exports. Except for Canadian exports of automobiles and auto parts to the United States, the main pattern of trade between the two countries is the exchange of Canadian raw and intermediate-state materials—primarily ores and metals, timber and newsprint, oil, and natural gas—for American manufactured goods.

Free-trade agreements designed to open markets and increase revenues for both countries have facilitated this exchange. Nevertheless, the countries have periodically tried to protect their industries from one another, often with poor results. In the 1990s, there was a so-called **wheat war** between the two countries as they struggled for greater market share in an environment of high yields and low prices. With agricultural prices for grains and pork dropping to the lowest levels in more than 50 years, many farmers both south and north of the U.S.-Canada border went bankrupt. U.S. producers claimed that Canada was **dumping** (selling the product for less than it cost to produce it) its durum wheat on the U.S. market.

There was also a **salmon war** between the two countries, marked by periodic skirmishes until a resolution was reached in 1999. Salmon hatch in Canadian and American rivers, spend most of their lives in the open seas, and return to their river birthplaces to spawn. The dispute centered on the issue of "interceptions," meaning that American fishers often intercepted fish that hatched in Canadian upstream waters but migrated into U.S. waters.

In 2003, a new dispute emerged when the United States, accusing Canada of dumping low-cost lumber in the United States, imposed heavy new tariffs of 27 percent on softwood lumber imported from Canada. An agreement reached in 2006 lowered the tariffs and required Canada to price its lumber more competitively. Other issues between the countries await resolution. There are dozens of small water disputes across the border, including one over North Dakota's plan to divert excess water into the Red River, which flows north into Manitoba's Lake Winnipeg.

Despite such volleys occasionally fired on both sides of the border, the trend has been toward more cooperation and free trade. In 1965, the two countries negotiated a free-trade agreement for the production, import, and export of automobiles. This sparked rapid expansion of the auto industry into southern Ontario (Canada's most industrialized province), on the border with the U.S. automobile-producing state of Michigan. When the auto industry suffered in Detroit, it did so in neighboring Windsor, Ontario. Ontario contributed to the "bailout" of the automakers General Motors and Chrysler. In this bailout, which was widely considered to be successful, the American government infused huge amounts of capital into the ailing auto industries.

The two countries entered into the comprehensive **Canada-U.S. Free Trade Agreement** in 1988, and in 1994 Canada, the United States, and Mexico enacted the North American Free Trade Agreement (NAFTA). Its goals and impacts are described in detail, especially as they affect Mexico, on page 371 in Chapter 10. NAFTA has brought a boom in trade between the three countries but also caused major industrial readjustments, generally characterized by the more labor-intensive firms (beginning with clothing) moving to Mexican locales from Canada's southeastern core region and from the industrial northeast of the United States.

Transportation Infrastructure

Both governments have sought to enhance national unity and economic strength with transportation networks spanning

• **Figure 11.31** There is a dense network of highways and railways in the United States and southern Canada.

their vast landscapes (•**Figure 11.31**). In both countries, east–west networks have been built over long distances, against the "grain" of the land, especially the great north–south-trending ranges of the Rocky Mountains. The coasts of both countries were first tied together effectively by heavily subsidized transcontinental railroads completed during the second half of the nineteenth century. Later, national highway and air networks improved transportation.

A landmark development in the United States was the growth of the **interstate highway system** beginning in the late 1950s. Today, this system serves as the primary network for the trucking of cargo across the country. Railways meet these needs in most countries. Although railways still carry the bulk of goods being transported across the country, few new lines are being built. Passenger rail traffic is particularly neglected, and once-popular routes are no longer used. In many cases, the old rails have been torn up and converted productively as "rails to trails," becoming bicycle and walking paths. There is discussion of building high-speed (90–200 mph) passenger rail service, using existing rails rather than building new ones as the Europeans do. A major problem with this plan is that more than 90 percent of the U.S. rail network is owned by freight companies, and in part for safety reasons they do not want to share their tracks with passenger traffic.

The stagnation or decline of U.S. passenger railways is a direct reflection of the American love affair with the car. Personal prosperity (or at least the ability to take out a loan) and the freedoms associated with being on the road in one's own automobile diminished appetites to travel by rail and bus. Public transportation remained popular only in the cities, where gridlock made it an attractive alternative, but rising energy prices could lead quickly to a reevaluation of public mass transport.

11.5 Geopolitical Issues

Even with the meteoric economic rise of China and the collective strength of the European Union, the United States is the most important economic, military, and diplomatic power in the world. This section first examines the country's strategic relations with its northern neighbor and then with the rest of the world.

Historical Relations between the United States and Canada

For many years after the American Revolution, which secured U.S. independence from Britain, the political division of Canada and the United States—between a group of British colonial possessions to the north and the independent United States to the south—was accompanied by serious friction between the peoples and governments.

There were several sources of antagonism. The northern colonies failed to join the Revolution, and the British used those colonies as bases during the war. A large segment of the northern population had come from **Tory** stock driven from U.S. homes during the Revolution (Tories were the "friends of the king," colonists who wanted to maintain political connections with the British government). And tensions were high over the issue of who would have ultimate control of the central and western reaches of the continent.

The **War of 1812** was fought largely as a U.S. effort to conquer Canada. Even after its failure, a series of border disputes occurred, and the United States openly expressed ambitions to possess this British territory in North America. Officials in the United States threatened annexation throughout the nineteenth century and even into the twentieth.

Canada's emergence as a unified nation came partly as a result of U. S. pressure. After the U.S. Civil War (1861–1865), during which more than 600,000 Americans died, the military power of the reunited nation seemed threatening to Canadian and British eyes. Recalling that the British had been hostile to the ultimately victorious Union (North) during the Civil War, some Americans argued that taking Canadian territory would be a just recompense. In 1867, the British Parliament passed the **British-North America Act**, bringing an independent Canada into existence by combining the separate colonies of Nova Scotia, New Brunswick, Quebec, and Ontario into one dominion named Canada. With that act, Great Britain sought to establish Canada as an independent nation capable of deterring U.S. conquest of the whole North American continent.

Although hostility between the United States and its northern neighbor did not immediately end with the establishment of an independent Canada, relations improved gradually. Canada and the United States today are strong allies with a connection only occasionally tested. But just as the United States overshadows Latin America in geopolitical affairs, it overshadows Canada as well. Canadians, like Mexicans, often feel overlooked and underappreciated by the neighbors across the border. Canada maintains a rather low profile in international affairs, sometimes disagreeing with the United States on such critical issues as the Iraq invasion of 2003. Canada exerts a policy of relative independence and neutrality despite membership with the United States in the North Atlantic Treaty Organization (NATO) and other international agreements. Canadian soldiers have died supporting the NATO mission in Afghanistan. Some American travelers in lands where Americans are often disliked have learned to exploit Canada's relative neutrality for their own protection by sewing badges of Canada's national symbol, the maple leaf, on backpacks and jackets.

The United States' Place in the World

Historically, the United States was not the same kind of empire-building power as many European countries were. The United States did have its overseas possessions—the Philippines, Puerto Rico, and Guam, for example—and still holds a few, but its power has been and continues to be projected more through military action and trade (•**Figure 11.32**).

The United States long enjoyed the geographic advantage of being something of an island situated far from the world's hot spots—especially Europe and the Middle East. In both world wars that ravaged Europe in the twentieth century, the United States initially clung to **isolationism**—the view that those conflicts were someone else's and that the United States would do best to stay out of them. Only late in World War I, at the unpopular insistence of President Woodrow Wilson, did the United States enter the stalemated war on the side of Britain and France. The United States lost 116,516 soldiers in that war, but in helping its allies secure victory, it also gained unprecedented influence and importance on the world stage.

Again in World War II, the United States managed to stay out of the conflict for more than two years after Hitler's troops invaded Poland and ignited conflagrations across Europe. But on the infamous day of December 7, 1941, Japan forced the United States into the war, with its surprise attack on Hawaii. And again, U.S. leadership and victory on both fronts of that war, which cost 405,399 American lives, helped the country achieve unprecedented strength in global affairs.

That war was succeeded by the Cold War, in which the two leading powers, the United States and the Soviet Union, faced off against one another, flanked by a host of often strong but

Pete Souza - White House via CNP/Newscom

• **Figure 11.32** Hanging in the Oval Office of the White House, this is probably one of the most consequential maps in the world.

88, 114, 139 always less powerful allies. U.S. concerns about the spread of communism from the Soviet Union and China into the newly independent countries of the postwar world—and the so-called domino theory that one after another of these countries might fall to communism—led to numerous and sometimes very costly American military interventions. The Vietnam War
258 was the most significant of these.

The peaceful conclusion of the Cold War around 1990 was followed by a decade in which the United States sought a new role for itself on the world stage. There were military interventions to quell the conflict in disintegrating Yugoslavia and to try to deliver lawless Somalia from famine, but there was a new sense of security—and, in hindsight, complacency. There was no longer any defining framework, such as east–west relations, in international affairs. Terrorism was thought to be a problem that existed far from home.

Not unlike Pearl Harbor, the attacks of September 11, 2001, brought an end to the notion that geographic distance protected the United States. The events of that day represented a great watershed in U.S. geopolitical history. The United States under President George W. Bush developed a policy of **preemptive engagement** (also known as the "Bush Doctrine"): whenever and wherever the country perceived a threat to its security, it would take military action if necessary to
201 defuse that threat. That action would be unilateral, if necessary; in its first term, the Bush administration downplayed the need for diplomacy and partnership. The preemptive engagement policy was founded mainly on the premise that such actions would prevent potentially devastating terrorist attacks on the U.S. homeland, perhaps with chemical, biological, or nuclear (CBN) weapons of mass destruction (WMD). Such threats could emerge not only from transnational groups like al-Qa'ida but also from "rogue states," most of whom are on the United States' list of **official state sponsors of terrorism**, which includes Iran.

As discussed earlier (page 178), the Bush administration justified the invasion of Iraq in 2003 on the grounds that Iraq was a state sponsor of terrorism and had WMD that it might use against the United States or its allies, especially Israel. Detractors wrote of a **new American imperialism**, proposing that U.S. post-9/11 actions marked the beginning of a new era of aggressive global involvement for the country. While continuing to prosecute America's wars, President Obama drew distinctions from his predecessor. In ousting Qaddafi from Libya, for example, his strategy was to "lead from behind," allowing America's European allies to do much of the heavy military "lifting."

The United States remains the world's sole **superpower**, with the strongest economy, military expenditures larger than those of the next 14 countries combined, dominance of global popular culture (from films to fast foods), the world's best universities for graduate studies, and headquarters to many of the world's leading international organizations (including the United Nations, the International Monetary Fund, and the World Bank). Its singular stature suggested a "unipolar" world in which its dominance is unchallenged. A shift appears to be in the works, however, as we have seen repeatedly in this book. China's surging economy is the principal challenge; it is expected to be the world's largest by 2027.

The following section takes a closer look at Canada and the United States.

11.6 Regional Issues and Landscapes

Canada

The Québec Separatist Movement

The core region of Canada (•**Figure 11.33**), with about two-thirds of Canada's total population, has developed in a narrow strip along Lakes Erie and Ontario and seaward along the Saint Lawrence River.

Ontario and Québec account for about 40 percent of products marketed from Canadian farms. Three-fourths of Canada's total manufacturing is in this core region, but in the pattern of deindustrialization also seen recently in Europe and the United States, China's output has caused factories here to shut down; household appliances, electrical equipment, plastics, rubber, and textiles have all been hit hard.

The two sections of this vital region differ in their agricultural, industrial, and cultural features. The Ontario Peninsula, which has strong British roots, mirrors the U.S. Midwest's crop belts, with corn and livestock production, dairy farming, and specialty crops like tobacco. Ontario, where about 85 percent of the people speak English as a first language and only 5 percent speak French as a first language, is the most populous of the nine mainly English-speaking provinces.

In Québec, 82 percent of the people speak French as a preferred language and 8 percent speak English (•**Figure 11.34**). This is the only province in which French speakers predominate and control the provincial government, and Québec's potential devolution or even independence from Canada casts a long shadow over Canadian affairs.

Citizens of more developed countries are generally unaccustomed to the prospect that a major region of their own country might break off to form an independent nation. This is a real prospect for Canadians, however, if the Québec separatist movement has its way. The modern separatist movement has its roots in early Canadian history: the French were early settlers in these lands, but after a major British military victory in 1763, the broad sweep of language, government, and authority became decisively British. A cultural dichotomy developed that gave the French dominant influence in Québec but established only modest influence for them beyond the Saint Lawrence Lowlands.

After the end of World War II, French Canadian dissatisfaction spread, partly as attention to the plight of minorities grew worldwide. In Québec, this led to the formation of a separatist political party, the ***Parti Québécois (PQ)***, dedicated to the full independence of Québec from Canada. In

•**Figure 11.33** Principal features of Canada and Greenland.

Source: Based on Getis, Arthur and Judith Getis. 1995. The United States and Canada: The Land and Its People. Dubuque, Iowa: William. C. Brown Publishers.

1976, the party gained control of Québec's provincial government and made French the language of commerce, requiring that all immigrants to the province be educated in French. Québec refused to ratify the country's 1982 constitution. In 1980 and again in 1985, the separatists forced the country to hold referenda on Québec's independence; both failed. In 1995, the issue was taken to provincial voters again, and those who favored staying a part of Canada again won, but by a very small margin (51%). In the referendum's wake, the federal government officially recognized Québec as a "distinct society" within Canada.

Although the PQ suffered a political setback in 2007, losing ground to a party favoring autonomy rather than independence for Québec, the issue of separatism is still very much alive and will likely stay on the national agenda. Separatists are frustrated by the steady influx of non-French-speaking immigrants seeking jobs and settlement in Québec City and other cities along the Saint Lawrence Seaway, who tend to vote with

It lies more than 2,000 miles (3,200 km) west of the European nation with which it is associated. In 1985, Greenland withdrew from the European Economic Community (EEC, the precursor to the European Union) because of its frustration over EEC fishing policies and the lack of any EEC development aid. Today, however, the European Union gives Greenland almost $50 million each year in exchange for Greenland's granting of fishing quotas to EU member states in its territorial waters. Denmark subsidizes the massive island with an average support of more than $8,000 per person per year. There is increasing talk of independence among the Greenlanders, but severing the aid pipeline from Denmark and the European Union would make that a very costly proposition. There is one way out of this: striking it rich with oil. Greenland may have done just that: there are estimates that Greenland may have oil reserves as great as Libya's, and a petroleum industry has started up in Greenland.

Greenland has nearly one-fourth the area of the United States, but about 80 percent of the island is covered by an icecap up to 10,000 feet (3,000 m) thick (•**Figure 11.37**). Until recently, Greenland seemed to have bucked the worldwide trend toward melting of glacial ice, but between 2003 and 2011 lost an average of about 50 cubic miles (c. 200 cu km) each year.
76, Scientists are working to determine whether the melting is re-
250, lated to overall climate change and what its impacts on the
358 "conveyer belt" of Atlantic oceanic waters and subsequent climates might be. Some models suggest that the ice sheet could shrivel completely over a period of about 1,000 years, which would raise the world's sea levels by 20 feet (6 m). With the ongoing melting, Greenland's geography is a work in progress. Maps of the coastline are outdated almost as quickly as they are drafted as new islands and inlets emerge from the thaw.

There was recently a cartographic brouhaha about Greenland's ice. The 2011 edition of Britain's revered *Times Atlas of the World* depicted significant portions of Greenland's coastline ice-free, representing a 15 percent loss of ice cover from 1999 to 2011. Critics insisted that the retreat was exaggerated. *The Times* promised a more accurate revision and called for an investigation.

Ninety percent of Greenland's population is distributed along the southern coast. Greenland has a total population of just over 50,000, giving it a population density of less than one-tenth of a person per square mile. The capital and largest city is Nuuk (originally called Godthab; population 15,000) on the southwestern coast. Under the auspices of NATO, joint Danish-American air bases are maintained there, and the United States has a giant radar installation at Thule in the remote northwest.

The great majority of inhabitants are Inuit, with Danes making up about 15 percent of the population. The leading means of livelihood are fishing, hunting, trapping, sheep grazing, and the mining of zinc and lead. As mentioned above, there may be oil in the island's future. This would make fishing seem like small fry. There is risk involved, however: Greenland is a pristine arctic wilderness, and an oil spill here would be infinitely more difficult to contain than the *Deepwater Horizon*'s.

Joe Hobbs

• **Figure 11.37** Greenland's spectacular glacial landscape.

The United States

Reference Map

The book concludes with United States, a country that is home to most of its readers (•**Figure 11.38**).

Adaptive Reuse and Gentrification

A characteristic urban landscape transformation, particularly in the Northeast and Midwest, is the **adaptive reuse** of older structures. In this process, abandoned commercial or factory buildings are being converted to new economic activities, generating jobs and attracting people back to the area. This shift is often associated with **gentrification**, the process by which a low-cost, run-down neighborhood undergoes physical renovation for residential or office space, resulting in increased property values and new homes and offices for more affluent urban professionals (•**Figure 11.39**). In Connecticut, for example, where the manufacturing economy shrank by nearly 25 percent in the past two decades, vacated mills and other factories were reborn as art shops and antique stores. Others were razed and replaced by big-box stores like Best Buy and discount department stores, including Kohl's and Target. There is still twice as much office space in suburbs as there is in central business districts, but the trend is unmistakable: during the Great Recession, suburban-based businesses like mortgage lenders and home lenders were hit hard, whereas the government offices and big banks of the downtown areas suffered less.

Adaptive reuse takes place most often in old central business districts, but the process has also been occurring in old residential neighborhoods. Adaptive reuse and gentrification give value to the past, especially when historic buildings can be modified and used rather than destroyed (•**Figure 11.40**). But although these processes revitalize run-down areas, attracting new wealth and revenue to them, they have a downside. People who have been living in these neighborhoods often can no longer afford to do so and are displaced to distant housing that is more affordable but far less convenient for work and services (•**Figure 11.41**). There is little they can do but protest—and people have been doing that in growing numbers in the United States as "the American Dream" retreats from their
419 grasp.

The Changing Geography of American Settlement

Across the United States, midsize cities with 750,000 to 2 million inhabitants have become more attractive communities for manufacturing and service firms. Traditional locational factors like proximity to coal or iron ore or rivers are no longer so important. Smaller cities have the infrastructure and capable workers that employers need. The Internet and other information technology allow many firms to locate almost anywhere they wish. Operating and living costs are often less than in the larger cities. Midsize cities often provide attractive incentives to new businesses, including low-cost land, reduced taxes, and subsidized transportation.

Although the midsize American city has blossomed, the small town has withered. Between 1900 and the most recent census year, 2010, while the American population more than tripled, 25 percent of U.S. counties lost population. There was a marked migration of people out of the rural Great Plains region in particular (•**Figures 11.42** and **11.43**). Another way to think of this is that an increasing share of America's wealth moved to cities, while the rural Great Plains and Mississippi Delta grew poorer.

A flight from small towns to larger cities was mainly responsible for this depopulation. Small towns had their heyday when they were vital clusters of merchants and townsfolk located on some avenue of transportation—highway, railway, or river—and serving a broad agricultural hinterland. The more fortunate towns became home to small industries like food processing, shoe, or textile factories.

Transportation gradually improved, and better roads led to ever-larger settlements that attracted most new businesses. By offering consistently lower prices and a wider selection of merchandise, Walmart and other large retailers drove out diverse businesses in small towns and in the middle-sized and larger towns in which they were opened; this **"Walmart effect"** has had a huge impact around the country. Smaller towns have lost their economic significance and their residents.

The past two decades have seen some reverse migration. Some small towns in rural states of the Great Plains began actively recruiting international immigrants from countries including Vietnam, Sudan, and India to take up residence and help revive local economies. Others are enticing outsiders with offers of tax breaks, low-interest loans, and land giveaways to those who would build homes, often playing up the virtues of small-town living and the distinctive attributes of these settlements.

Another trend changing American settlement patterns has been the process of **exurbanization**. This represents a kind of re-embrace of rural life, but in response to urban conditions. City dwellers who are either priced out of closer-in suburbs or drawn to the wider spaces of more distant ones move to the "exurb," an outer suburb of the city, typically so far out as to stand almost alone. The more affluent may even buy a rural "ranchette"—a plot of often about 35 acres (c. 14 ha)—well away from the big city. American settlement patterns continue to evolve: there are signs that rising gas prices would lure faraway families back to the cities and their closest suburbs.

The Big Apple

The United States' largest city, New York City, is centrally located on the Northeastern Seaboard's urban strip midway between Boston and Washington. An enormous harbor, an active business enterprise, and a central location within the economy of the colonies and the young United States made New York City the country's leading seaport. Superior access to the developing Midwest also gave New York unchallenged leadership among American cities in size, commerce, and economic impact.

• **Figure 11.38** Principal features of the United States. Alaska is depicted in Figure 11.51.

KEY TO NUMBERED STATES
1 Vermont
2 New Hampshire
3 Massachusetts
4 Rhode Island
5 Connecticut
6 New Jersey
7 Delaware
8 Maryland
9 Virginia
URBAN AREAS
Name Over 10 million
Name 1 million to 10 million
Name Under 1 million*
* selected cities
0 100 200 mi.
0 100 200 km.
ATLANTIC OCEAN
GULF OF MEXICO
BAHAMAS
CANADA
MINNESOTA
WISCONSIN
MICHIGAN
IOWA
ILLINOIS
INDIANA
OHIO
PENNSYLVANIA
NEW YORK
MAINE
MISSOURI
KENTUCKY
WEST VIRGINIA
TENNESSEE
NORTH CAROLINA
SOUTH CAROLINA
GEORGIA
ALABAMA
MISSISSIPPI
ARKANSAS
LOUISIANA
FLORIDA
LAKE SUPERIOR
LAKE HURON
L. MICHIGAN
L. ERIE
L. ONTARIO
APPALACHIAN MTS.
New York
Chicago
Detroit
Washington
Philadelphia
Boston
Atlanta
Miami
Houston
Tropic of Cancer
95°W
90°W
85°W
80°W
40°N
35°N
30°N

Jerry Watcher/Photo Researchers, Inc

• **Figure 11.39** The Baltimore Inner Harbor Development, called Harborplace, is a successful reclamation of an old warehouse and wharf area. It boasts a shopping mall with outdoor restaurants and cafés, an aquarium, power plant buildings converted into elegant townhouses and condominiums, and access to Camden Yards, the home of baseball's Baltimore Orioles.

Joe Hobbs

• **Figure 11.40** Rather than being torn down, an old shoe warehouse in downtown St. Louis has been remade into the interesting, family-friendly and historic City Museum.

New York City itself has a population of 8 million, and the metropolitan area (which includes much of northern and central New Jersey, all of Long Island, southernmost New York State just north and west of the Bronx, and southwestern Connecticut) is home to 22 million people. The city is centered on the less than 24 square miles (62 sq km) of Manhattan Island (•**Figure 11.44**). The island is narrowly separated from the southern mainland of New York State on the north by the Harlem River, from Long Island on the east by the East River, and from New Jersey on the west by the Hudson River. Manhattan is one of five **boroughs** (administrative divisions) of the city proper. The others are the Bronx, on the mainland to the north; Brooklyn and Queens, on the western end of Long Island; and Staten Island, to the south across Upper New York Bay.

Expressways and mass-transit lines converge on bridges, tunnels, and ferries linking the boroughs to each other and to northern New Jersey. Port activity is concentrated in New Jersey along the Upper Bay and Newark Bay. The whole area is held together in part by an amazing 722 miles (1,155 km) of subway system, parts of which are more than a century old (•**Figure 11.45**).

Of the city's total population today, more than 35 percent are foreign-born, with the largest contingents coming from the Dominican Republic, China, and Jamaica. Their growing numbers more than offset the out-migration of longer-resident New Yorkers, with the trends of whites and Asians moving elsewhere in the greater New York region, Asians moving to the West Coast, retiring whites to Florida, blacks whose families originated in the South moving back to the South, and some Puerto Ricans returning to their home island. Immigrants make up more than 40 percent of New York City's labor force, including more than half in restaurants and hotels, about 60 percent in construction, and nearly 70 percent in manufacturing. Most live in the outer boroughs, with two-thirds of them in Queens or Brooklyn. Whereas New York historically was an ethnic checkerboard of Irish, Italian, Jewish, Chinese, and other groups, Queens and Brooklyn today are ethnic stewpots where immigrants from numerous countries live cheek by jowl. Restaurants and other services in these areas are extraordinarily diverse.

New York Mayor Michael Bloomberg promised a "sustainable city" for New Yorkers of the future. His plan, unveiled in 2006, had 10 goals, including a huge increase in affordable housing, a pledge that every New Yorker would live within 10 minutes' walk of a public park, an overhaul of the subways and other transport systems, and a target of having the cleanest air of all American cities. Not to be outdone by efforts in California, the mayor insisted in this plan that the Big Apple should reduce greenhouse gas emissions by 30 percent by 2030. 416

• **Figure 11.41** Chicago's Pilsen area has been spiffed up in recent years. As the area becomes more expensive and attractive to urban professionals, its poorer, longtime, mainly Latino residents are driven out. With this mural, the less affluent residents condemn Pilsen's gentrification. Note on the left the sign reading "STOP GENTRIFICATION IN PILSEN!" Eastern European immigrants once lived in Pilsen, and moved out during an earlier cycle of urban change.

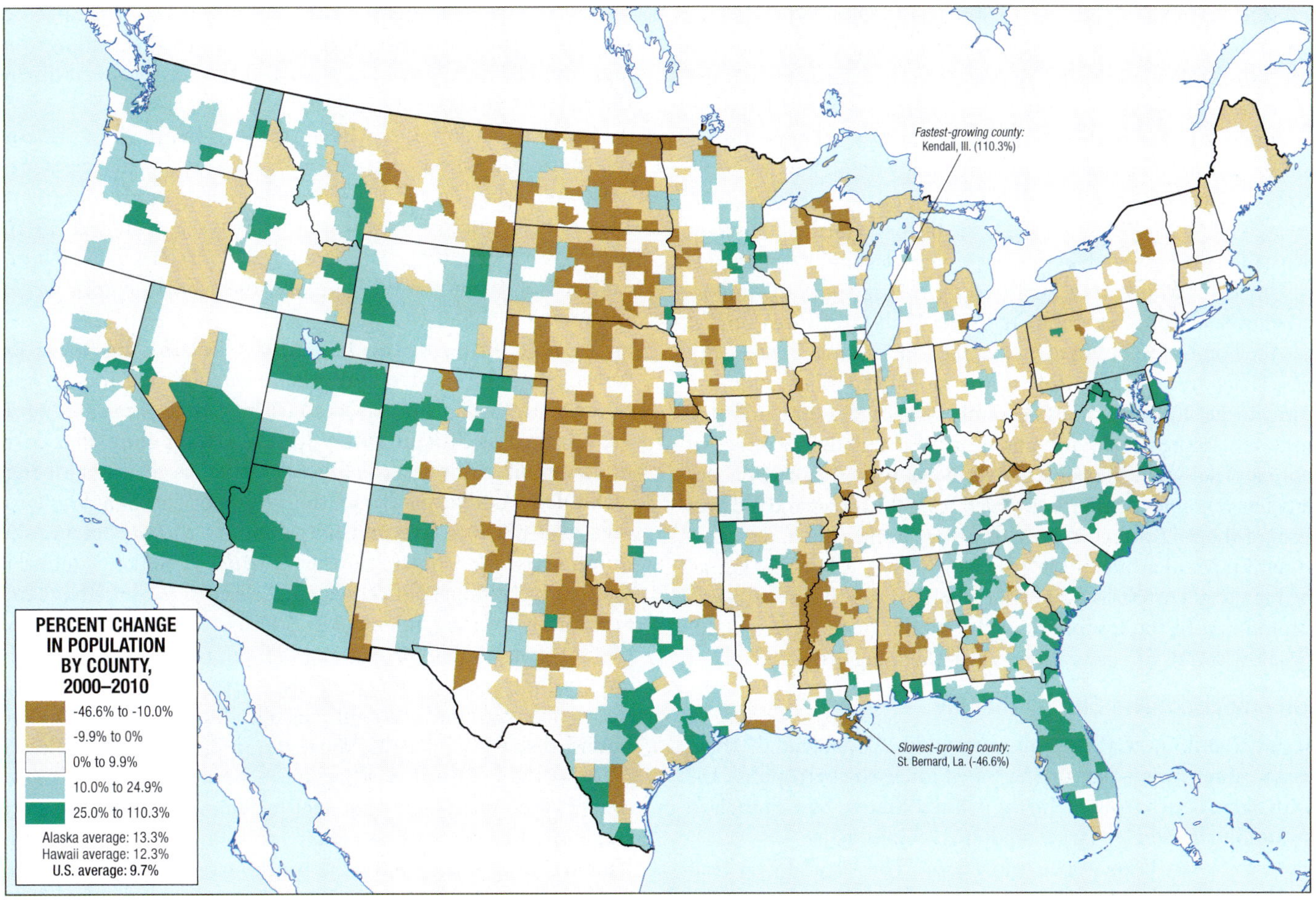

• **Figure 11.42** This map clearly shows population movement trends in the U.S. Note in particular the depopulation of the Great Plains.

Source: From Timothy Egan, "Vanishing Point: Amid Dying Towns of Rural Plains, One Makes a Stand," From the New York Times, December 1, 2003, p. A18. Copyright © 2003 The New York Times Co. Used by permission and protected by the Copyright Laws of the United States. The printing, copying, redistribution, or retransmission of this Content without express written permission is prohibited.

Joe Hobbs

• **Figure 11.43** Many small towns across the Great Plains are dying—or even dead.

Size, financial importance, cultural complexity and variety, and national history have given New York City an uncontested position of urban prominence in the United States. The heart of this extraordinary city is roughly the lower two-thirds of Manhattan. The island's Midtown hotel, shopping, theater, and office district includes most of the city's best-known hotels, restaurants, department stores, specialty shops, theaters, concert halls, and museums. The Midtown area also has numerous office buildings, including the Empire State and Chrysler buildings. At the southern end of this area are Times Square and Broadway, formerly notorious for seedy nightclubs and strip joints but spiffed up in the 1980s to be a family-friendly, must-see area (•Figure 11.46). Here too is the historic Garment District, also called the Fashion District, to present a less industrial image.

The lower Manhattan Financial District includes the imposing cluster of office skyscrapers in the Wall Street area, and it was the home of the twin towers of the World Trade Center until 9/11. This area's critical role in the global economy made it a prime target for al-Qa'ida's September 11 attacks (and an often forgotten, more feeble attempt in 1993), and the prospect of
183 more attacks will always loom over New York. The city's economy has rebounded remarkably well from the direct and indirect costs of the 9/11 attack. Many firms lost in the rubble relocated to central Manhattan and nearby New Jersey. While rebuilding, the Financial District is also being reborn as an affluent bedroom community. In the footprints of the World Trade Center buildings now stands an inspirational, heart-rending memorial.

San Francisco: The City by the Bay

About 7 million Californians—roughly 20 percent of them Asian Americans—live in the San Francisco metropolis, the sixth largest urban area in the United States. San Francisco is our illustration of large- and small-scale maps—see Figure 1.6, page 9. San Francisco—beloved as "The City" by its residents and neighbors—is located on a hilly peninsula between the Pacific Ocean and San Francisco Bay, but the metropolitan area includes counties on all sides of the bay (•Figure 11.47). The climate is unusual. Marine conditions moderate San Francisco's temperatures from an average of 48°F (9°C) in December to only 64°F (18°C) in September. Local marine and topographic circumstances give rise to San Francisco's trademark fog, which can produce very chilly days even in the height of summer. Mark Twain is said to have quipped, "The coldest winter I ever spent was a summer in San Francisco."

The area is in the border zone between the marine west coast and Mediterranean climates. Characteristic of the Mediterranean influence, about 80 percent of the area's rainfall comes in five months from November through March. Mean annual precipitation is 20 inches (50 cm) in San Francisco, enough to produce a naturally forested landscape. There are even large groves of redwood trees within a short distance both north and south of the Golden Gate Strait. East of San Francisco Bay, warmer temperatures and lower precipitation have produced a scrubby forest (chaparral and scattered oak trees) and grassland typical of the Mediterranean biome.

San Francisco began as a Spanish fort (*presidio*) and a mission on the south side of the Golden Gate Strait. It grew as a port for the gold-mining industry and then for the expanding agricultural economy of central California. The **Gold Rush of 1849** centered on the gold-bearing gravels of Sierra Nevada streams pouring into the Central Valley north and south of Sacramento. San Francisco, with its spacious natural harbor, was the nearest port to receive immigrants and supplies for the gold camps, and the city boomed during the race for gold.

Today's San Francisco Bay Area is a major domestic and international tourist destination, famed for its scenery, wines, arts, and other cultural attractions. Along with the greater Los Angeles metropolitan area, it plays a vital role in the California's economy. If considered apart from the United States, California would have had the world's eighth largest economy.

Manufacturing in the Bay Area is varied. In recent decades, computer-related and other electronics businesses have boomed from San Francisco south to San Jose, center of an area known as **Silicon Valley** for the silicon chips that carry microcircuits. The rise of Silicon Valley as a center of high technology is due partly to the allure of the San Francisco area as a place to live and its proximity to two outstanding universities: the University of California at Berkeley (close to Oakland, an industrial center on the east side of the bay), and Stanford University, in Palo Alto south of San Francisco. When the **dot-com bubble** (or tech bubble) popped at the turn of the
twenty-first century, much of the software capital fled, and 421
the region suffered a downturn. The local economy had rebounded sharply by 2005, however, in part by reaffirming its place as the country's leading biotechnology hub, with more than 800 companies in that sector. Among the most recently established is the leading center for stem cell research in the United States. But perhaps the best illustration of Silicon Valley's place in global economies and cultures is that it is home to Apple. When the company's CEO Steve Jobs passed away in 2011, many of us paused to reflect on American leadership in innovation and technology.

• Figure 11.44 New York from three perspectives and scales. **(a)** New York is located at the center of a strip of highly urbanized land christened the "megalopolis" by the French geographer Jean Gottmann in 1961. Many waterfronts and other areas shown on the maps of New York City **(b, c)** are being redeveloped for new commercial, residential, and recreational uses. Note the complex of bridges, tunnels, and ferries tying Manhattan Island to the rest of metropolitan New York.

Joe Hobbs

• **Figure 11.45** Lower Manhattan. For anyone familiar with pre-9/11 New York, the twin towers of the World Trade Center are always conspicuously absent.

Joe Hobbs

• **Figure 11.46** Manhattan's Times Square, one of the world's most famed places.

The Thirsty West

The subregion of the American West has some environmental challenges: dry climates, rough topography, and a lack of water transportation except along the Pacific Coast. Because settlement is clustered in places where water is adequate, the population pattern is oasis-like, with oases separated from each other by steppe, desert, and mountains. Prevailing dry conditions often menace the region with wildfire and droughts, ironically punctuated by floods and landslides.

Water is a therefore a critical issue throughout the western United States. Historically, settlements were small and self-sufficient in water. But with explosive suburbanization following World War II, huge numbers of people moved west, especially to California, in search of the American dream. Populations quickly surpassed local water supplies, and new sources had to be found or redirected from elsewhere. The region's older cities, such as Denver and Phoenix, were able to grow because of federal support of giant dam, reservoir, and irrigation projects during the twentieth century. But the realities of living in an arid environment have set in, and newer communities cannot meet their needs with large-scale manipulation of the environment. They must rely far more on water conservation and other ways of adapting to local conditions. Business as usual when it comes to water is out of the

Joe Hobbs

• **Figure 11.47** San Francisco: the top of "The Crookedest Street in the World" (Lombard Street), with Coit Tower and the hills of the East Bay visible in the distance.

• **Figure 11.48** The Colorado River Basin.

• **Figure 11.49** The Glen Canyon Dam, just upstream from the Grand Canyon, was constructed in the 1950s. Note the low water level in this 2005 photo. A full reservoir would cover the white strip of bank that extends behind the dam.

question. There is a book about Phoenix entitled *Bird on Fire: Lesson's from the World's Least Sustainable City.*

Large-scale twentieth-century waterworks included canals dug to bring water from distant Mono Lake and Owens Lake, in the shadow of the Sierra Nevada, to the Los Angeles metropolitan region (see Figure 11.12). These controversial projects produced decades of discord between northern and southern California. Environmentalists finally succeeded in efforts to restore these debilitated lakes with projects that have resumed water flow to both watersheds.

The region's largest water source is the Colorado River, carrying snowmelt from mountain peaks and flanks in north central Colorado and the Green River in Utah (•**Figure 11.48**). It has been highly manipulated to suit human needs. Its waters are impounded behind two major dams: the Glen Canyon in Arizona (•**Figure 11.49**) and the Hoover on the Arizona-Nevada border. Hoover Dam's Lake Mead (with the largest volume capacity of all American reservoirs) supplies a major portion of the water needs for about 25 million people in the three lower basin states of California, Arizona, and Nevada. Almost 20 million people in southern California drink these waters. Lake Powell (behind Glen Canyon Dam) and Lake Mead also generate hydroelectricity for the region. Below the dams, a canal network known as the **Central Arizona Project (CAP)** siphons water from the Colorado River to sustain large, booming, and very thirsty Phoenix and Tucson.

Colorado River water has been a subject of litigation and federal and state disputes ever since the Hoover Dam was completed in the mid-1930s. For decades, California has used more water from the river than was allocated to it in a 1922 agreement known as the **Colorado River Compact**, whereas the six upstream river basin states (Arizona, Nevada, New Mexico, Utah, Colorado, and Wyoming) have drawn less. Meanwhile, the upstream states, especially Nevada and Arizona, have had explosive urban growth and demand for water. It took threats from the upstream states to actually cut off water to California to bring that powerful state to the negotiating table. In 2003, California agreed to reduce its use of Colorado River water over a period of 14 years, to allow the six upstream basin states to use their share.

There is unhappiness in California; water allocation is a classic zero-sum game in which any benefit to one party equals a loss to another. Meanwhile, booming cities search desperately for new water sources. Las Vegas wants to divert water by pipeline from northern Nevada—a project opposed by people there and in neighboring Utah. Saint George, Utah, a rapidly growing retirement and recreation community, wants a new pipeline from Lake Powell—much to the dismay of Nevada.

Hurricane Katrina and Its Aftermath

Katrina was an unforgettable storm. What do you remember about it? On August 29, 2005, Hurricane Katrina made landfall as a Category 3 storm, with maximum sustained winds of 125 miles per hour (205 kph), along the vulnerable Gulf of Mexico coast from southeastern Louisiana to Alabama. New Orleans was the storm's epicenter of tragedy. Known as the "Big Easy" and the "Crescent City," New Orleans had enjoyed a reputation of carefree soulfulness, but Katrina changed that.

Hurricane Katrina was the worst natural disaster in U.S. history. Yet the city's tragedy was not unforeseeable. Geographically, New Orleans is extremely vulnerable to the hurricane hazard. It sits in a bowl between the waters of Lake Pontchartrain to the north and the Mississippi River to the south, only 15 miles (25 km) from the waters of the Gulf of Mexico (•**Figure 11.50**). The city already lies an average of 9 feet (2.75 m) below sea level. It continues to slowly sink at a rate of 3 feet (0.9 m) each century as loose sediments are consolidated and cannot

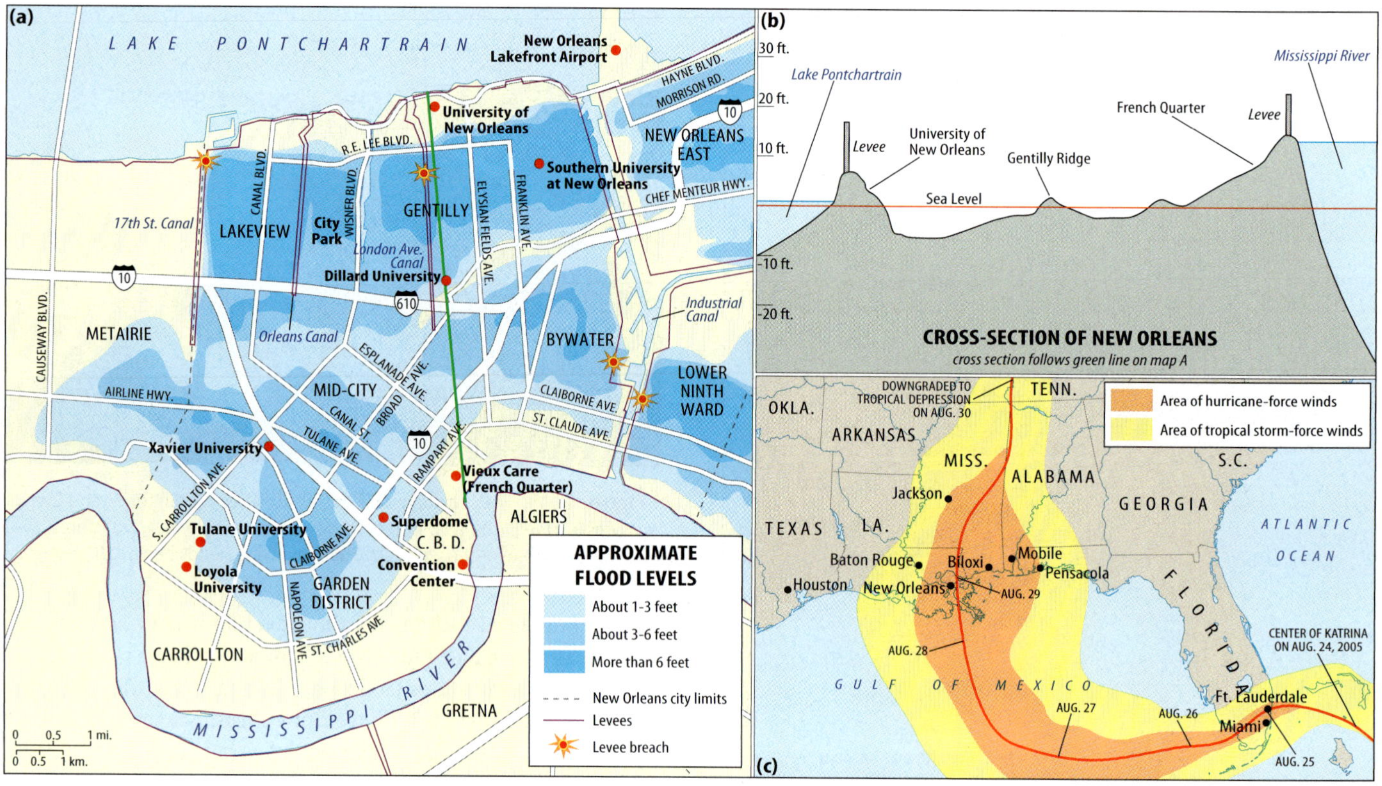

• **Figure 11.50** A monster storm, Hurricane Katrina overwhelmed the inadequate levees defending the vulnerable city of New Orleans.

be replenished: far upstream, dams on Missouri River tributaries hold back 80 percent of the sediment that would naturally reach New Orleans, while the very dikes that protect the city also keep available sediment in the Mississippi at bay.

Hurricane Katrina's **storm surge**—a powerful wave of water—penetrated the city's defenses, a 350-mile (560 km) system of elevated levees and 22 pumping stations. Before Katrina, the U.S. Army Corps of Engineers received less than half of the funds it had asked for to shore up this network, leaving New Orleans buffered by levees that were thought to be able to withstand a Category 3 hurricane. Katrina's storm surge put tremendous pressure on the levees, and they gave way in several places, notably along the 17th Street Canal levee near Lake Pontchartrain. Power supplies failed, and the city's pumps were unable to prevent water from inundating fully 80 percent of the city.

Preparations for the storm were inadequate. The city's mayor issued a mandatory evacuation order well in advance of the storm but did not make transportation available to tens of thousands of people who did not own automobiles. The disaster hit minorities and the disadvantaged especially hard: of the most affected, 4 in 10 were black, 2 in 10 poor, and 1 in 10 elderly.

The emergency response to Katrina was tragically inept. Missteps and poor communication among city, state, and federal officials led to a delay of several days in deployment of security personnel and relief supplies. An estimated 25,000 people sought refuge in the giant Superdome sports stadium and an equal number in the city's Convention Center, but both sites were short on medical supplies, food, and emergency personnel. Many New Orleans police fled the city, and a few joined in the widespread looting that followed the evacuation. The appointed head of the government body responsible for disaster relief, the Federal Emergency Management Agency (FEMA), had no disaster management experience. Thousands of National Guard troops from Louisiana and neighboring states who might have been rapidly deployed to deal with the crisis were instead serving in Iraq.

New Orleans's population three months after the storm was about 280,000, or just 58 percent of its pre-Katrina level. The poorest and most vulnerable residents of the city had fled to Houston and other distant cities, and it still appears that most will not return to New Orleans. Those who returned as the city rebuilt were more affluent and more predominantly white, but large sections of the city remained wastelands.

The economic costs of repair and cleanup from Katrina were enormous: over $80 billion, making this by far the costliest natural disaster in U.S. history (in lives, Katrina took 1,577, well below the 8,000 lost to a hurricane in Galveston in 1900). Nearly 1 million people along the Gulf Coast were made homeless, and 300,000, jobless. Most of New Orleans's housing stock was made of wood and was irreparably damaged by floodwaters.

As the city continues to be rebuilt, questions persist about whether there should even be a new New Orleans (as we remember from page 70 in Chapter 4, its *situation* makes perfect sense, but its *site* is a nightmare). Not far from the sinking city, the Gulf of Mexico is eating away at the coastline: Louisiana has lost more than 1,560 square miles (4,040 sq km) of coastal land since 1930 and continues to do so at the rate of about one-tenth of a square mile (0.25 sq km) every day. There are concerns that climate change may be providing fuel for stronger hurricanes. Gulf of Mexico waters are about 1°F (0.56°C) warmer now than they were a century ago. The warmer these waters are, the greater their potential to support catastrophic hurricanes. Both Katrina and, later in the 2005 hurricane season, hurricanes Rita and Wilma spent part of their time in the Gulf as Category 5 storms, the strongest possible.

Do you recall the discussion of adaptation and mitigation in dealing with climate change (page 34)? New Orleans will have to adapt to anticipated climate change. How can the city be protected in the future? Work to repair and strengthen existing levees began in Katrina's wake, but engineers agree that these will not prevent another disaster. The U.S. Army Corps has a larger strategy, still in the planning stage, with three legs. One is to rebuild the narrow, steadily eroded barrier islands that form a punctuated chain between Louisiana's mainland and the Gulf of Mexico. Second, restored wetlands along the Gulf of Mexico coast would serve as a natural buffer against storm surge. Finally, federal and other funding would build the Morganza Project, a semipermeable levee system that would allow engineers to control the flow of both freshwater and tidal water across a wide swath of southeastern Louisiana. This **"great wall of Louisiana"** would not protect New Orleans directly but would reduce the losses of much of the state's coastal marshlands. These help buffer the city, serve as a nursery for Louisiana's valuable fishing industry, and are central to Cajun culture.

New Orleans is not the only city calculating and constructing adaptations to climate change. City planners in Chicago expect their city's climate to be similar to that of Baton Rouge, Louisiana, by the end of the century. Conditions would be hotter and wetter, with more frequent snowstorms and extreme events including tornadoes. Workers have already replaced native Illinois species with southern species that thrive in higher temperatures. They are widening sidewalks to allow these trees to produce maximum shade cover. Light-reflecting pavement is being put down to replace the dark surfaces that contribute to the urban "heat island" effect.

Although these cities may be the "canaries in the coal mines" of climate change in the United States, fewer and fewer Americans seem to be concerned. In 2006, 79 percent of Americans believed the Earth was warming. In 2011, that figure was 59 percent. We might assume that Chinese and Indians would put concerns about economic growth ahead of climate change, but 2010 polls showed that 70 percent of both were willing to pay more for energy in order to address climate change. In the United States, just 38 percent were willing. 35

The Arctic National Wildlife Refuge

Alaska (from *Alyeska,* an Aleut word meaning "great land") (•**Figure 11.51**) makes up about 17 percent of the United States by area, but it is so rugged, cold, and remote that its total population was only 722,000 in 2011.

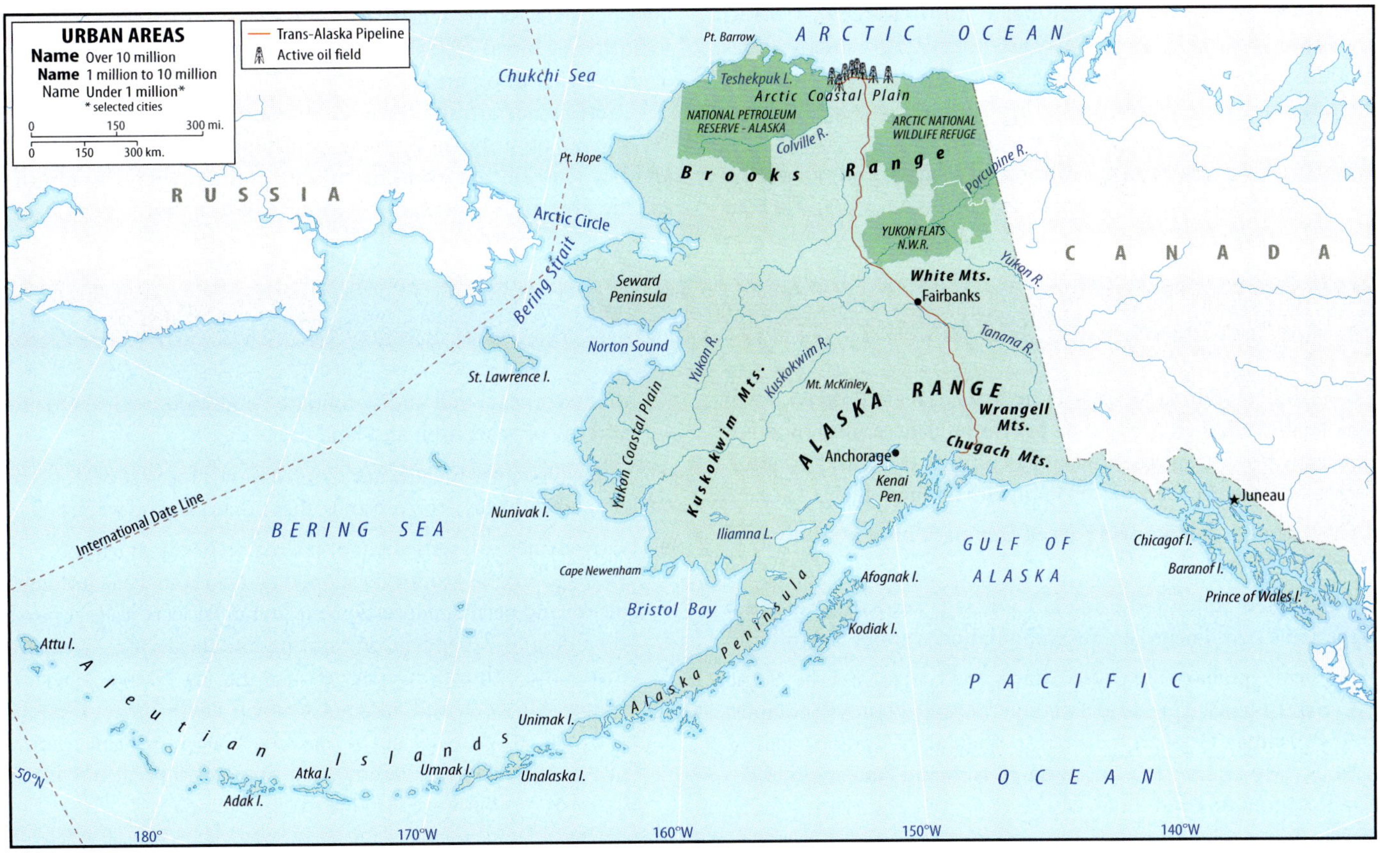

• **Figure 11.51** Alaska.

Alaska's Arctic Coastal Plain is an area of tundra along the Arctic Ocean north of the Brooks Range. Known as the North Slope, it holds the reserves of oil and natural gas that make Alaska the biggest fossil fuel–producing state in the United States. Alaska hopes to pump natural gas from the North Slope to the Lower 48. To date, the focus has been on oil, which supplies more than 80 percent of the state's revenues. Oil income has built a hefty surplus known as the **Permanent Fund**. Interest accrued in this account provides every resident about $1,100 in a typical year just to live in the Union's harsh but beautiful northernmost state.

The 1973 energy crisis delivered a profound shock to U.S. and
176 other industrialized economies. In its wake, there was a push to develop more domestic sources of energy, particularly oil. The most significant reserves were in the North Slope of Alaska, where Prudhoe Bay and nearby oilfields were established. Since 1977, their oil has been transported southward through the Trans-Alaska Pipeline, an extraordinary feat of engineering snaking 800 miles (1,300 km) across some of the most challenging environments on earth, all the way to the port of Valdez on the south coast. The pipeline route may be seen in Figure 11.51.

Originally holding about 10 billion barrels of proven reserves (just under 4% of Saudi Arabia's estimated reserves in 2011), Prudhoe Bay and nearby fields now produce about 11 percent of total U.S. domestic oil production. The fields' output peaked in 1988, when 20 percent of the country's oil was produced there, and has declined steadily since then. Attention has inevitably turned eastward and southward to the estimated 3 to 8 billion barrels beneath the **Arctic National Wildlife Refuge** (**ANWR**, often spoken of as "Anwar"). When the 12,500 square miles (32,000sq km) of the refuge were set aside as wilderness in 1980, the fate of the oil beneath this tundra ecosystem was left for Congress to decide at some later date.

ANWR's oil has been the subject of a bitter and polarizing debate ever since. On the one side are the oil interests, Alaskan politicians, and almost all Republicans in Congress, who have argued that the United States must develop these reserves to give the country more independence from Middle Eastern oil. On the other side is a coalition of environmentalists, many Native Americans, and congressional Democrats who argue that oil production in ANWR will do irreparable damage to the coastal plain's unique environment and the indigenous people who depend on it.

The main Native American group involved is the **Gwich'in**, a 7,000-strong Athabascan tribe that derives much of its livelihood from exploiting caribou (New World reindeer). The Gwich'in live mainly south of ANWR, but the 180,000-strong population of caribou known as the Porcupine Herd spends the winter in Gwich'in territory. The Gwich'in insist that oil drilling and related activity in ANWR will disrupt caribou calving, decimate the herd, and thus endanger the Gwich'in way of life.

The indigenous perspective on ANWR is not united, however; an Inuit group called the **Inupiat** supports oil production there, especially because of the jobs it would bring them. Environmentalists argue that habitats and native cultures would be despoiled to provide just six months of U.S. oil needs, whereas the oil industry says that 20 years of critical U.S. supply is in ANWR. Meanwhile, attention is turning to two other potential oil bonanza sites in Alaska: the North Slope's Teshekpuk Lake, a particularly oil-rich section of the **National Petroleum Reserve**, estimated to have more reserves than ANWR—and one that successive administrations have avoided developing because of its sensitive populations of geese and other wildlife—central Alaska's Yukon Flats National Wildlife Refuge, rich in oil and natural gas.

Our journey around the world has ended in northern Alaska. May your own journeys bring you happiness and knowledge. And when you are thinking about courses, majors, minors, certificates and careers, don't forget Geography!

summary

- The United States and Canada make up a region often known as Anglo America because of British influences. Both had numerous pre-European indigenous cultures and through immigration have been shaped by many others since 1492.
- Greenland is politically a part of Denmark but physically part of this region.
- The United States and Canada have large land areas of about the same size. Their combined population is about 335 million.
- Physical regions of the United States and Canada are the Arctic Coastal Plains, Gulf and Atlantic Coastal Plains, Piedmont, Appalachian Highlands, Interior Highlands, Interior Plains, Rocky Mountains, Intermountain Basins and Plateaus, and Pacific Mountains and Valleys.
- Climate types of the region are tundra, subarctic, humid continental with short and long summers, tropical savanna, Mediterranean, and undifferentiated highland, with associated patterns of vegetation and land use.
- Major resources include agricultural land, forests, coal, petroleum, natural gas, iron ore and other minerals, and waterpower.
- Native American settlement of the region began at least 12,500 years ago. Some groups developed civilizations based on agriculture and permanent settlement, but most were at least partly nomadic and dependent on hunting and gathering. Great linguistic and cultural diversity existed among Native American groups. All were well adapted to local environmental conditions, and all were disrupted profoundly by European contact. Native Americans are among the poorest populations in the United States but have fared better in Canada.
- Early European settlement displaced native populations in both countries, where waves of immigration continually changed

demographic and ethnic characteristics. The French are the most significant minority in Canada. Hispanics and African Americans are the largest minority groups in the United States. Asians are prominent minorities in both countries.

- The United States built a land empire through warfare and purchases, including the Louisiana Purchase (from France) of 1803; the acquisition of Florida from Spain in 1819; the addition of Texas, the Southwest, and California as spoils of war with Mexico in 1845 and 1848; the Gadsden Purchase from Mexico of a strip of Arizona and New Mexico in 1853; and Alaska, bought from Russia in 1867.
- Both the United States and Canada industrialized in the nineteenth and twentieth centuries. The economies have changed as manufacturing became less important while employment in service industries and information technology grew.
- Canada and the United States are among the world's richest countries because of many factors. Both have large endowments of natural assets; both are large in area and population; and both industrialized fairly early. Innovations have increased their efficiency and productivity. Peace and stability at home have stimulated investment and economic development. Both have strong internal unity and political stability.
- The United States has the world's largest appetite for fossil fuel. Its energy efficiency has improved since 1973, but energy consumption and the percentage of energy imported have both increased. Nuclear energy production has stalled because of environmental and health considerations. Production of ethanol from corn has surged in the United States, as has production of oil from tar sands in Canada.
- The two countries have had generally good economic and political relations. Each is a major consumer of exports from the other. About 90 percent of the Canadian population lives within 100 miles of the international border.
- The United States is the world's strongest military and economic power. Its influence in world affairs grew in the wake of the two world wars of the twentieth century. The terrorist attacks of September 2001 put U.S. foreign policy on a new course that led to a more aggressively interventionist foreign policy, but in the recent Libyan conflict the United States "led from behind."
- Atlantic Canada consists of the provinces of Newfoundland and Labrador, New Brunswick, Nova Scotia, and Prince Edward Island. Cod fishing was long the mainstay of its economy, but the Grand Banks have become so overfished that government regulation has restricted the harvest, causing much unemployment.
- The Ontario Peninsula is strongly British in origin and character, and Québec is strongly French, with 82 percent of the Québec population of French origin. Québec separatists have been a political presence for the past 40 years, and in 1995 the referendum for an independent Québec was defeated by less than 1 percent. But Québec has gained more provincial autonomy.
- As temperatures warm in the Arctic region, Canada is one of several nations capitalizing on new opportunities for trade and mineral exploitation. Several countries have claimed parts of the Arctic Ocean as sovereign territory, and some have conflicting claims to various areas, including the North Pole.
- Greenland is geologically a part of North America but is a self-governing property of Denmark. Greenland's foreign affairs are controlled by Denmark, which, along with the European Union, spends heavily on subsidies for the 50,000 inhabitants of the world's largest island. Scientists are watchful of the potential melting of the massive icecap that covers much of the island.
- From the fall-line cities, early industrialization, beginning with steel production, spread. The greatest center of major industry, finance, textile manufacturing, publishing, and many other enterprises was New York City.
- Major demographic changes have taken place in recent decades in the United States. Rural areas in the Midwest have been depopulated, whereas southern and western cities have exploded in population.
- Settlement in the West is concentrated in well-watered areas that are, in effect, oases. Topography has a strong role in the region's climates.
- California's cities and farms have huge water requirements that cannot be met from local resources. California has long exceeded its allotment of Colorado River waters, and states upstream in the Colorado Basin have won court actions to increase their share and reduce California's.
- The San Francisco Bay Area is home to Silicon Valley, the dominant software development center in the United States. Like Los Angeles and other Californian cities, the Bay Area is very multicultural.
- Alaska stretches north of the Arctic Circle and has only about 685,000 people. Recent decades have seen a pronounced warming that has affected natural and human environments. There is an ongoing dispute among Congress, the oil industry, environmentalists, and indigenous peoples about whether the oil reserves of northeastern Alaska's Arctic National Wildlife Refuge (ANWR) should be exploited.

Key Terms + Concepts

adaptive reuse (p. 429)
African Americans (p. 409)
Afro-Americans (p. 409)
Aleuts (p. 404)
Algic language family (p. 404)
American dream (p. 419)
amnesty (p. 396)
"anchor baby" (p. 396)
Arctic Bridge (p. 427)
Arctic National Wildlife Refuge (ANWR) (p. 440)
Asian Americans (p. 410)
bank (p. 426)
birth tourism (p. 395)
blacks (p. 409)
blizzard (p. 398)
borough (p. 432)
"Boswash" (p. 391)
British–North America Act (p. 423)
California Gold Rush (p. 409)
California Trail (p. 409)

Glossary

1948–1949 Arab-Israeli War A war fought between Israel and the Arab countries of Egypt, Iraq, Syria, Lebanon, and Transjordan after Israel declared its independence as a Jewish country. Hundreds of thousands of Arab Palestinians chose or were forced to flee Israel during this war, many settling in the Gaza Strip and the West Bank.

1973 Arab-Israeli War A war launched against Israel by Egypt and Syria in an attempt to reverse the Arab losses of the 1967 war. Egypt succeeded in winning back the eastern side of the Suez Canal.

Abkhaz-Adyghean language family A collection of languages spoken in the Caucasus region.

Aboriginal languages Indigenous languages to Australia that are not clearly related to any languages outside the continent.

Aborigines People indigenous to Australia that likely arrived in a single major migration 50,000 years ago. Europeans colonizing Australia in the eighteenth century slaughtered large numbers of Aborigines and drove them into marginal areas of the continent. Aborigines have long been Australia's most disadvantaged group and were not allowed to vote until 1962.

absolute (mathematical) location Determined by the intersection of lines, such as latitude and longitude, providing an exact point expressed in degrees, minutes, and seconds.

Adamawan A branch of Bantu languages spoken in West Africa.

adaptation Measures designed to cope with and reduce the unavoidable impacts of climate change.

adaptive reuse New uses for older buildings and stores, often accompanied by a shift from decline to steady renewal in an urban neighborhood.

Adena A mound-builder Native American culture that lived in what is now the northeastern United States.

Africa's First World War A widespread war involving nine African countries and 20 rebel movements between 1996 and 2001 that resulted in the deaths of more than 5 million people.

African Americans Americans of African ancestry.

African Development Bank (AfDB) A multi-government-owned bank, based in the Ivory Coast and dedicated to social and economic development (especially in loans for infrastructure) in African countries.

African Growth and Opportunity Act (AGOA) A bill passed by the U.S. Congress in 2000 that reduced or ended tariffs and quotas on many manufactured, mineral, and food items that could be imported from Africa.

African National Congress (ANC) The governing political party in South Africa since 1994. During the apartheid era, the ANC was dedicated to the nonviolent resolution of conflicts between ruling whites and the black majority. The ANC was banned until 1990. ANC leader Nelson Mandela became South Africa's first post-apartheid president in 1994.

African Standby Force A continental military force established by the African Union.

African Union (AU) An organization of most African countries formed to promote democracy, good governance, human rights, gender equity, and development.

Afrikaners Dutch-speaking descendants of Dutch, French Huguenot, and German settlers in South Africa. They are the largest white population in South Africa; also known as Boers.

Afro-Americans See *African Americans.*

Afro-Asiatic language family A language family that includes Semitic languages such as Arabic and Hebrew.

Age of Discovery (Age of Exploration) The three to four centuries of European exploration, colonization, and global resource exploitation and trading led largely by European mercantile powers, beginning with Columbus at the end of the fifteenth century and continuing into the nineteenth century.

Age of Exploration See *Age of Discovery.*

age structure diagram (population pyramid) The graphic representation of a country's population by gender and 5-year age increments.

Agricultural (Neolithic or **New Stone Age** or **Food-Producing) Revolution** The domestication of plants and animals that began about 10,000 years ago.

Agricultural Triangle See *Slavic core.*

AIDS (acquired immunodeficiency syndrome) The disease caused by HIV. AIDS is a global pandemic that has killed over 25 million people since the first case was reported in Africa in 1959.

Ainu An indigenous people and language to Japan.

Al-Aqsa *Intifada* A Palestinian uprising, beginning in 2000, triggered by Israeli Prime Minister Ariel Sharon's visit to the Temple Mount in Jerusalem.

Al-Asqa Mosque A sacred Islamic site on the Temple Mount in Jerusalem.

al-Haraam ash-Shariif See "Noble Sanctuary."

Al-Qa'ida A terrorist group formerly led by Osama bin Laden that was behind several attacks against U.S. interests, including the September 11, 2001, attacks.

Al-Qa'ida in Iraq An insurgent group arising in Iraq and affiliated with al-Qa'ida during the Iraq War, led by Abu Mus'ab al-Zarqawi until 2006.

Alawite A minority Shiite Muslim group.

albedo The amount of the sun's energy reflected by the ground. Less vegetation cover correlates with high albedo, and vice versa.

Aleuts An indigenous people of the Aleutian Islands in Alaska.

Algic language family A formerly widespread Native American language family across what is now Canada and the United States.

Allah God in Islam.

Alliance of Small Island States A group of 39 island countries arguing for significant reductions in global greenhouse gas emissions.

Altaic languages A language family that includes Turkish, Korean, and Japanese.

alternative development An overall effort to enhance the education, health, and livelihoods of Latin American villages in an attempt to reduce the amount of cropland devoted to coca and poppies.

American Dream The concept of upward mobility in American society, associated in part with material possessions such as homes and cars.

American Invasion Mexican term for the Mexican War.

Amerindians See *Native Americans.*

Amhara The politically dominant ethnic group of Ethiopia.

amnesty In the U.S. immigration debate, not persecuting illegal immigrants if they meet certain conditions and granting them permanent resident status.

Anasazi A Native American civilization living in what is now the southwestern United States up until about the year 1500. The

Anasazi gradually abandoned the pueblo dwellings they inhabited for unknown reasons.

ancestor veneration A widespread custom in eastern Asia, with shrines and offerings dedicated to preserving the memory of one's ancestors.

anchor baby A baby born in the United States to foreign parents.

animism The belief that natural processes and objects possess souls.

Attila Line See *Green Line* (Cyprus).

Antarctic Circle A line of longitude (65.56° S) commonly used to divide the southern middle latitudes to the north from the southern high latitudes to the south.

Antarctic Treaty An agreement signed by 45 countries forbidding any exploitation of Antarctica's natural resources until 2048.

Anthropocene Epoch A proposed new name for the present geological epoch, currently termed the Holocene.

antiretroviral (ARV) drugs A medication for treatment of HIV infection.

anti-Semitism Anti-Jewish sentiments and activities.

apartheid The Republic of South Africa's former official policy of "separate development of the races," designed to ensure the racial integrity and political supremacy of the white minority.

aquaculture The cultivation of aquatic organisms such as fish for food.

Arab A person of Semitic Arab ethnicity whose ancestral language is Arabic.

Arab-Israeli War of 1956 See *Suez Crisis*.

archipelago A chain or group of islands.

Arctic Bridge A potential shipping route between Churchill, Canada, and Murmansk, Russia, across the Arctic Ocean.

Arctic Circle A line of longitude (65.56° N) commonly used to divide the northern middle latitudes to the south from the northern high latitudes to the north.

Arctic National Wildlife Refuge (ANWR) A wildlife refuge in Alaska that contains 3 to 8 billion barrels of oil.

Armenian genocide The deaths of an estimated 1.5 million Armenians between 1915 and 1918. Turkey and Azerbaijan have not accepted responsibility for their roles in the genocide.

arms race Usually associated with the competition between the United States and the Soviet Union, it refers to rival and potential enemy powers increasing their military arsenals—each in an open-ended effort to stay ahead of the other.

Ashkenazi Jews Jewish immigrants to Israel from Europe after Israel's independence.

Asia-Pacific Economic Cooperation (APEC) group A 21-member organization (including China, Australia, and the United States) that has agreed to establish free and open trade and investment among member countries by 2020.

Asian Americans Americans of Asian ancestry.

Asian Century A term for the twenty-first century, anticipating the large role many Asian countries may play in changing the globe.

Asian Tigers Several Asian regions including South Korea, Taiwan, Hong Kong, and Singapore that had rapidly growing economies in the late twentieth century.

Association of Southeast Asian Nations (ASEAN) A regional economic association of ten Southeast Asian countries.

asylum People given permission to immigrate on the grounds that they would be harmed or persecuted in their country of origin.

atmosphere The layer of gases surrounding the Earth.

atoll Low islands made of coral and usually having an irregular ring shape around a lagoon.

Attila Line See *Green Line* (Cyprus).

Australia New Zealand United States (ANZUS) Security Alliance A defense and security agreement among the three countries, created in 1951. New Zealand has not been an active member since 1987 because of the country's anti-nuclear stance.

Austric language family A language family of Southeast Asia.

Austronesian subfamily The family of languages of most indigenous peoples of the Pacific Islands.

Austronesians Original settlers of many locations in Southeast Asia and the Pacific beginning after 5000 B.C.E. Many of today's Micronesians and Polynesians are their descendants.

autonomous regions A region of a country granted a measure of self-rule without being fully independent of that country.

autonomy Self-rule, generally with reference to Palestinians' rights to run their own civil (and some security) affairs in portions of the West Bank and Gaza Strip allocated to them in the 1993–2000 peace agreements.

ayatollah A revered and powerful cleric at the head of the religious establishment in Iran.

azimuthal A type of map projection in which the developable surface is based on a geometric plane.

Aztec-Tanoan language family A language family that included the Nahuatl language the Aztecs spoke.

Aztecs A Native American people that created an empire across central Mexico in the fifteenth century, which was destroyed by the Spanish in 1521.

baby bounty A financial incentive provided by the Italian government to families to have more than one child, to combat declining population.

Balkanization The fragmentation of a political area into many smaller independent units, as in former Yugoslavia in the Balkan Peninsula.

balloon effect A term referring to the fact that if a balloon is pressed in one place, the air shifts elsewhere. This term is often used to describe areas where drugs are produced and shipped through; when drug supply or shipment is eradicated in one area, it will tend to surge in previously marginal locations.

Baltic languages A language group of Europe that includes Latvian and Lithuanian.

banana war Reference to the incidence in the late 1990s, when the United States wanted to sell bananas grown in Latin America by U.S. corporations to countries of the European Union, but the EU members refused to buy them, as they wanted to protect their investments in banana farming in former European colonies in Africa and the Pacific.

bank In physical geography, an elevated area of the sea bottom.

Bantustans See *homelands*.

barrio A densely settled neighborhood in city space that is characteristically inhabited by migrants of Latino or Hispanic origin.

barter The exchange of goods or services in the absence of cash.

Basques A people, living primarily in Spain and France, who are ethnically and culturally unrelated to those of their host country majorities.

Bay of Pigs Invasion A failed attempt by the John F. Kennedy administration to overthrow Fidel Castro by military force.

***bazaar* (*suq*)** The central market of the traditional Middle Eastern city, characterized by twisting, close-set lanes and merchant stalls.

Benue-Congo A language family found in central and southern Africa.

Benelux The name used to collectively refer to Belgium, the Netherlands, and Luxembourg.

Berber An ethno-linguistic group of people living in northern Africa.

big bang The addition of ten primarily Eastern European countries to the European Union in 2004.

biodiversity See *biological diversity.*

biodiversity hot spots A ranked list of places scientists believe deserve immediate attention for flora and fauna study and conservation.

biogeography The distribution of living things on Earth.

biological diversity (biodiversity) The number of plant and animal species and the variety of genetic materials these organisms contain.

biomass The collective dried weight of organisms in an ecosystem.

biome A terrestrial ecosystem type categorized by a dominant type of natural vegetation.

biosphere The global ecological system, including all the relationships played out among the lithosphere, hydrosphere, and atmosphere. Also called the ecosphere.

birth dearth The declining population of many European countries from low birth rates.

birth rate The annual number of live births per 1,000 people in a population.

birth rate solution Intentionally lowering birth rates to prevent an overpopulation crisis.

birth tourism A practice in which an expectant mother from an LDC comes to the United States for the purpose of having her baby on U.S. soil, making the baby an automatic U.S. citizen.

black-earth belt An important area of crop and livestock production spanning parts of Russia, Ukraine, Moldova, and Kazakhstan. The main soils of this belt are mollisols.

blacks See *African Americans.*

blizzard A severe winter storm with heavy snowfall and high winds.

blood diamonds See *dirty diamonds.*

blowback Term associated with attacks on U.S. forces by former *mujahadiin* armed and trained by the United States to fight the Soviet occupation in Afghanistan.

Bolivarian Revolution Hugo Chávez's term for his policy agendas after taking control of Venezuela.

Bolshevik Revolution The seizure of the Russian government by Vladimir Lenin's faction of the Communist Party in 1917.

boreal forest A type of forest (also called taiga) primarily composed of coniferous trees that thrive in subarctic climates.

borough In New York City, one of the five administrative divisions: Manhattan, Brooklyn, Queens, Staten Island, or the Bronx.

Boswash A term for the megalopolis stretching from Boston to Washington, D.C.

bourgeoisie In Marxist doctrine, the capitalist class.

brain drain The exodus of educated or skilled persons from a poor to a rich country or from a poor to a rich region within a country.

branded diamonds Diamonds produced by the De Beers company, certified as coming from nonconflict areas.

break-of-bulk point A classic geographic term describing a point in transit when bulk goods must be removed from one mode of transport and installed on another. A trainload of grain carried to a port for transshipment on cargo boats or barges is a common example.

BRIC economy An acronym for Brazil, Russia, India, and China, four countries that may have the world's largest economies by 2050.

British–North America Act An act of the British Parliament in 1867 uniting several British North American colonies to create the independent country of Canada.

bubble economy See *economic bubble.*

Buddhism A religion based on the life and teachings of Siddhartha Gautama, known as the Buddha (or Enlightened One), who lived in the sixth century B.C.E. in India.

Burmese Way to Socialism Rigid state control of the economy in Myanmar.

California Gold Rush The discovery of gold in California, leading to a population boom in that state.

California Trail One of the major westward routes that American pioneers took in the middle of the nineteenth century, leading to what is now Sacramento, California.

Camp David Accords A 1979 agreement between Israel and Egypt that returned the Israeli-occupied Sinai Peninsula to Egypt in exchange for Egypt's recognition of Israel as a sovereign state.

Canada-U.S. Free Trade Agreement A free trade agreement between Canada and the United States in 1988 that was the forerunner to NAFTA.

Canadian Shield The ancient geological core of North America.

Cao Dai A faith originating in Vietnam that blends elements of Buddhism, Daoism, Confucianism, Judaism, Christianity, and Islam.

cap-and-trade system An international system designed to reduce carbon emissions by having richer countries that produce lots of greenhouse gases pay poorer, energy-inefficient countries to cut their emissions. This allows the richer countries to offset their carbon emissions with the cuts in emissions made by the poorer countries. The poorer country is obligated to use the income from cap-and-trade to invest in energy-efficient and nonpolluting technologies.

carbon dioxide (CO_2) A greenhouse gas, the human production of which is very likely the cause of global warming.

carbon sequestration The capture and long-term storage of carbon dioxide. Usually refers to natural processes, but there are also artificial means.

carbon sink An area such as forest, farmland, or ocean that sequesters a large amount of carbon dioxide.

cargo cults Belief systems that appeared in New Guinea and on various Pacific islands during World War II when "Stone Age" peoples were exposed for the first time to Western technology and consumer goods.

carrying capacity The size of a population of any organism that an ecosystem can support.

Carter Doctrine President Jimmy Carter's declaration, following the Soviet Union's invasion of Afghanistan in 1979, that the United States would use any means necessary to defend its vital interests in the Persian Gulf Region. The "vital interests" were interpreted to mean oil, and "any means necessary" interpreted to mean that the United States would go to war with the Soviet Union if oil supplies were threatened.

cartography The craft of designing and making maps, the basic language of geography. In recent years, this traditional manual art has been changed profoundly through the use of computers and geographic information systems (GIS) and through major improvements in machine capacity to produce detailed, colored map products.

cash (commercial) crops Crops produced generally for export.

caste The hierarchy in the Hindu religion that determines a person's social rank. It is established by birth and cannot be changed.

cataracts Areas along the Nile River where its valley narrows and rapids form.

Catholic Republicans One side of "The Troubles" in Northern Ireland, those who wished the British Army to leave the area and for Ireland to be unified.

Celtic Tiger A nickname for Ireland in the 1990s for the country's rapidly growing economy.

Central Arizona Project (CAP) A canal network bringing water from the Colorado River to the populous cities of Arizona.

Chadic A family of languages spoken mainly in northern Nigeria.

chaebols A selection of small companies in South Korea that were heavily funded by the government, eventually growing into giant conglomerates.

Chang Jiang Water Transfer Project A massive engineering project in China that will move water from China's wet south into its populated but drier north.

Chari-Nile A language family spoken in east-central Africa.

Charney Effect Observed by an atmospheric scientist named Charney, this states that the less plant cover there is on the ground, the higher the albedo—solar energy deflected back into the atmosphere—and therefore the lower the humidity and precipitation.

Chernobyl The site in the Ukraine where, in April 1986, the worst nuclear power plant accident in history occurred. It is thought that approximately 5,000 people died, and a zone with a 20-mile (32 km) radius is still virtually uninhabitable; 116,000 people were moved from the area, and cleanup continues to this day.

chernozem A Russian term meaning "black earth." It is a grassland soil that is exceptionally thick, productive, and durable.

Chibcha A small, pre-Columbian civilization in South America renowned for their exquisite gold works.

chokepoint A strategic narrow passageway on land or sea that may be closed off by force or threat of force.

choropleth maps Maps that are drawn to show the differing distribution of goods or geographic characteristics (including population) across a broad area. Such maps are good for generalizations but often mask significant local variations in the presence of the item being mapped.

Christianity One of the world's major faiths, predominant in Europe and many former European colonial areas. Christianity has several major denominations, including Roman Catholicism, Protestantism and Eastern Orthodox.

Chukotko-Kamchatkan A language family of far northeastern Siberia.

Church of the Holy Sepulcher Located in Jerusalem, this is the center of the Christian world, containing the places where tradition says Jesus Christ was crucified and buried.

City of Light A nickname for Paris, France.

civilization The complex culture of urban life.

climate The average weather conditions, including temperature, precipitation, and winds, of an area over an extended period of time.

Closer Economic Relations Agreement A free-trade pact between Australia and New Zealand.

coal A fossil fuel, used for creating heat and electricity, which has many industrial uses. The Industrial Revolution began in European locations situated near large deposits of coal.

coconut civilization A term characterizing the importance of coconuts in the economies and diets of several low-island Pacific countries.

coke Coal with most of its volatile constituents burned off.

Cold War The tense but generally peaceful political and military competition between the United States and the Soviet Union, and their respective allies, from the end of World War II until the collapse of the Soviet Union in 1991.

collective farm *(kolkhoz)* A large-scale farm in the former Soviet Union that usually incorporated several villages. Workers received shares of the income after the obligations of the collective had been met.

collectivization The process of forming collective farms in Communist countries.

collectivized agriculture See *collective farm.*

colonization The European pattern of establishing dependencies abroad to enhance economic development in the home country.

Colorado River Compact A 1922 agreement between seven western U.S. states that the Colorado River flows through, which determined how much river water each state may take.

Coloreds A South African term for persons of mixed racial ancestry.

Columbian Exchange The transfer of plants and animals between Europe and the Americas after Christopher Columbus's voyages.

comanagement A joint system of national park management in Australia between the Aborigines and the Australian government.

command economy A centrally planned economy typical of the Soviet Union and its Communist allies, in which the government rather than free enterprise determines the production, distribution, and sale of economic goods and services.

commercial crops See *cash crops.*

Common Market An earlier name given to the (current) 25 countries that make up the European Union. In 1957, an initial six countries combined to form the European Economic Community (EEC), and this supranational community has grown to have considerable economic and political importance in Europe. See *European Union.*

Commonwealth of Independent States (CIS) A loose political and economic organization of former Soviet republics (except for Estonia, Latvia, and Lithuania) formed in 1991.

Commonwealth of Nations A voluntary association of 54 countries around the world, most of them former British colonies.

communal violence Term used in India to refer to sectarian violence.

communism A type of national economy characterized by abolition of trade unions, abolition of private ownership in fields such as manufacturing and commerce, varying degrees of state ownership of agriculture, centrally planned and directed economies, and one-party dictatorial government.

Compact of Free Association A 1986 agreement between the United States and the Marshall Islands and the Federated States of Micronesia granting independence and subsidies to the two Pacific countries in exchange for renting their military sovereignty to the United States.

Comprehensive Test Ban Treaty A treaty signed by 149 nations in 1996 that prohibits all nuclear tests.

compromise projections A map projection that does not preserve any one metric but are designed to distort all metrics roughly equally to achieve a map that "looks right."

conflict diamonds See *dirty diamonds.*

conformal projections A map projection that preserves shape.

Confucianism A Chinese sociopolitical philosophy based upon the writings of Confucius, either on its own or blended with other religions.

conic A type of map projection in which the developable surface is based on a cone.

coniferous forest Needleleaf evergreen trees; most bear seed cones.

consumption overpopulation The concept that a few persons, each using a large quantity of natural resources from ecosystems across the world, add up to too many people for the environment to support.

continental drift See *plate tectonics.*

continental islands Once attached to nearby continents, these are the islands north and northeast of Australia, including New Guinea.

contingent sovereignty An agreement in Myanmar between the government and several minority ethnic groups, granting them more civil rights and economic opportunities.

Convention on the Law of the Sea A 1970s United Nations treaty permitting a sovereign power to have greater access to surrounding marine resources.

coordinate system A grid consisting of horizontal and vertical lines used to establish absolute location. On the Earth's surface, latitude and longitude lines form the coordinate system.

Copts A Christian sect in Egypt.

core location An area of the world with significant importance due to a central location relative to others.

Cornucopians People who argue that humans have always been able to conquer food shortages in the past, thanks to technological innovations, and therefore will be able to in the future as well. Also called *technocentrists*.

Corporate Average Fuel Economy (CAFÉ) standard A 1975 U.S. law stating that passenger cars produced by all auto companies must achieve an average of 27.5 miles per gallon.

Cossacks Peasant soldiers of the Russian steppes who were among the first Russian settlers in Siberia in the sixteenth century.

credit crisis A financial crisis originating in the United States in 2008, after the housing bubble burst, that quickly spread worldwide, causing global selloffs in stock exchanges and a widespread economic recession.

Creole Tongues that developed among black slaves and indentured servants in the Americas, consisting mainly of vocabularies from the European colonizers' languages.

Creole languages See *Creole*.

crop substitution Efforts to get Latin American farmers to switch from growing illegal coca or poppies to legal flowers or coffee.

Crusades A series of European Christian military campaigns between the eleventh and fourteenth centuries aimed at recapturing Jerusalem and the rest of the Holy Land from the Muslims.

Cuban Americans Cubans who fled Castro's regime for the United States, and their descendants.

Cuban Missile Crisis The debacle between the United States and the Union of Soviet Socialist Republics in 1962 that put both countries on the brink of nuclear war. Soviet rocket launch facilities were being built in Cuba, presumably for possible nuclear sites on the United States. President Kennedy demanded that the Soviets withdraw the missiles; they did.

cultural landscape The landscape modified by human transformation, thereby reflecting the cultural patterns of the resident culture.

culture The values, beliefs, aspirations, modes of behavior, social institutions, knowledge, and skills that are transmitted and learned within a group of people.

culture hearth An area where innovations develop, with subsequent diffusion to other areas.

Cushitic A family of languages spoken in Ethiopia and Somalia.

cylindrical A type of map projection in which the developable surface is based on a cylinder.

Dakota A Native American people living on the Great Plains.

Dalai Lama The spiritual and political leader of Tibetan Buddhism (Lamaism).

Dalits The bottom of the social hierarchy in India, not part of the caste system. Once known as "untouchables."

Daoism (Taoism) A Chinese school of thought, second only to Confucianism, originating in the *Tao-te Ching* in the sixth century B.C.E. Daoism philosophy is commonly blended with elements of Buddhism.

Dayton Accord A peace agreement creating a joint multiethnic, democratic national government in Bosnia and Herzegovina, responsible for foreign and economic policies, while also recognizing a second tier of government charged with overseeing internal functions composed of the Federation of Bosnia and Herzegovina and the Republika Srpska.

death rate The annual number of deaths per 1,000 people in a population.

death rate solution A catastrophic increase in death rates, seen as "nature's solution" to overpopulation.

debt-for-nature swap An arrangement in which a certain portion of international debt is forgiven in return for the borrower's pledge to invest that amount in nature conservation.

decertification A country's loss of foreign aid from the United States if that country is deemed to be uncooperative in the war on drugs.

decommissioning (of dams) A major modification or removal of a dam primarily for environmental reasons.

deforestation The removal of trees by people or their livestock.

degrees Unit of measurement for longitude and latitude. There are 180 degrees of latitude (90 degrees north of the equator, and 90 degrees south) and 360 degrees of longitude (180 degrees east of the Prime Meridian, 180 degrees west).

deindustrialization An economy shifting away from industries requiring large amounts of energy and labor toward an economy based on services and the production of high-tech goods.

delta The usually triangular-shaped alluvial area at the mouth of a river.

demilitarized zone (DMZ) The border between North and South Korea.

demographic transition A model describing population change within a country. The country initially has a high birth rate, a high death rate, and a low rate of natural increase; then moves through a middle stage of high birth rate, low death rate, and high rate of natural increase; and ultimately reaches a third stage of low birth rate, low or medium death rate, and low or negative rate of population increase.

demography The study of population.

denying the countryside to the enemy The use of explosives and defoliants such as Agent Orange by U.S. troops to create enormous damage to the farms and forests of Vietnam.

dependency theory A theory arguing that the world's more developed countries continue to prosper by dominating their former colonies, the now independent less developed countries.

desert An area too dry to support a continuous cover of trees or grass. A desert generally receives less than 10 inches (25 cm) of precipitation per year.

desert shrub vegetation Scant, bushy plant life occurring in deserts of the middle and low latitudes where there is not enough rain for trees or grasslands. The plants are generally xerophytic.

desertification The expansion of a desert brought about by changing environmental conditions or unwise human use.

development A process of improvement in the material conditions of people often linked to the diffusion of knowledge and technology.

devolution The process by which a sovereign country releases or loses more political and economic control to its constituent elements, such as states and provinces.

Diaspora The scattering of the Jews outside Palestine beginning in the Roman Era.

digital divide The divide between the handful of countries that are the technology innovators and users and the majority of nations that have little ability to create, purchase, or use new technologies.

direct rule The policy of administering Northern Ireland directly by the central government of the United Kingdom.

dirty diamonds Rough diamonds used by rebel movements or their allies to finance conflict aimed at undermining legitimate governments.

distance-decay relationship The geographical concept that the interactions between places (and the people that occupy them) decline as the distance between them grows.

divine rule by clerics Iranian principle of the Guardian Theologian, an infallible supreme leader whose authority is based on Shiite interpretations of Islam.

Dome of the Rock A Muslim shrine built on the Temple Mount in Jerusalem, on the site where two Jewish temples had previously stood.

domestication The controlled breeding and cultivation of plants and animals.

geographic information systems (GIS) The growing field of computer-assisted geographic analysis and graphic representation of spatial data. It is based on superimposing various data layers that may include everything from soils to hydrology, to transportation networks, to elevation. Computer software and hardware are steadily improving, enabling GIS to produce ever more detailed and exact output.

geography The study of the spatial order and associations of things. Also defined as the study of places, the study of relationships between people and environment, and the study of spatial organization.

geologic hot spot A small area of Earth's mantle where molten magma is relatively close to the crust. Hot spots are associated with island chains and thermal features.

geomancy The arrangement of homes and villages for maximum harmony with natural and spiritual environments.

geospatial A term pertaining to the geographic location and characteristics of natural or constructed features and boundaries on, above, or below the Earth's surface, especially referring to data that is geographic and spatial in nature.

Germanic languages A language family of northern and central Europe including German, Dutch, Swedish, and English.

ghetto Impoverished areas of central cities where there are many problems, including high crime and lack of jobs.

ghurba The worldwide diaspora of Palestinians.

gigantomania A Soviet preoccupation with fulfilling huge quotas or implementing grandiose schemes.

glacial deposition In the process of continental and valley glaciation, the deposition of moraines that become lateral or terminal in the act of glacial retreat. This same process also leads to glacial scouring as moving ice picks up loose rock and reshapes the landscape as the glacier moves forward or retreats.

glacial scouring See *glacial deposition*.

glaciation A geologic and climatologic process in which great ice sheets form in the Arctic and Antarctic regions and advance toward the equator.

glasnost Policies introduced by Russian leader Mikhail Gorbachev in the 1980s to allow a more democratic political system, more freedom of expression, and a more market-oriented economy.

globalization The spread of free trade, free markets, investments, and ideas across borders and the political and cultural adjustments that accompany this diffusion.

Global Warming Solutions Act of 2006 A California law requiring the state's greenhouse gas emissions to be cut to 1990 levels by 2020.

Gold Rush of 1849 See *California Gold Rush*.

Golden Crescent A term for portions of Iran, Afghanistan, and Pakistan, referring to the large amounts of the world's opium supply being produced there.

Golden Horde Russian term for the Kazan Tatars in the thirteenth century.

Golden Triangle A region in the hinterlands of Myanmar, Laos, and Thailand that is the center of Southeast Asia's illegal drug production.

Good Friday Agreement An agreement between the Unionists and the IRA in 1998 to replace direct rule of Northern Ireland with the power-sharing Northern Ireland Assembly.

Gosplan (Committee for State Planning) The agency in Moscow that formulated the Soviet Union's national economic plans.

GNI PPP Per capita gross national income purchasing power parity (per capita GNI PPP), a method of comparing the real value of output between countries' economies by combining gross national income (GNI) and purchasing power parity (PPP).

graduated symbol map A type of map that uses simple symbols such as circles or bar graphs scaled proportionally to the quantity of the attribute being mapped.

Great Depression The worst economic downturn in U.S. history, beginning in 1929 and lasting through most of the 1930s.

Great Game A rivalry between Russia and the United Kingdom in the nineteenth and early twentieth centuries over strategic influence in Afghanistan and other areas of Central Asia.

Great Migration The migration of about 6 million blacks from the rural South to the cities of the Northeast and Midwest between 1915 and 1970.

Great Rift Valley As the result of tectonic processes, a broad, steep-walled trough extending from the Zambezi Valley in southern Africa northward to the Red Sea and the valley of the Jordan River in southwestern Asia.

Great Volga Scheme The transformation of the Volga River in Russia in the twentieth century to make the river more navigable and to supply hydroelectric power and water for irrigation.

Great Wall of Louisiana Several projects in the planning stage to minimize flooding in Louisiana, including rebuilding the eroded barrier islands in the Gulf of Mexico, restoring the wetlands along the coast, and building a semipermeable levee system.

Greek The ethnolinguistic group inhabiting Greece.

Green Line (Cyprus) A buffer zone dividing Cyprus between the Greek Cypriots and the Turkish Cypriots.

Green Line (Israel) A term referring to Israel's border with the West Bank as of 1967.

Green Revolution The introduction and transfer of high-yielding seeds, mechanization, irrigation, and massive application of chemical fertilizers to areas where traditional agriculture has been practiced.

greenhouse effect The observation that increased concentrations of carbon dioxide and other gases in Earth's atmosphere cause a warmer atmosphere.

greenhouse gases Gases such as carbon dioxide and methane that trap heat from the Sun in the atmosphere, increasing the Earth's temperature.

Greenwich Meridien See *prime meridien*.

gross domestic product (GDP) The value of goods and services produced in a country in a given year. It does not include net income earned outside the country. The value is normally given in current prices for the stated year.

gross national income (GNI) A measure of a country's wealth, counting gross domestic production plus income from abroad such as rents, profits and labor.

gross national product (GNP) The value of goods and services produced internally in a given country during a stated year plus the value resulting from transactions abroad. The value is normally expressed in current prices of the stated year. Such data must be used with caution in regard to developing countries because of the broad variance in patterns of data collection and the fact that many people consume a large share of what they produce.

Group of Eight (G-8) The world's eight most economically powerful and politically influential countries: Canada, France, Germany, Italy, Japan, Russia, the United Kingdom, and the United States.

GUAM A regional grouping of Georgia, Ukraine, Azerbaijan, and Moldova, promoting economic integration, democratic reforms, and orientation toward Europe and away from Russia.

guano Seabird excrement that is also found in phosphate deposits.

guest workers Migrants (mainly young and male) from less developed countries, who are employed, sometimes illegally, in more developed countries.

Gulf Stream The strong ocean current originating in the tropical Atlantic Ocean that skirts the eastern shore of the United States, curves eastward, and reaches Europe as a part of the broader current called the North Atlantic Drift.

Gulf War The 1990–1991 confrontation between a large military coalition, spearheaded by the United States, and the Iraqi forces that had invaded Kuwait. The allies successfully drove Iraqi forces from Kuwait. Earlier, the Gulf War had referred to the 1980–1988 war between Iraq and Iran.

Gwich'in A Native American people in Alaska opposed to oil development in ANWR.

Gypsies See *Roma.*

hacienda A Spanish term for large rural estates owned by the aristocracy in Latin America.

hajj Pilgrimage to Mecca.

Hamas Palestinian Islamist party long dedicated to the destruction of Israel and opposed to the Palestinian-Israeli peace process. A major rival to the moderate al-Fatah party within the Palestine Liberation Organization.

Han Chinese The original Chinese peoples who settled in North China on the margins of the Yellow River (Huang He) and who were central to the development of Chinese culture. Han Chinese make up more than 90 percent of China's population.

Hanification Efforts by the Chinese government to increase ethnic Han presence in China's outlying areas, particularly Tibet.

Hapa The mixed-race offspring between Asian immigrants to the United States and Canada and other ethnic groups.

Hawaiian language The native Hawaiian language, a member of the Malayo-Polynesian subfamily of the Austronesian language family.

Hebrews A term for Jews in Biblical times.

hemisphere Half of the Earth's surface.

Hermit Kingdom A term for North Korea.

hero projects The construction of dams, railways, factories, and the like, encouraged by the Soviet government.

high islands Generally the result of volcanic eruptions, these are the higher and more agriculturally productive and densely populated islands of the Pacific World.

high latitudes See *latitude.*

Hinduism A regionally varied, complex belief system that lacks a definite creed or theology but includes a belief in reincarnation, the reverence of many living things, an unlimited pantheon of deities, and an infinite range of types of permissible worship.

hinterland The region of a country that lies away from the capital and largest cities and is most often rural or even unsettled; it is often seen in the minds of economic planners as an area with development potential.

Hizbullah Iranian-inspired and supported Shiite Lebanese political party holding many seats in Lebanon's Parliament. Broadly regarded as a terrorist organization, and opposed to the existence of Israel.

Hmong-Mien subfamily A language family of Southeast Asia and southern China.

Hokan-Siouan language family A Native American language family in North America.

Holocaust Nazi Germany's attempted extermination of Jews, Roma (Gypsies), homosexuals, and other minorities during World War II.

homelands Ten former territorial units in South Africa reserved for native Africans (blacks). The homelands had elected African governments, and some were designated as "independent" republics, although they were not recognized outside South Africa. Formerly known as *Bantustans*, they were abolished in 1994. Also known as *native reserves* and *national states.*

Homestead Act An 1862 U.S. law that opened up much of the interior of the country for settlement. A pioneer family could claim up to 160 acres of farming land for a $10 fee.

Hopewell A mound-builder Native American civilization that lived in what is now the U.S. Midwest.

homogenous region See *region.*

Human Development Index (HDI) A United Nations–devised ranked index of countries' development that evaluates quality of life issues (such as gender equality, literacy, and human rights) in addition to economic performance.

human–environment interaction Study of the ways human beings use and change the natural environment.

human geography One of the two major branches of geography, primarily studying the world's peoples, their spatial relationships, and their usage of the natural environment.

human immunodeficiency virus (HIV) The virus that causes AIDS.

human trafficking Movement of people against their will or intention, largely for the sex trade and manual labor.

humid continental A climate with cold winters, warm to hot summers, and sufficient rainfall for agriculture, with the greater part of the precipitation in the summer half-year.

humid subtropical A climate that generally occupies the southeastern margins of continents, with hot summers, mild to cool winters, and ample precipitation for agriculture.

humus Decomposed organic soil material. Grasslands characteristically provide more humus than forests do.

hunting and gathering A mode of livelihood, based on collecting wild plants and hunting wild animals, generally practiced by preagricultural peoples. Also known as *foraging.*

hurricane Large tropical storms that form over the warm waters of the Atlantic and Pacific Oceans.

Hutu (Bahutu) The majority population of Rwanda and Burundi. Ethnically, culturally, and linguistically the same as the minority Tutsi; the differences between the groups are socioeconomic, with the Tutsi a ruling class that dominated the Hutu majority.

hydraulic fracturing (fracking) A method of extracting natural gas by pumping water, sand and chemicals underground at high pressure, creating fissures in the rock that allow the natural gas to rise to the surface.

hydrologic cycle The process by which the sun evaporates seawater into water vapor that is later released as freshwater precipitation.

hydropolitics Political leverage and control over water.

hydrosphere All of the world's water features.

Ibo (Igbo) A major ethnic group in southern Nigeria.

Ice Age A period of widespread glaciation from about 2 million to 10,000 years ago.

ice cap A climate and biome type characterized by permanent ice cover on the ground, no vegetation (except where limited melting occurs), and a severely long, cold winter. Summers are short and cool.

Ijaw A major ethnic group in southern Nigeria.

illegal alien A migrant moving to another country illegally.

immigrant A person moving into a place.

Inca An indigenous people of South America who created an empire running from modern-day Ecuador to Chile. The Inca Empire was crushed by the Spanish in 1532.

Indians See *Native Americans.*

Indo-European languages A major language group made up of smaller language groups such as Germanic, Romance, and Slavic.

Indo-Fijians Residents of Fiji who are descended from indentured laborers from India, brought to the islands by the British in the early 1900s.

Indo-Iranian subfamily A language family that contains many of the languages spoken in India, Pakistan, Afghanistan, and Iran.

Indus Waters Treaty A 1960 treaty between India and Pakistan, describing how much water from the Indus River each country can use.

Industrial Revolution A period beginning in mid-eighteenth-century Britain that saw rapid advances in technology and the use of inanimate power.

informal settlements Settlements by people who do not hold legal claims to the land their homes are built upon.

information technology (IT) The Internet, wireless telephones, fiber optics, and other technologies characteristic of more developed countries. Generally seen as beneficial for a country's economic prospects, IT is also spreading in the less developed countries.

Inkatha Freedom Party (IFP) South African political party with mainly Zulu membership. Founded by Zulu chief Mangosuthu Buthelezi, the IFP is a rival of African National Congress (ANC), long led by Nelson Mandela.

insolation Heat from the sun.

insourcing Keeping certain jobs within a country instead of offshoring those jobs to another country.

intensive land use A livelihood such as farming, requiring use of small land areas.

intensive subsistence agriculture Agricultural practices built around the growing of cereals (especially rice) that require a steady output of arduous manual labor.

Intergovernmental Panel on Climate Change (IPCC) A United Nations panel, composed of 2,500 atmospheric scientists from 130 countries, that has concluded that the evidence for global warming is "unequivocal."

internally displaced persons (IDPs) People dislodged and impoverished by strife in their home countries but who have little prospect of emigrating from that country.

International Atomic Energy Agency (IAEA) A United Nations agency that carries out inspections of nuclear facilities.

International Criminal Court A United Nations forum for crimes against humanity, located in the Netherlands.

International Date Line A line roughly concurrent with the 180-degree line of longitude (with several deviations for political and practical reasons) where the beginning of one day and the end of another day meet. The date west of the International Date Line is one day ahead of the date east of the line.

interstate highway system A massive infrastructure development in the United States that started in the 1950s and serves as the primary network for trucking cargo across the country.

Intifada Palestinian uprising against Israeli occupation.

Inuit A people living in Greenland and far northern Canada and Alaska, sometimes known as Eskimos to others.

Inupiat A Native American people in favor of oil production in ANWR.

Iranian A subgroup of the Indo-European language family. Persian and Tajik are Iranian languages.

Irish Republican Army (IRA) A terrorist group based in Northern Ireland that launched many attacks in the United Kingdom, which were designed to drive out the British army from Northern Ireland.

Iron Curtain A term used by Winston Churchill to describe the dividing line in Europe between capitalist democracies to the west and Communist authoritarian regimes to the east.

irredentism Movement by an ethnic group in one country to revive or reinforce kindred ethnicity in another country—often in an effort to promote succession there.

irrigation The artificial placement of water to produce crops, generally in arid locations.

isarithmic map A type of map that uses lines or bands of color to join the points of equal value of some phenomenon across a mapped area.

Islam A monotheistic faith built upon the foundations of Judaism and Christianity. One of the world's major religions, it originated in the seventh century in what is now Saudi Arabia.

Islamic law (sharia) Islamic interpretations of divine intention in legal matters, based on the Qur'an, the Hadith, and Islamic scholars' opinions in matters of jurisprudence.

Islamic Movement of Uzbekistan (IMU) A group dedicated to the overflow of Uzbekistan's president and the creation of an Islamic state in the Fergana Valley of Central Asia.

Islamists Islamic militants.

Islamophobia A fear of Islam and the Muslims that practice it.

isolationism The tendency of a country to stay out of the conflicts of other countries.

isostatic rebound The slow uplift of land that had been pressed down by the great weight of glaciers during the Ice Age.

Israelites A term for Jews in Biblical times.

Jainism A small but influential religion of India renowned for its respect for geographic features and animal life.

Jain A practitioner of Jainism.

Jasmine Revolution The uprising in Tunisia against that country's leader in early 2011.

Jemaah Islamiah (JI, or the Islamic Group) A militant Islamist group in Southeast Asia.

Jewish settlements Israeli citizens settling in the West Bank to strengthen Israel's claims to the area.

Jews Practitioners of Judaism.

jihad A term that for Muslims can have a range of meanings, from expressing one's commitment to a personal conviction to all-out war against nonbelievers.

juche North Korean principle of self-reliance.

Judaism The first significant monotheistic faith, originating at least 4,000 years ago. It does not have a fixed creed or doctrine, and it is not a proselytizing religion.

Judea and Samaria Biblical kingdoms that are now part of the West Bank.

Ka'aba A Muslim shrine located in Mecca.

Kanaka Maoli The term native Hawaiians use for themselves.

karma An important Buddhist concept of a person's acts and the consequences of those acts.

karoshi Japanese term for death by overwork.

Kartvelian (South Caucasian) language A collection of languages spoken in the Caucasus region.

keystone species A species that affects many other organisms in an ecosystem.

Khalistan A desired homeland for India's Sikhs.

Khoisan language family The languages of the San people in southern Africa. These languages, nearly extinct, are among the world's oldest.

Kimberley Process A United Nations–backed certification plan for ensuring that only legally mined rough diamonds, untainted by violence, reach the market.

kleptocracy An economic and political system based on crime.

knowledge economy An economy based on innovation and services.

Kurdistan Workers Party (PKK) The main Kurdish resistance to Turkish rule.

Kurds The largest minority group in Turkey and the largest non-Arab minority in Iraq. Kurds are the world's largest ethnic group without a country.

Kyoto Protocol A treaty on climate change signed by 160 countries in Kyoto, Japan, in 1997 and put into force in 2005. It requires MDCs to reduce their greenhouse gas emissions by more than 5 percent below their 1990 levels by the year 2012.

Labor Party The liberal political party of Israel.

Lakota Native American people living on the Great Plains.

land empire The establishment of colonies in a continental hinterland as opposed to overseas.

Land for Peace Part of the peace process between Israel and Arabs, with Israel swapping Arab lands it occupied in 1967 in exchange for peace with its neighbors.

land hemisphere See *hemisphere*.

land reform In Latin America, the reallocation of land owned by the wealthy to small farmers.

landscape A portion of the earth's land surface. Geographers are interested in the transformation of natural landscapes into cultural landscapes.

landscape perspective The understanding of the changes people have made on the natural environment over time.

language death When the last speaker of a certain language dies

lapse rate The decrease in temperature over a corresponding increase in elevation. The lapse rate is (on average) about 3.6°F (2.0°C) for each increase of 1,000 feet (305 m) in elevation.

large-scale map A map constructed to show considerable detail in a small area.

Laskar Jihad A militant Islamist group in Indonesia.

laterite A material found in tropical regions with highly leached soils; it is composed mostly of iron and aluminum oxides that harden when exposed and make cultivation difficult.

lateritic soils A soil characteristic of many tropical areas. Excessive exposure to the sun and repeated wetting and drying turns the soil into a bricklike substance nearly impossible to cultivate.

***latifundia* (sing. *latifundio*)** Large agricultural Latin American estates with strong commercial orientations.

Latino An American of Hispanic background.

latitude A measurement that denotes position with respect to the equator and the poles. Latitude is measured in degrees, minutes, and seconds, which are described as parallels. Places near the equator are said to be in low latitudes; places near the poles are in high latitudes. The Tropic of Cancer and the Tropic of Capricorn, at 23.5°N and 23.5°S, respectively, and the Arctic Circle and Antarctic Circle, at 66.5°N and 66.5°S, respectively, form the most commonly recognized boundaries of the low and high latitudes. Places occupying an intermediate position with respect to the poles and the equator are said to be in the middle latitudes. There are no universally accepted definitions for the boundaries of the high, middle, and low latitudes.

Law of Return An Israeli law permitting all Jews living in Israel to have Israeli citizenship.

less developed countries (LDCs) The world's poorer countries.

levee A raised riverbank designed to prevent flooding.

liberal autonomy See *autonomous regions*.

Liberal Democratic Party (LDP) The dominant political party of Japan since 1952.

lifeboat ethics Ecologist Garrett Hardin's argument that, for ecological reasons, rich countries should not assist poor countries.

Liga Veneta An Italian organization wishing to create an autonomous region centered on the city of Venice.

Likud Party The conservative political party in Israel.

lithosphere The rocky portion of the Earth's surface.

location A concept central to all geographic analysis. Where something is relates to all manner of influences, from climate to migration routes, and is a crucial component in trying to understand patterns of historic and economic development.

lock Devices that allow ships to move from one water level to another in a canal.

loess A fine-grained material that has been picked up, transported, and deposited in its present location by wind; it forms an unusually productive soil.

longitude A measurement that denotes a position east or west of the prime meridian (Greenwich, England). Longitude is measured in degrees, minutes, and seconds, and meridians of longitude extend from pole to pole and intersect parallels of latitude.

loose nukes Stolen radioactive materials.

Louisiana Purchase A large purchase of land by the United States from France in 1803 that significantly added to the size of the United States.

low islands Made of coral, these are the generally flat, drier, less agriculturally productive, and less densely populated islands of the Pacific World.

low latitudes See *latitude*.

Ma'adan (Marsh Arabs) People living on marshlands in Iraq, creating a unique culture found nowhere else in the Middle East. Most of their habitat was destroyed by Saddam Hussein after the first Gulf War.

Mahayana Buddhism One of the two branches of Buddhism, originating in India and spreading northward to Tibet and China.

mainstream European American culture A culture based on the English language and British educational principles brought by British colonists combined with core American values that developed on U.S. soil.

major cluster of continuous settlement An area where all habitations lie no more than 3 miles (5 km) from other habitations in at least six different directions, and roads or railroads lie no more than 10–20 miles (16–32 km) away in at least three directions.

Malayo-Polynesian language subfamilies A subfamily of the Austronesian language family.

Mandé A language family of west Africa.

Malthusian scenario The model forecasting that human population growth will outpace growth in food and other resources, with a resulting population die-off.

Manifest Destiny The opening and settlement of new frontiers across the United States from the Atlantic to the Pacific Oceans.

map A representation of various phenomena over all or a part of the Earth, either on a globe or a flat surface.

map projection A way to minimize distortion in one or more properties of a map (direction, distance, shape, or area).

maquiladoras Operations dedicated to the assembly of manufactured goods, generally in Mexico and Central America, from components initially produced in the United States or other places. With the enactment of NAFTA in 1994, there was a massive expansion of *maquiladora* operations.

marginalization A process by which poor subsistence farmers are pushed onto fragile, inferior, or marginal lands that cannot support crops for long and that are degraded by cultivation.

Maria Lionza A syncretic faith in Venezuela that blends Catholic, West African, and indigenous beliefs.

marine west coast A climate occupying the western sides of continents in the higher middle latitudes; it is greatly moderated by the effects of ocean currents that are warm in winter and cool in summer relative to the land.

Maronites A Christian sect in Lebanon.

Marshall Plan The plan designed largely by the United States, after the conclusion of World War II, by which U.S. aid was focused on the rebuilding of the very Germany that had been its enemy in the war just concluded. Secretary of State George Marshall (who had been Chief of Staff of U.S. Army from 1939–1945) was central to the plan's design and implementation.

Masdar A carbon-neutral city under construction in the United Arab Emirates.

mathematical location See *absolute location.*

Maya A Native American ethnic group inhabiting southern Mexico and northern Central America.

Mayan language family A language family of southern Mexico and Guatemala.

medina An urban pattern typical of the Middle Eastern city before the twentieth century.

Mediterranean (dry-summer subtropical) A climate that occupies an intermediate location between a marine west coast climate on the poleward side and a steppe or desert climate on the equatorward side. During the high-sun period, it is rainless; in the low-sun period, it receives precipitation of cyclonic or orographic origin.

Mediterranean scrub forest The xeroyphytic vegetation typical of hot, dry summer, Mediterranean climate regions. Local names for this vegetation type include *maquis* and *chaparral.*

megacity A city with more than 10 million inhabitants.

megadiversity country A country containing a large percentage of the world's animal and plant species.

megalanguage One of the world's most-spoken languages, including Chinese, English, Spanish, Hindi, and Arabic.

megalopolis A chain of metropolitan areas with a population over 10 million.

Mekong River Commission A coalition of Vietnam, Laos, Cambodia, and Thailand, to determine rules to ensure water quality and quantity of the Mekong River.

Melanesia (black islands) A group of relatively large islands in the Pacific Ocean bordering Australia.

mental map A term used to define the geographies encompassed in every individual's mind as a series of locations, access routes, physical and cultural characteristics of places, and often a general sense of the good or bad of locales.

mercantile colonialism The historical pattern by which Europeans extracted primary products from colonies abroad, particularly in the tropics.

Mercator Projection A cylindrical, conformal map projection intended to be used for navigation developed by Gerardus Mercator in 1569.

Mercosur Acronym for Southern Cone Common Market, an economic association among Brazil, Argentina, Paraguay, and Uruguay.

meridian See *longitude.*

mestizo In Latin America, a person of mixed European and Native American ancestry.

Methane Greenhouse gas resulting from human activities (from rice paddies and the guts of ruminating animals like cattle).

Mexican War A war between the United States and Mexico in the 1840s, in which Mexico lost a significant portion of its territory to the United States.

Mezzogiorno A term for southern Italy.

microcredit The lending of small sums to poor people to set up or expand small businesses.

Micronesia (tiny islands) Thousands of small and scattered islands in the central and western Pacific Ocean, mainly north of the equator.

microstate A political entity that is tiny in area and population and is independent or semi-independent.

Middle Eastern ecological trilogy The model of mostly symbiotic relations among villagers, pastoral nomads, and urbanites in the Middle East.

middle latitudes See *latitude.*

migration A temporary, periodic, or permanent move to a new location.

millenarian movements See *cargo cults.*

***minifundia* (sing. *minifundio*)** Small Latin American agricultural landholdings, usually with a strong subsistence component.

minutes A unit of measurement equal to 1/60th of a degree.

Mississippian A mound-builder Native American culture that lived in what is now the southeastern United States.

misdeveloped country A term sometimes applied to post-Soviet Russia because of the rampant criminality and corruption in its society and its demographic trends of high death rates and low birth rates.

Mistake of 1914 Nigerian term for the creation of the country's modern borders by the British.

mitigation Measures taken to avoid the adverse impacts of climate change in the long term.

Mogollon A Native American civilization that existed in what is now the southwestern United States.

Mohajirs Muslim immigrants to Pakistan from India after the two countries were partitioned in 1947.

mollisol soils Thick, productive, and durable soils, such as the *chernozem,* whose fertility comes from abundant humus in the top layer.

money laundering Ways of turning money earned in illegal ways into money that looks legal.

Mongols Conquerors of most of Russia in the thirteenth century.

Mon-Khmer A branch of the Austro-Asiatic language subfamily that includes the Khmer and Vietnamese languages.

monoculture The single-species cultivation of food or tree crops, usually very economical and productive but threatening to natural diversity and change.

Monroe Doctrine A policy of U.S. President James Monroe stating the United States would prevent European countries from establishing any new colonies in the Western Hemisphere.

monsoon A current of air blowing fairly steadily from a given direction for several weeks or months at a time. Characteristics of a monsoonal climate are a seasonal reversal of wind direction, a strong summer maximum of rainfall, and a long dry season lasting for most or all of the winter months.

Montreal Protocol A 1989 international treaty to ban chlorofluorocarbons.

moor A rainy, deforested upland, covered with grass or heather and often underlain by water-soaked peat. Also, Muslim inhabitants of Spain.

more developed countries (MDCs) The world's wealthier countries.

Mosan language family A Native American language family along the Pacific coast.

Mound Builder civilizations A collection of Native American culture groups named for their creation of large earthen mounds. They created extensive trading networks and urban settlement patterns.

Movement for the Emancipation of the Niger Delta (MEND) A mainly ethnically Ijaw group in Nigeria that has expressed dissatisfaction with the redistribution of Nigeria's oil wealth by kidnapping foreign oil workers.

Movement for the Survival of the Ogoni People (MOSOP) An ethnic Ogoni organization in Nigeria that petitions the government for a safe environment and more federal support for their people.

mujahidin Anti-Soviet rebel bands in Afghanistan during the Soviet invasion of the 1980s.

mulatto A person of mixed European and black ancestry.

mullah Religious leaders in Iran.

Multiculturalism Act A 1988 Canadian law that officially established Canada as a multicultural society, recognizing and encouraging traditional identities.

multi-Latina companies Companies, established in Latin America by Latin American workers who gained experience in the United States, that invest in the United States.

multinational companies Companies that operate, at least in part, outside their home countries.

Muslims Adherents of Islam.

mutually assured destruction (MAD) The concept of nuclear deterrence: if one side in a conflict were to launch nuclear weapons, the other would. Both countries would be obliterated.

Na-Dene language family A Native American language family of western North America, including Apache and Navajo.

nagana See *trypanosomiasis.*

Nakh-Dagestanian A collection of languages spoken in the Caucasus region.

narcoterrorist organization Armed insurgent groups that are funded by the sale of illegal drugs.

National Petroleum Reserve An area along the North Slope of Alaska estimated to have more petroleum resources than ANWR.

nationalism The drive to expand the identity and strength of a political unit that serves as home to a population interested in greater cohesion and often expanded political power.

national minorities A substantial population of ethnic groups originating in other nations but now living in foreign countries.

Native Americans The indigenous peoples of North and South America who originally migrated to the New World from Asia over a land bridge during the Ice Age. Christopher Columbus misidentified them as "Indians" in 1492.

native reserves See *homelands.*

Native Title Bill Legislation passed by the Australian government in 1993, addressing many issues relating to Aborigines' rights to claim and buy land.

natural replacement rate The highest rate at which a renewable resource can be used without decreasing its potential for renewal. Also known as *sustainable yield.*

natural landscape An environment without (or before) transformations caused by humans.

natural resource A product of the natural environment that can be used to benefit people. Resources are human appraisals.

Naxalites Maoist insurgents in India that claim to champion the causes of the rural poor.

Nazca An extinct ethnic group from Peru that left massive carvings in the desert two thousand years ago.

Near Abroad Russia's name for the now-independent former republics, other than Russia, of the Union of Soviet Socialist Republics.

Near North Australian term for neighboring Asian countries.

Negroes An archaic term for African Americans.

neocolonialism The perpetuation of a colonial economic pattern in which developing countries export raw materials to, and buy finished goods from, developed countries. This relationship is more profitable for the developed countries.

neo-Europes Areas colonized by Europeans in lands similar climatically and geographically to Europe, creating "new Europes" with cultural similarities and ties to Europe.

Neolithic Revolution See *Agricultural Revolution.*

Neo-Malthusians Supporters of forecasts that resources will not be able to keep pace with the needs of growing human populations.

new American imperialism The belief that post-9/11 actions by the United States marked the beginning of a new era of aggressive global involvement.

New Asian Tigers A collection of Southeast Asian countries that have had rapid economic growth.

new lands See *virgin and idle lands.*

New Partnership for Africa's Development (NEPAD) An agency of the African Union responsible for monitoring improvements in human rights across Africa.

New Rome A term referring to Constantinople as a center of Christendom in the late Roman period.

newly industrializing countries (NICs) The more prosperous of the world's less developed countries.

Niger-Congo language family A language family of central Africa.

Night Journey In Islam, Muhammad's brief ascension into heaven from the great rock beneath the Dome of the Rock.

Nile Water Agreement A treaty guaranteeing Egypt 75 percent of the Nile River's flow; it forbids any projects downstream that might threaten the volume of water reaching Egypt.

Nilo-Saharan language family A language family spoken in parts of Africa around the Sahara Desert.

NIMBY An acronym for "not in my back yard." Refers to the opposition of local residents to new developments in their area.

nirvana In Buddhism, a transcendent state in which one is able to escape the cycle of birth and rebirth.

Nitrous oxide Greenhouse gas resulting from human activities (from the breakdown of nitrogen fertilizers).

Noble Sanctuary (al-Haraam ash-Shariif) What Muslims call the Temple Mount.

nodal region See *region.*

Non-Pama-Nyungan A branch of the Aboriginal language family spoken in northern Australia.

nonselective migrants People responding to push factors.

nontimber forest products (NTFPs) Goods, other than trees, made from products found in forests, to prevent deforestation.

North American Free Trade Agreement (NAFTA) A free-trade agreement among Canada, the United States, and Mexico.

North Atlantic Drift A warm current, originating in tropical parts of the Atlantic Ocean, that drifts north and east, moderating temperatures of Western Europe. Also known as the *Gulf Stream.*

North Atlantic Treaty Organization (NATO) A military alliance formed in 1949 that included the United States, Canada, many European nations, and Turkey.

North European Plain The level to rolling lowlands that extend from the low countries on the west, through Germany to Poland. These are areas rich in agricultural development, dense human settlements, and a number of major industrial centers. The plain is broken by a number of rivers flowing from the Alps and other mountain systems in central Europe into the North and Baltic Seas.

north latitude See *latitude.*

North Pole A point at 90 degrees north latitude. The highest latitude in the northern hemisphere.

Northern Alliance The main domestic opponent of the Taliban in Afghanistan.

Northern Hemisphere See *hemisphere.*

Northern League An Italian organization calling for the creation of Padania as an autonomous area or independent country.

Northern Sea Route A waterway developed by the Soviets to provide a connection between the Pacific and Arctic oceans.

Northwest Passage A shipping channel through Canada's Arctic islands if global warming decreases sea ice enough to make a reliable route.

Nuclear Nonproliferation Treaty (NPT) See *Comprehensive Test Ban Treaty.*

Nunavut A Canadian territory created in 1999 to be the home of the Inuit.

Oceania A term collectively referring to Australia, New Zealand, and the islands of the Pacific Ocean.

Occupied Territories The territories captured by Israel in the 1967 War: the Gaza Strip and West Bank, captured, respectively, from Egypt and Jordan, and the Golan Heights, captured from Syria.

official state sponsors of terrorism A U.S. list of countries that sponsor terrorism.

offshore outsourcing (offshoring) See *outsourcing.*

Ogoni An ethnic group in Nigeria.

oil embargo The 1973 embargo on oil exports imposed by Arab members of the Organization of Petroleum Exporting Countries (OPEC) against the United States and the Netherlands.

oligarchs Literally "rule of the few," where a small number of wealthy businessmen control much of a state's production.

one-child campaign See *one-child policy.*

one-child policy Official policy of the Chinese government, limiting families to one child.

Operation Barbarossa A German onslaught against the Soviet Union during World War II.

Operation Crossroads A series of 67 nuclear experiments by the United States in the Marshall Islands of the Pacific Ocean after World War II.

Operation Enduring Freedom The U.S. military strike against Afghanistan that began in October 2001.

Operation Iraqi Freedom The U.S.-led ground assault on Iraq that began in March 2003.

Orange Revolution The 2005 Ukrainian presidential election that brought pro-Western Victor Yushchenko to power.

Oregon Trail One of the major routes leading westward that American pioneers took in the middle of the nineteenth century, leading to what is now the Willamette Valley of Oregon.

Organization of American States (OAS) A political organization of every country in North and South America, except Cuba.

Organization of Petroleum Exporting Countries (OPEC) An international organization created to get higher profits from oil sales to its member countries.

orientation The relationship between direction on a map and the corresponding compass direction in reality.

original affluent society A description of hunter-gatherers, who are thought by some scholars to have enjoyed harmony with the natural world and suffered few social and psychological problems.

Oslo I Accord An agreement signed by Israeli prime minister Yitzhak Rabin and Palestinian leader Yassir Arafat, designed to be a pathway to peace between Israel and the Palestinians.

Oslo II Accord Another agreement, not yet fully implemented, between Israel and the Palestinian Authority that would transfer more of the West Bank to Palestinian control in exchange for increased Palestinian efforts to crack down on terrorists and guarantee Israel's security.

Oto-Manguean language family A family of Native American languages spoken in northern Mexico.

outsourcing The flight of jobs and technology from countries with high manufacturing and service costs to countries with lower costs.

overseas foreign workers (OFW) A term referring to the one-quarter of the workforce of the Philippines that works outside that country and sends back remittances that make up more than 10 percent of the Philippines' GDP.

Padania A term for northern Italy.

paddy (padi) The field in which rice is grown.

Palestine Liberation Organization (PLO) The Palestinian military and civilian organization created in the 1960s to resist Israel and recognized in the 1990s as the sole legitimate organization representing official Palestinian interests internationally.

Palestinian Authority (PA) The Palestinian government created as a result of the Oslo I Accord.

Palestinians Arabs who formed the majority of inhabitants of Palestine prior to the creation of the modern country of Israel.

pampa Subtropical grasslands in South America.

Pama-Nyungan branch of the Aboriginal language family spoken over most of Australia.

Pan-African Parliament A lawmaking body of the African Union.

pan-Turkism The goal of uniting all Turkic peoples of Asia.

Pancasila A pan-Indonesian nationalist ideology designed to neutralize all ethnic identities.

pandemic A very widespread epidemic.

Papuan languages A collection of the 750 languages spoken on New Guinea and surrounding islands.

Papuans People of mostly Melanesian origin that live on New Guinea.

parallel A latitude line running parallel to the equator.

páramos Alpine meadows in Latin America.

Parsis An ethnic group in India that practices Zoroastrianism.

Parti Quebecois (PQ) A separatist political party in Quebec.

pastoral nomads Desert peoples of the Middle East and northern Africa who migrate through arid lands with their livestock, following patterns of vegetation and rainfall.

Penutian language family A Native American language family of Mexico and California.

People of the Book Term used by Muslims to refer to Jews and Christians.

people overpopulation The concept that many persons, each using a small quantity of natural resources to sustain life, add up to too many people for the environment to support.

PEPFAR An acronym for President's Emergency Plan for AIDS Relief, in which the George W. Bush administration pledged over $45 billion to combat AIDS in Africa.

per capita GDP A measure of economic well-being found by dividing a country's gross domestic product by its population.

per capita gross national income A measure of economic well-being found by dividing a country's gross domestic product plus any foreign output by domestically-owned producers by its population.

per capita gross domestic product purchasing power parity (GDP PPP) In this figure, annual per capita gross domestic product (GDP)—the total output of goods and services a country produces for home use in a year—is divided by the country's population. For comparative purposes, that figure is adjusted for purchasing power parity (PPP), which involves the use of standardized international dollar price weights that are applied to the quantities of final goods and services produced in a given economy. The resulting measure, per capita GDP PPP, provides the best available starting point for comparisons of economic strength and well-being between countries.

perceptual region See *region.*

perestroika See *glasnost.*

peripheral location An area of the world with less importance due to a distant location relative to others.

permafrost Permanently frozen subsoil.

Permanent Fund A government surplus in Alaska, created from oil revenues.

Persian The language spoken by the majority of Iranians.

physical geography The subdiscipline of geography most concerned with the climate, landforms, soils, and physiography of the earth's surface.

Pidgin The official language of Papua New Guinea, a combination of English and foreign words mixed with indigenous vocabulary and grammar.

PIGS An acronym for Portgual, Ireland, Greece, and Spain, used in reference to the weaker economies of those countries as opposed to those of France and Germany.

Pillars of Islam The five fundamental tenets of the faith of Islam, which are *profession of faith, prayer, almsgiving, fasting,* and the *hajj*.

pivotal countries Those countries whose collapse would cause international refugee migration, war, pollution, disease epidemics, or other international security problems.

place In geographic analysis of a given locale, the nature of place identity becomes a means of understanding people's response to that particular place. Determination of the environmental and cultural characteristics that are most frequently associated with a certain place helps to establish that "sense of place" for a given location.

Plan Colombia A U.S.-funded effort to reduce the amount of coca production in Colombia.

Plan Patriot An operation in Colombia involving U.S. military personnel and civilian contractors targeting FARC to reduce drugs and secure oil reserves.

plates Giant sections of rock the lithosphere is split into, which slowly drift over time.

plate tectonics The dominant force in the creation of the continents, mountain systems, and ocean deeps. The steady, but slow, movement of these massive plates of the earth's mantle and crust has created the positions of the continents and major patterns of volcanic and seismic activity. Areas where plates are being pulled under other plates are called subduction zones.

platform economy Common in high-tech industries, with a corporation headquartered in the United States but most of the work on its products done overseas.

Pleistocene overkill hypothesis A hypothesis stating that hunters and gatherers of the Pleistocene Era hunted many species to extinction.

plinthite The bricklike substance into which lateritic soils can turn.

podzol Soil with a grayish, bleached appearance when plowed, lacking in well-decomposed organic matter, poorly structured, and very low in natural fertility. Podzols are the dominant soils of the taiga.

Polynesia (many islands) A Pacific island region that roughly resembles a triangle with its corners at New Zealand, the Hawaiian Islands, and Easter Island.

population change rate The birth rate minus the death rate in a population.

population explosion The surge in Earth's human population that has occurred since the beginning of the Industrial Revolution.

population implosion A dramatic loss of population, caused by low fertility rates or death rates that are much higher than birth rates.

population pyramid See *age structure diagram*.

population replacement level The number of new births required to keep a population steady. Generally calculated as 2.1 children per woman in MDCs.

postindustrial Description of an economy or society characterized by the transformation from manufacturing to information management, financial services, and the service sector.

potato famine The starvation of 10 percent of Ireland's population in the late 1840s, caused by a crop disease called blight.

potlatch A Native American system in which rank and status were determined by the quantity and quality of material goods one could give away.

Poverty Point A mound-builder Native American culture that lived in what is now the southeastern United States.

power projection hub In geopolitical terms, a strategic outpost from which a country could launch naval, air, and other forces against an enemy.

prairie An area of tall grass in the middle latitudes, composed of rich soils that have been cleared for agriculture. The original lack of trees may have been due to repeated burnings or periodic drought conditions.

Pre-1967 borders The borders of Israel after 1949 and before the Six-Day War of 1967 in which Israel captured and occupied the Gaza Strip, West Bank, and Golan Heights.

precautionary principle The notion that is it appropriate to attempt to reduce or eliminate any risky practice.

Pre-Columbian times North and South America before Christopher Columbus's first voyage in 1492.

precipitation Water falling from the sky to the surface in the form of rain, snow, sleet or hail.

preemptive engagement A U.S. policy developed after 9/11 that allowed the country to take unilateral military action if necessary to defuse threats to its security whenever and wherever those threats occur.

primate city A city that dominates a country's urban scene and is usually defined as being larger than the country's second and third largest cities combined. Primate cities are generally found in developing countries, although some developed countries, such as France, have them.

prime meridian The line of zero degrees longitude, which passes through Greenwich, England. It separates the Western Hemisphere from the Eastern.

privatization The shift of ownership of economic activities from the state to nongovernmental.

proletariat In Marxism, the industrial working class.

Promised Land Canaan, for the Israelites in Biblical times.

Protestant Reformation The schism in Western Christianity that began in 1517 when Martin Luther registered his objections to much of the dogma, rituals, and political organization of the Roman Catholic Church.

Protestant Unionists One side of "The Troubles" in Northern Ireland, those who wished Northern Ireland to remain unified with the United Kingdom and not unify with the rest of Ireland.

Protestant work ethic In the United States, the emphasis on work, productivity, and economic success in an individual's life. Despite its name, this ethic is shared by people of many faiths and ethnicities.

Protestantism One of the denominations of Christianity, which began in the Protestant Reformation of the sixteenth century.

Proto-Asiatic A family of languages spoken largely in northeastern Russia.

proxy war During the Cold War, the situation created when the United States and the Union of Soviet Socialist Republics played African countries against one another, arming them with weapons, to boost their competing aims.

pueblo An interconnected series of residences and ceremonial centers composed of adobe, used by the Hohokam culture of New Mexico.

purchasing power parity (PPP) A method of comparing the real value of output between different countries' economies, considering factors such as differences in relative prices of goods and services.

push and pull forces of migration Emigration caused by so-called push factors, as when hunger or lack of land "pushes" peasants out of rural areas into cities, or by pull factors, as when an educated villager responds to a job opportunity in the city. People responding to push factors are often referred to as nonselective migrants, whereas those reacting to pull factors are called selective migrants. Both push and pull forces are behind the rural-to-urban migration that is characteristic of most countries.

Quechua See *Quechua-Aymaran language family.*

Quechua-Aymaran language family A language family of central South America, to which the language of the Inca belonged. Many residents of Peru and Bolivia still speak Quechua as a first language.

Rastafarianism A religion originating in Jamaica in 1930 that holds that Ethiopian emperor Haile Selaisse was the earthly incarnation of God.

Red-Dead Peace Conduit A joint venture between Israel and Jordan that will carry seawater from the Gulf of Aqaba northward to replenish the shrinking Dead Sea.

Reducing Emissions from Deforestation and Forest Degradation (REDD) A United Nations effort to create a financial value for the carbon stored in forests and offer incentives for developing countries to reduce carbon emissions from forested areas and invest in low-carbon development.

reference map A type of map primarily depicting the locations of various features on the Earth's surface.

refugees Victims of severe push factors such as persecution, political repression, and war.

region A "human construct" that is often of considerable size, that has substantial internal unity or homogeneity, and that differs in significant respects from adjoining areas. Regions can be classed as formal (homogeneous), functional, or vernacular. The formal region, also known as a uniform region, has a unitary quality that derives from a homogeneous characteristic. The United States is an example of a formal region. The functional region, also called the nodal region, is a coherent structure of areal units organized into a functioning system by lines of movement or influence that converge on a central node or trunk. A major example would be the trading territory served by a large city and bound together by the flow of people, goods, and information over an organized network of transportation and communication lines. Vernacular or perceptual regions are areas that possess regional identity, such as the Sun Belt, but share less objective criteria in the use of this regional name. General regions, such as the major world regions in this text, are recognized on the basis of overall distinctiveness.

Regional Initiative for the Infrastructure Integration of South America An international effort to improve the economies of South American countries by linking them together by road networks and other infrastructure projects.

relative location The location of a place defined by relationship to other places.

remittances Money earned by workers in a foreign country sent back to that worker's home country.

remote sensing Through the use of aerial and satellite imagery, geographers and other scientists have been able to get vast amounts of data describing places all over the face of the earth. Remote sensing is the science of acquiring and analyzing data without being in contact with the subject. It is used in the study of patterns of land use, seasonal change, agricultural activity, and even human movement along transport lines. This process relates closely to GIS. See also *geographic information systems (GIS).*

renationalization A process of reversing the trend toward private ownership of economic production in Russia during the presidency of Vladimir Putin.

renewable resource A resource, such as timber, that is grown or renewed so that a continual supply is available. A finite resource is one that, once consumed, cannot be easily used again. Petroleum products are a good example of such a resource and because it takes too much time to go through the process of creation, they are not seen as renewable.

reservations Areas of land reserved for Native American tribes by the U.S. government.

resilience An ecosystem that can recover from the stress of drought and has mechanisms to cope with a natural cycle that includes extended periods of dryness and rain.

resource curse A paradox of plenty, when a country with a great abundance of a valuable natural resource experiences lower economic growth than countries without such abundance.

Richter scale A system for measuring the strength of earthquakes.

right of return The Palestinian Arab principle that refugees (and their descendants) displaced from Israel and the Occupied Territories in the 1948 and 1967 wars be allowed to return to the region.

risk minimization The exploitation of multiple resources so some resources will support a population if others fail.

River Sea A term referring to the enormity of the Amazon River.

road map for peace An effort led by the United States, the European Union, Russia, and the United Nations to reach a peace settlement between Israel and the Palestinians, ultimately leading to an independent Palestinian state.

Roma (Gypsies) Originating in India, they are one of Europe's largest ethnic minorities. Commonly referred to as Gypsies, they often move from place to place in caravans and have been discriminated against for centuries.

Roman Catholic Church The largest denomination of Christianity, led by the Pope in Vatican City.

Romance languages A language family primarily of southern Europe, including Italian, Spanish and French.

Roosevelt Corollary A 1904 policy in which the United States declared it had the right to supervise the internal affairs of Latin American countries to ensure U.S. national security.

Rose Revolution The peaceful transition of power in the 2003 presidential election in Georgia.

rules of origin A NAFTA requirement that half or more of all the components of any manufactured good must originate in Canada, the United States, or Mexico.

rural-to-urban migration See *urbanization.*

Rus Viking adventurers from Scandinavia that settled in Slavic regions during the Middle Ages.

Russian cross The dramatic rise in death rates and fall of birth rates in Russia after the breakup of the Soviet Union.

Russian Revolution The overthrow of Czar Nicholas II of Russia in 1917.

Russification The effort, particularly under the Soviets, to implant Russian culture in non-Russian regions of the former Soviet Union and its Eastern European neighbors.

Rust Belt An area across the Midwest United States that was home to many heavy industries.

sacred space (sacred place) Any locale that people hold in reverence, such as places of worship, cemeteries, and battlefields.

Saharan A subfamily of the Nilo-Saharan language family spoken in north-central Africa.

salinization The deposition of salts on, and subsequent fertility loss in, soils experiencing a combination of overwatering and high evaporation.

salmon war A dispute regarding salmon fishing between the United States and Canada until it was resolved in 1999.

San The proper name for Bushmen, a major group of the Khoisan ethnolinguistic cultures living mainly in Botswana and Namibia. They were among the last people on Earth to practice hunting and gathering.

sand sea A virtual "ocean" of sand characteristic of parts of the Middle East, where people, plants, and animals are all but nonexistent.

Santa Fe Trail One of the major routes leading westward that American pioneers took in the middle of the nineteenth century, leading to what is now New Mexico.

Santeria A syncretic religion originating in the Caribbean, blending Catholicism, indigenous, and Yoruba beliefs.

savanna A low-latitude grassland in an area with marked wet and dry seasons.

scale The size ratio represented by a map; for example, a map with a scale of 1:12,500 is portrayed as 1/12,500 of the actual size.

Schengen Agreement A framework among signatory countries allowing for free movement of people without need for visas, passports, or other border controls.

Schengenland A term for the countries participating in the Schengen Agreement.

scorched earth The wartime practice of destroying one's own assets to prevent them from falling into enemy hands.

seafloor spreading The creation of new crust from molten material rising from the Earth's mantle through ridges on the ocean floor.

seamount An underwater volcanic mountain.

Second Kuwait A term describing the Russian island of Sakhalin and the potential large reserves of oil and natural gas it has offshore.

second law of thermodynamics A natural law stating that high-quality, concentrated energy is increasingly degraded as it passes through the food chain.

Second Russian Revolution The collapse of the Soviet Union in 1991.

Second Temple The second Jewish temple to stand on the Temple Mount, which was destroyed by the Romans.

seconds A unit of measurement equal to 1/60 of a minute, or 1/3,600 of a degree.

sectarian violence Conflict between people of different religions.

security fence A physical barrier constructed by Israel in the West Bank designed to prevent Palestinian suicide bombers from reaching Israel.

sedentarization Voluntary or coerced settling down, particularly by pastoral nomads in the Middle East.

segregation A policy of separation of races.

seismic activity Usually refers to tectonic forces that result in earthquakes.

selective migrants People reacting to pull factors.

semiarid A mainly dry climate, with sparse precipitation.

Semitic language A language family that includes Hebrew and Arabic.

Sephardic Jews Jews who had inhabited Palestine and other parts of the Middle East.

settler colonization The historical pattern by which Europeans sought to create new or "neo-Europes" abroad.

Sephardic Jews Jews who had inhabited Palestine and other parts of the Middle East since early times.

Shanghai Cooperation Organization (SCO) An organization composed of China, Russia, and several Central Asian countries designed to combat terrorism, separatism, and extremism in the region.

shantytown Areas of often squalid slums on the outskirts of large cities throughout the developing world.

shatter belt A large, strategically located region composed of conflicting states caught between the conflicting interests of great powers.

shell banks A bank existing only on paper that specializes in money laundering.

Shia, or Shiite The branch of Islam regarding male descendants of Ali, the cousin and son-in-law of the Prophet Muhammad, as the only rightful successors to the Prophet Muhammad.

shifting cultivation A cycle of land use between crop and fallow years that is needed to work around the infertility of tropical soils. After clearing tropical forest, a farmer may get only a few years of crops before there is no fertility in the soil and must move on to clear more land. In the meantime, forest reclaims the previously farmed plots. Where new lands are not available, fallow periods on old fields are reduced or eliminated, resulting in soil deterioration. Also known as *swidden* or *slash-and-burn cultivation*.

shifting ranching A destructive process in which ranchers raise cattle on areas of former rainforest originally cleared by farmers. When the land is exhausted, ranchers move their cattle on to the next plot of cleared land.

Shintoism The strongly nationalistic religion of Japan, a blend of nature reverence, Japanese folklore, and Chinese ritual.

Silicon Valley A region of the San Francisco Bay area known for its concentration of high-tech companies and workers.

Silk Road A mostly overland trade route that stretched from Venice to Xi'an in China.

Sinhalese The majority ethnic group of Sri Lanka, predominantly Buddhist.

Sinn Fein The political counterpart to the Irish Republican Army.

Sino-Tibetan language family A major language family of Asia, including Chinese and its variants.

Sioux Native American people living on the Great Plains.

site The specific geographic location of a given place.

situation The accessibility of that site and the nature of the economic and population characteristics of that locale.

Six-Day War A 1967 war between Egypt, Syria, and Jordan against Israel, resulting in a dramatic Israeli victory and the Israeli occupation of the Gaza Strip, West Bank, and Golan Heights.

six essential elements of geography 1. The world in spatial terms; 2. Places and regions; 3. Physical systems; 4. Human systems; 5. Environment and society; 6. Uses of geography.

slash-and-burn cultivation See *shifting cultivation*.

Slavic Core (Fertile Triangle or Agricultural Triangle) The large area of the western former Soviet region containing most of the region's cities, industries, and cultivated lands.

Slavic languages A language family in eastern Europe including Russian, Polish and Czech.

sleeping sickness See *trypanosomiasis*.

small-scale map A map constructed to give a highly generalized view of a large area.

socialization State ownership of economic production.

Solidarity An independent trade union in Poland, the creation of which was one of the conditions resulting in the withdrawal of Soviet control in Poland in 1989.

songlines See *Footprints of the Ancestors*.

south latitude See *latitude*.

South Pole A point at 90 degrees south latitude. The highest latitude in the southern hemisphere.

Southeast Anatolia Project (GAP) An agricultural effort in Turkey aiming to double the amount of the country's irrigable farmland.

Southern African Customs Union (SACU) A trade organization including South Africa, Lesotho, Swaziland, Botswana, and Namibia.

Southern Cone Common Market (Mercosur) See *Mercosur.*

Southern Hemisphere See *hemisphere.*

Soviet satellites Communist countries in Eastern Europe between World War II and 1989.

space In geography, the exact placement of locations on the face of the Earth.

Spanglish A hybrid tongue of Spanish and English.

spatial Geographers recognize spatial distributions and patterns in Earth's physical and human characteristics. The term *spatial* comes from the noun *space*, and it relates to the distribution of various phenomena on Earth's surface. Geographers portray spatial data cartographically—that is, with maps.

special economic zones (SEZs) In China, the urban areas designated after the 1970s to attract investment and boost production through tax breaks and other incentives.

spodosols Acidic soils that have a grayish, bleached appearance when plowed, lack well-decomposed organic matter, and are low in natural fertility. Also known as *podzols.*

state farm (*sovkhoz*) A type of collectivized state-owned agricultural unit in the former Soviet Union; workers receive cash wages in the same manner as industrial workers.

State Law and Order Restoration Council (SLORC) The military government that seized control of Burma in 1988 and changed the country's name to Myanmar.

steppe (temperate grassland) A biome composed mainly of short grasses. It occurs in areas of steppe climate, which is a transitional zone between very arid deserts and humid areas.

storm surge A powerful wave in the ocean that is associated with landfalling hurricanes.

strategic hamlet Fortified communities in Vietnam that the U.S. military forced millions of South Vietnamese farming families into, in a failed attempt to prevent Viet Cong forces from infiltrating South Vietnamese villages.

String of Pearls China's strategy of gaining critical allies along the Indian Ocean shoreline.

subarctic A high-latitude climate characterized by short, mild summers and long, severe winters.

subduction The process of one tectonic plate descending below another.

subduction zone See *plate tectonics.*

subsidies Payments made by governments to domestic producers.

Suez Crisis A British, French, and Israeli invasion of Egypt in 1956 after Egyptian president Nasser nationalized the Suez Canal. International pressure caused the withdrawal of the invading forces, and the canal stayed in Egyptian control.

summer monsoon See "Monsoon."

Sunni The branch of Islam regarding successorship to the Prophet Muhammad as a matter of consensus among religious elders.

superpower A country with a bigger economy, stronger military, and pervasive popular culture than most other countries. Currently the United States is the only global superpower.

supranational organization An international organization in which member countries are united beyond the authority of any single national government and are planned and controlled by a group of nations.

suq A large commercial zone adjoining the ceremonial and administrative heart of a Middle Eastern city. See *bazaar.*

sustainable development (ecodevelopment) Concepts and efforts to improve the quality of human life while living within the carrying capacity of supporting ecosystems.

sustainable yield (natural replacement rate) The highest rate at which a renewable resource can be used without decreasing its potential for renewal.

sweatshops Factories in the LDCs infamous for low wages and long working hours. Part of the dark side of globalization: Chinese and other Asian companies in particular can undercut competition by underpaying and overworking employees.

swidden cultivation See *shifting cultivation.*

syncretism The incorporation of some animistic beliefs into mainstream religions.

taiga A northern coniferous (needleleaf) forest.

Tai-Kadai sufamily

Taliban One of the mujahidiin factions in Afghanistan fighting the Soviet invasion in the 1980s. The Taliban gained control over most of the country in the 1990s and instituted a very strict code of Islamic law. The Taliban were driven from power during the U.S.-led Operation Enduring Freedom but remain a significant insurgent force.

Tamils The minority ethnic group of Sri Lanka, primarily Hindu.

Tamil Tigers Tamil insurgents in Sri Lanka that wanted to establish an independent Tamil homeland. They were defeated by the Sri Lanka government in 2009.

Taoism See *Daoism.*

tar sands Mineral deposits in which liquid petroleum is mixed with sand and clay. Extracting the oil from tar sands is expensive and consumes large amounts of energy and water.

tariffs Tax penalties imposed on imports.

Tatars Nomads of Central Asia that conquered much of Russia in the thirteenth century along with the Mongols.

tech bubble An economic bubble in the United States in the late 1990s, centering on high-tech products and services.

technocentrists (cornucopians) Supporters of forecasts that resources will keep pace with or exceed the needs of growing human populations.

tectonic forces Processes that derive their energy from within the earth's crust and serve to create landforms by elevating, disrupting, and roughening the earth's surface.

temperate grasslands See *steppe.*

temperate mixed forest Dominant forest type in areas with humid subtropical *and* humid continental climates.

Temple Mount A site in Jerusalem sacred to both Jews and Muslims. Two Jewish temples stood on the Temple Mount, and currently the Muslim Dome of the Rock occupies the site.

Teotihuacanos A Native American ethnic group that lived in central Mexico two thousand years ago, who were later replaced by the Aztecs.

terra nullius A legal doctrine declaring a location "no one's land," allowing for claims to that land.

terraces Carefully constructed rice paddies arranged in a stair-step fashion.

terrorism Premeditated, politically motivated violence perpetrated against noncombatant targets by subnational groups or clandestine agents, usually intended to influence an audience.

terrorism hot spot An area regarded by the United States as real or potential training grounds for al-Qa'ida-affiliated groups.

thematic map A type of map showing the spatial distribution of one or more attributes across a given area.

theory of Himalayan environmental degradation The concept that overpopulation on steep slopes in the middle and high elevations of the Himalaya leads to excessive deforestation and erosion through shifting cultivation. This has many knock-on effects downstream, including flooding and siltation of hydroelectric reservoirs.

theory of island biogeography A theoretical calculation of the relationship between habitat loss and natural species loss, in which a 90 percent loss of natural forest cover results in a loss of half of the resident species.

Theravada Buddhism One of the two branches of Buddhism, found largely in Sri Lanka and Southeast Asia. Therevada Buddhists claim to follow the true teachings and practices of the Buddha.

Third Revolution The belief that sustainable development will bring about a marked shift in human ways of interacting with the natural environment, so dramatic that it will be compared with the origins of agriculture and industry.

Third Rome A term used to describe Moscow's importance in Christian affairs (after Rome and Constantinople) in the eleventh century.

Thirteenth Amendment An amendment to the U.S. Constitution in 1865, ending the practice of slavery.

Tibeto-Burman subfamily A language subfamily spoken in Southeast Asia and western China.

tierra caliente A Latin American climatic zone reaching from sea level upward to approximately 3,000 feet (914 m). Crops such as rice, sugarcane, and cacao grow in this hot, wet environment.

tierra fría A Latin American climatic zone found between 6,000 feet (1,800 m) and 12,000 feet (3,600 m) above sea level. Frost occurs and the upper limit of agriculture and tree growth is reached.

tierra helada A Latin American climatic zone above 10,000 feet (3,048 m) that has little vegetation and frequent snow cover.

tierra templada A Latin American climatic zone extending from approximately 3,000 to 6,000 feet (914–1,829 m) above sea level. It is a prominent zone of European-induced settlement and commercial agriculture such as coffee growing.

Tiger Cubs See *New Asian Tigers.*

tipping point In climate change, a time beyond which some changes are irreversible, likely bringing rapid acceleration in temperature and its related effects.

Tornado Alley An area of the Midwestern United States where tornado activity is very common.

Torres Strait Islanders A distinct indigenous people of Australia with origins in Papua New Guinea.

tortilla wars Increases in the cost of corn in Mexico from U.S. imports that threatened to make it too expensive for the poor to buy. Also refers to the controversial appearance of genetically modified corn in Mexican fields at the same time.

Tory In the American Revolution, colonists who wished to remain politically connected with the British government.

Totonac language family A small language family of eastern Mexico.

trade barriers Restrictions on international trade, especially those created by MDCs to protect their industries from competition from cheaper goods produced in LDCs.

trade wars The use of barriers to restrict trade between countries to protect their interests and industries.

Trading with the Enemies Act A U.S. law forbidding Americans from spending more than $300 in Cuba.

Trans-Amazon Highway A highway linking the Atlantic coast of Brazil with the Pacific coast of Peru.

Trans-Pacific Partnership A free-trade initiative between the United States, Australia, New Zealand, and several countries in Asia and South America.

tree line The upper limit (or northernmost limit) of natural tree growth.

Trench Deep linear Earth feature (for example, the Mariana Trench off Japan).

triangular trade The sixteenth- to ninteenth-century trading links between West Africa, Europe, and the Americas, involving guns, alcohol, and manufactured goods from Europe to West Africa exchanged for slaves. Slaves brought to the Americans were exchanged for the gold, silver, tobacco, sugar, and rum carried back to Europe.

Tropic of Cancer A line of latitude located at 23.44°N. The general dividing line in the northern hemisphere between the tropics to the south and the mid-latitudes to the north.

Tropic of Capricorn A line of latitude located at 23.44°S. The general dividing line in the southern hemisphere between the tropics to the north and the mid-latitudes to the south.

tropical deciduous forest The vegetation typical of some tropical areas with a dry season. Here, broadleaf trees lose their leaves and are dormant during the dry season and then add foliage and resume their growth during the wet season.

tropical rain forest A low-latitude broadleaf evergreen forest found where heat and moisture are continuously, or almost continuously, available.

tropical savanna A relatively moist low-latitude climate that has a pronounced dry season.

The Troubles A period of strife in Northern Ireland marked by violence between Catholic Republicans and Protestant Unionists.

Truth and Reconciliation Commission A South African commission that allows those who were party to racial violence during the apartheid era to confess their misdeeds and, in a sense, be absolved of them.

trypanosomiasis A disease spread by the tsetse fly of Africa. Known as sleeping sickness in humans, and nagana in domestic cattle.

tsunami Large waves in the ocean created by undersea earthquakes.

Tuareg An ethnolinguistic group in northern Africa.

tundra A region with a long, cold winter, when moisture is unobtainable because it is frozen, and a very short, cool summer. Vegetation includes mosses, lichens, shrubs, dwarf trees, and some grass.

Turkic A subfamily of the Altaic language family, spoken mainly in western and central Asia.

Turks The predominant ethnic group of Turkey.

Tutsi (Watusi) See *Hutu.*

two-state solution A 1947 United Nations proposal to establish separate Jewish and Arab states in Palestine.

underemployment The underutilization of labor.

underground economy The "black market" typical of the former Soviet region and many LDCs.

undifferentiated highland A climate that varies with latitude, altitude, and exposure to the sun and moisture-bearing winds.

undocumented workers People who are in the United States illegally but have obtained temporary or long-term employment.

uniform region See *region.*

union state The integration of economic, political, and defense systems between two independent countries.

United Nations Resolutions 242 and 338 Resolutions passed by the United Nations after the Middle Eastern wars of 1967 and 1973, respectively, calling on Israel to withdraw from the Occupied Territories.

untouchables See *Dalits.*

upper limit of agriculture The highest elevation at which agricultural activities can be performed, at about 12,000 feet (3,600 m).

upstream countries Countries where a major river originates that have a large amount of control over that river's water, at the expense of countries farther down the river's course.

Uralic languages A language group that includes Finnish, Hungarian, and many Siberian languages.

urbanites In the Middle Eastern ecological trilogy, the educated people who provide manufactured goods and innovations, secular and religious educational instruction, and cultural amenities.

urbanization The population growth of cities, mostly from the movement of people from rural regions to built-up areas.

Ushahidi A nonprofit software firm and platform founded in Kenya. The software uses "crowdsourcing" for activist mapping of political, environmental, and other problems. It is a fine example of local people in an LDC developing a geospatial tool kit to address problems in development.

usufruct Nondestructive mutual use of resources by multiple clans or tribes located within the territory of another.

value-added manufacturing The process of refining and fabricating more valuable goods from raw or semi-processed materials.

value-added products Finished products, worth much more than the raw materials they are created from.

Varangians See *Rus.*

vernacular region See *region.*

Viet Cong An insurrectionist Communist force in South Vietnam, supported by the North Vietnamese government, that aided the North's conquest of the South and reunification of the country during the Vietnam War.

villagers In the Middle Eastern ecological trilogy, the peasant farmers that form the cornerstone of the trilogy.

virgin and idle lands (new lands) Steppe areas of Kazakhstan and Siberia brought into grain production in the 1950s.

virtual fence A high-tech surveillance system for parts of the U.S.-Mexico border.

volcanism Movement of molten material from the Earth's mantle, usually released through volcanoes.

wa A Japanese concept of harmony based upon the principle of economic equality.

walkabout A ritual journey for Australian Aborigines in which a boy on the verge of adulthood follows the pathway of his creator-ancestors.

Wallace's Line A marked biogeographic division located in Indonesia.

Walmart effect The loss of independent or small retail establishments in towns after being unable to compete with lower prices at Walmart and similar large stores.

War of 1812 A war between the United States and the United Kingdom, fought largely as a result of the United States attempting to annex Canada.

war on drugs A long-running U.S. policy of reducing drug usage within the United States, mostly by attempting to eradicate drugs at their sources, especially in South America.

War on Terror An open-ended, unconventional war declared by President George W. Bush in 2001 against the Islamist groups that carried out terrorist attacks against the United States.

Warsaw Pact A military alliance, now dissolved, consisting of the Soviet Union and the European countries of Poland, East Germany, Czechoslovakia, Hungary, Romania, and Bulgaria.

Washington Consensus A political-economic philosophy by which the United States would pressure democratic governments of Latin America to open their markets to international trade, reduce tarriffs, sell off state companies to private owners, and make other economic changes to bring prosperity to the region.

water hemisphere See *hemisphere.*

Water Towers of Asia The glaciers and snowpack of the Himalaya, Karakorum, and other great mountain ranges.

Way of the Law See *Footprints of the Ancestors.*

weapons of mass destruction (WMD) Weapons designed to kill large numbers of people by chemical, biological, or nuclear means.

weather The atmospheric conditions prevailing at one time and place.

welfare state A national government that provides generous social services to citizens, paid for by high taxation rates.

west longitude See *longitude.*

westerly winds An airstream located in the middle latitudes that blows from west to east. Also known as *westerlies.*

Western Big Development Project A Chinese project to improve infrastructure and the economic livelihoods of residents of China's autonomous regions.

Western Hemisphere See *hemisphere.*

Western Wall The most sacred site in the world accessible to Jews, the only remnant of the Second Temple.

wet rice cultivation Rice harvests producing two or three crops per year that can sustain large populations over long periods of time.

wheat war An economic dispute between the United States and Canada in the 1990s, as wheat yields were high and prices were very low.

White Australia policy An official Australian immigration policy until 1972 that excluded Asian and black immigrants.

white gold A term indicating how valuable cotton in Central Asia was to the Soviet economy.

Wik case A 1996 Australian high court ruling that Aborigines can claim title to public lands held by farmers and ranchers under long-term pastoral leases granted by state governments.

winter monsoon See "Monsoon."

Window on the West Russia's city of St. Petersburg, located on the Baltic Sea, giving Russia marine access to Europe.

world city A city that is a center of economic and political power in the global economy.

Xhosa One of the most populous ethnic groups of South Africa.

yang In *feng shui*, the masculine, bright, and positive.

yin In *feng shui*, the feminine, dark, and negative.

Yoruba A major ethnic group in southern Nigeria.

youth bulge A demographic situation in which a large percentage of a country's population is young.

zero population growth (ZPG) The condition of equal birth rates and death rates in a population.

zero-sum game A term referring to a gain by one party resulting in an equivalent loss to another.

Zion A synonym for Jerusalem.

Zionist movement

Zomia A term for the upland areas of Southeast Asia.

Zoroastrianism Religion practiced by the Parsis of India.

Zulu One of the most populous ethnic groups of South Africa.

Index